SPACE EXPLORATION
TRIUMPHS AND TRAGEDIES

ISSN 1551-210X

SPACE EXPLORATION

TRIUMPHS AND TRAGEDIES

Kim Masters Evans

INFORMATION PLUS® REFERENCE SERIES
Formerly Published by Information Plus, Wylie, Texas

Detroit • New York • San Francisco • New Haven, Conn. • Waterville, Maine • London

THOMSON

GALE

Space Exploration: Triumphs and Tragedies
Kim Masters Evans
Paula Kepos, Series Editor

Project Editor
John McCoy

Permissions
Shalice Caldwell-Shah, Emma Hull,
Jackie Jones

Composition and Electronic Prepress
Evi Seoud

Manufacturing
Drew Kalasky

ISBN-13 978-0-7876-5103-9 (set)
ISBN-10 0-7876-5103-6 (set)
ISBN-13 978-1-4144-0426-4
ISBN-10 1-4144-0426-3
ISSN 1551-210X

This title is also available as an e-book.
ISBN-13 978-1-4144-1045-6 (set)
ISBN-10 1-4144-1045-X (set)
Contact your Thomson Gale sales representative for ordering information.

Printed in the United States of America
10 9 8 7 6 5 4 3 2 1

TABLE OF CONTENTS

PREFACE vii

CHAPTER 1

Introduction to Space Exploration 1

Humankind has long had a fascination with the mysterious heavens. This chapter provides an overview of human attempts to understand and reach out into the universe from ancient times to the present.

CHAPTER 2

Space Organizations, Part 1: NASA 21

From its origins in the early days of powered flight, NASA has grown to be one of the leading forces in space science and exploration, despite its flaws. This chapter examines NASA's beginnings in the early twentieth century and recounts its achievements and setbacks to the present day.

CHAPTER 3

Space Organizations, Part 2: U.S. Military, Foreign, and Private 43

Many other groups besides NASA are actively pursuing a variety of endeavors in space, including human exploration. This chapter details the various types of organizations that have a stake in space.

CHAPTER 4

The Space Shuttle Program 61

NASA proved that it could put human beings on the Moon and bring them back safely. Bringing their spaceships back in one piece was another question. This chapter tells the story of the development and use of the world's first reusable "space plane," which revolutionized space science, and the terrible price at which its advances were achieved.

CHAPTER 5

The *International Space Station* 81

An idea that began during the cold war battle of wills between the United States and the Soviet Union ended as a project that brought many nations together with a common goal. This chapter recounts the effort to build a continuously inhabited scientific space station and describes difficulties that jeopardize the station's future.

CHAPTER 6

Robotic Missions in Sun-Earth Space 101

Human exploration of space is expensive and dangerous. A great deal more can be achieved by sending out a robotic ambassador. This chapter discusses the various machines that have gone into space to perform missions that humans cannot.

CHAPTER 7

Mars 123

There may be no little green men there, but humanity is still fascinated by its next-door neighbor, Mars, and in January 2004 President George W. Bush proposed that astronauts travel to the planet. While that mission is still some years away, this chapter recounts efforts thus far to understand and explore the Red Planet.

CHAPTER 8

The Far Planets 141

Pluto wobbles in an irregular orbit, trading places with Neptune; Uranus orbits on its side. Saturn wears rings, and Jupiter stares back at Earthlings with its big red eye. The strange planets at the far end of our solar system and the mechanical ambassadors we have sent to study them are discussed in this chapter.

CHAPTER 9

Public Opinion about Space Exploration 155

The space program has given us heroes and incredible advances in science, but is it right to spend billions of dollars in space when the money could be used to improve life on Earth? This chapter examines the attitudes of scientists and ordinary citizens toward the U.S. space program, presents the arguments on both sides of the space question, and describes NASA's efforts to win over a wary public.

IMPORTANT NAMES AND ADDRESSES 169

RESOURCES 171

INDEX 173

PREFACE

Space Exploration: Triumphs and Tragedies is part of the *Information Plus Reference Series*. The purpose of each volume of the series is to present the latest facts on a topic of pressing concern in modern American life. These topics include today's most controversial and most studied social issues: abortion, capital punishment, care for the elderly, crime, health care, the environment, immigration, minorities, social welfare, women, youth, and many more. Although written especially for the high school and under-graduate student, this series is an excellent resource for anyone in need of factual information on current affairs.

By presenting the facts, it is Thomson Gale's intention to provide its readers with everything they need to reach an informed opinion on current issues. To that end, there is a particular emphasis in this series on the presentation of scientific studies, surveys, and statistics. These data are generally presented in the form of tables, charts, and other graphics placed within the text of each book. Every graphic is directly referred to and carefully explained in the text. The source of each graphic is presented within the graphic itself. The data used in these graphics are drawn from the most reputable and reliable sources, in particular from the various branches of the U.S. government and from major independent polling organizations. Every effort has been made to secure the most recent information available. The reader should bear in mind that many major studies take years to conduct, and that additional years often pass before the data from these studies are made available to the public. Therefore, in many cases the most recent information available in 2006 dated from 2003 or 2004. Older statistics are some-times presented as well, if they are of particular interest and no more recent information exists.

Although statistics are a major focus of the *Information Plus Reference Series*, they are by no means its only content. Each book also presents the widely held positions and important ideas that shape how the book's subject is discussed in the United States. These positions are explained in detail and, where possible, in the words of their proponents. Some of the other material to be found in these books includes: historical background; descriptions of major events related to the subject; relevant laws and court cases; and examples of how these issues play out in American life. Some books also feature primary documents, or have pro and con debate sections giving the words and opinions of prominent Americans on both sides of a controversial topic. All material is presented in an even-handed and unbiased manner; the reader will never be encouraged to accept one view of an issue over another.

HOW TO USE THIS BOOK

The achievements of the National Aeronautics and Space Administration (NASA) and its counterparts in other nations are widely admired. Yet mankind's space exploration efforts have been expensive and include some disturbing failures and tragic accidents. As a result, space exploration is at the center of numerous controversies. This volume presents the facts on space exploration's successes and failures and the questions that surround them. How can tragedies like the *Columbia* disaster be prevented? Why did they happen in the first place? Are manned missions necessary or are robotic probes more cost effective? What do we gain from space exploration? Where should our priorities lie? Should we continue to explore space at all?

Space Exploration: Triumphs and Tragedies consists of nine chapters and three appendices. Each chapter is devoted to a particular aspect of space exploration. For a summary of the information covered in each chapter, please see the synopses provided in the Table of Contents at the front of the book. Chapters generally begin with an overview of the basic facts and background information on the chapter's topic, then proceed to examine subtopics of particular interest. For example, Chapter 4: The Space Shuttle Program begins with an examination of the goals and issues that led to the decision to build an entirely new kind of spacecraft in the 1970s. The factors that determined the space shuttle's final design are discussed. This is followed by a description of how the shuttle operates when launching, orbiting, and landing. The accomplishments of the shuttle program are discussed, and its two great failures—the *Challenger* and *Columbia* disasters—are examined in detail. The chapter concludes with information on plans to retire the space shuttles from service. Readers can find their way through a chapter by looking for the section and subsection headings, which are clearly set off from the text. Or, they can refer to the book's extensive Index if they already know what they are looking for.

Statistical Information

The tables and figures featured throughout *Space Exploration: Triumphs and Tragedies* will be of particular use to the reader in learning about this issue. The tables and figures represent an extensive collection of the most recent and important statistics on space exploration, as well as related issues—for example, graphics in the book depict breakdowns of NASA's budget and workforce; public opinion about NASA and space exploration in general; a listing of every mission mankind has sent to Mars and its outcome; the plaque mounted on the *Pioneer* spacecraft, a map of the trajectory followed by *Galileo*; and diagrams of dozens of spacecraft and other equipment. Thomson Gale believes that making this information available to the reader is the most important way in which we fulfill the goal of this book: to help readers understand the issues and controversies surrounding space exploration and reach their own conclusions about them.

Each table or figure has a unique identifier appearing above it, for ease of identification and reference. Titles for the tables and figures explain their purpose. At the end of each table or figure, the original source of the data is provided.

In order to help readers understand these often complicated statistics, all tables and figures are explained in the text. References in the text direct the reader to the relevant statistics. Furthermore, the contents of all tables and figures are fully indexed. Please see the opening section of the Index at the back of this volume for a description of how to find tables and figures within it.

Appendices

In addition to the main body text and images, *Space Exploration: Triumphs and Tragedies* has three appendices. The first is the Important Names and Addresses directory. Here the reader will find contact information for a number of government and private organizations that can provide further information on aspects of space exploration. The second appendix is the Resources section, which can also assist the reader in conducting his or her own research. In this section the author and editors of *Space Exploration: Triumphs and Tragedies* describe some of the sources that were most useful during the compilation of this book. The final appendix is the Index.

ADVISORY BOARD CONTRIBUTIONS

The staff of Information Plus would like to extend its heartfelt appreciation to the Information Plus Advisory Board. This dedicated group of media professionals provides feedback on the series on an ongoing basis. Their comments allow the editorial staff who work on the project to continually make the series better and more user-friendly. Our top priorities are to produce the highest-quality and most useful books possible, and the Advisory Board's contributions to this process are invaluable.

The members of the Information Plus Advisory Board are:

- Kathleen R. Bonn, Librarian, Newbury Park High School, Newbury Park, California
- Madelyn Garner, Librarian, San Jacinto College—North Campus, Houston, Texas
- Anne Oxenrider, Media Specialist, Dundee High School, Dundee, Michigan
- Charles R. Rodgers, Director of Libraries, Pasco-Hernando Community College, Dade City, Florida
- James N. Zitzelsberger, Library Media Department Chairman, Oshkosh West High School, Oshkosh, Wisconsin

COMMENTS AND SUGGESTIONS

The editors of the *Information Plus Reference Series* welcome your feedback on *Space Exploration: Triumphs and Tragedies*. Please direct all correspondence to:

Editors
Information Plus Reference Series
27500 Drake Rd.
Farmington Hills, MI 48331-3535

CHAPTER 1
INTRODUCTION TO SPACE EXPLORATION

Mankind will migrate into space, and will cross the airless Saharas which separate planet from planet and sun from sun.

—Winwood Reade, 1872

Humans have always been explorers. When ancient peoples stumbled upon unknown lands or seas they were compelled to explore them. They were driven by a desire to dare and conquer new frontiers and a thirst for knowledge, wealth, and prestige. These are the same motivations that drove people of the twentieth century to venture into space.

By definition space begins at the edge of Earth's atmosphere, just beyond the protective blanket of air and heat that surrounds our planet. This blanket is thick and dense near the Earth's surface and light and wispy farther away from the planet. About sixty-two miles above Earth the atmosphere becomes quite thin. This altitude is considered the first feathery edge of outer space.

The very idea of space exploration has a sense of mystery and excitement about it. Americans call their space explorers astronauts. Astronaut is a combination of two Greek words, *astron* (meaning star) and *nautes* (meaning sailor). Thus, astronauts are those that sail amongst the stars. This romantic imagery adds to the allure of space travel.

The truth is that space holds many dangers to humans. Space is an inhospitable environment, devoid of air, food, or water. Everywhere it is either too hot or too cold for human life. Potentially harmful radiation flows in the form of cosmic rays from deep space and electromagnetic waves that emanate from the Sun and other stars. Tiny bits of rock and ice hurtle around in space at high velocities, like miniature missiles.

Space is not readily accessible. It takes a tremendous amount of power and thrust to hurl something off the surface of Earth. It is a fight against the force of Earth's gravity and the heavy drag of an air-filled atmosphere.

Getting into space is not easy, and getting back to Earth safely is even tougher. Returning to Earth from space requires conquering another mighty force—friction. Any object penetrating Earth's atmosphere from space encounters layers and layers of dense air molecules. Traveling at high speed and rubbing against those molecules produces a fiery blaze that can rip apart most objects.

It was not until the 1950s that the proper combination of skills and technology existed to overcome the obstacles of space travel. The political climate was also just right. Two rich and powerful nations (the Union of Soviet Socialist Republics and the United States) devoted their resources to besting one another in space instead of on the battlefield. It was this spirit of competition that pushed humans off the planet and onto the Moon in 1969.

Once that race was over, space priorities changed. Today, computerized machines do most of the exploring. They investigate planets, asteroids, comets, and the Sun. Human explorers stay much closer to Earth. They visit and live aboard a space station in orbit 200 miles above the planet. On Earth people dream of longer journeys because most of space is still an unknown sea, just waiting to be explored.

ANCIENT PERSPECTIVES ON SPACE

Since the earliest days people have looked up at the heavens and dreamed of flying there. In ancient Greek and Roman mythology gods and goddesses rode chariots through the skies or had wings of their own. These included Eros (the god of love), Nike (the goddess of victory), Hermes (the messenger to the gods), and Apollo (god of the arts). In Roman mythology they were called Cupid, Victoria, Mercury, and Apollo, respectively.

One famous Greek tale concerned a young man named Icarus and his father, Daedalus. Imprisoned on an island they decide to escape by building wings of feathers and wax for themselves and flying to freedom. However, Icarus disregards his father's warning against flying too close to the Sun, and the heat melts the wax in his wings. When Icarus's wings fall apart, he plunges to his death in the sea.

In about 160 AD a Greek writer named Lucian of Samosata (c. 120–180) wrote the story *True History* about a sailing ship whisked to the Moon by a giant waterspout. The sailors find the Moon inhabited by strange creatures that are at war with beings living on the Sun. In a later story, *Icaro-Menippus*, an adventurer more successful than Icarus uses eagle and vulture wings to fly to the Moon.

ENLIGHTENED OBSERVATIONS

Centuries later the great Italian painter and engineer Leonardo da Vinci (1452–1519) foresaw the day that humans would fly. He wrote "There shall be wings! If the accomplishment be not for me, 'tis for some other." Leonardo made several sketches of human-powered flying machines and gliders with bird-like wings.

At the time astronomical knowledge was limited, and it was widely believed that the Earth was the center of the universe and everything else revolved around it. The Polish astronomer Nicolaus Copernicus (1473–1543) studied the motion of the heavens and drew different conclusions. In 1543 he published the famous book *De revolutionibus orbium coelestium* (*On the Revolutions of the Celestial Orbs*). Copernicus insisted that Earth and the other planets orbit the Sun. He said "At the middle of all things lies the sun."

During the early 1600s the first telescopes appeared in Europe. Although historians are not sure who invented the telescope, they know that Italian astronomer Galileo Galilei (1564–1642) popularized their use. Galileo also improved the design and power of the telescope. He used several to study the cosmos and published his findings in the 1610 book *Sidereus Nuncius* (*Starry Messenger*).

At about the same time the German astronomer Johannes Kepler (1571–1630) was also studying the solar system. He discovered that planets move according to mathematical rhythms, and he derived laws of planetary motion from these rhythms that are still studied today. In 1593 Kepler wrote Galileo a letter in which he said "Provide ships or sails adapted to the heavenly breezes, and there will be some who will not fear even that void."

The mechanics of spaceflight were explored by English physicist Sir Isaac Newton (1642–1727). Newton first unraveled the mysteries of gravity on Earth and then extended his findings into space. He was the first to explain how a satellite (an orbiting body) could be put into orbit around Earth.

Newton's thought experiment, as it was called, proposed a cannon atop a tall mountain as a theoretical means for putting an object into orbit around Earth. The cannon shoots out projectiles one after another, using more gunpowder with each successive firing. Newton said each projectile would travel farther horizontally than the previous one before falling. Finally, there would be a projectile shot with enough gunpowder that it would travel very far horizontally and when it began to fall toward Earth, its path would have the same curvature as the Earth's surface. The projectile would not fall to Earth's surface but continue to circle around the planet. It would be another two centuries before humans could prove Newton's theory.

It was only a few decades after Newton's death that humans began to fly upon earthly breezes. During the late 1700s two French brothers named Joseph and Etienne Montgolfier built the first hot-air balloons. On November 21, 1783, two noblemen from the French court ascended to 500 feet and sailed across the city of Paris. They landed safely miles from the city and offered champagne to the terrified villagers there. The age of flight had begun.

SPACE TRAVEL IN EARLY SCIENCE FICTION

Science fiction is a category of literature in which an imaginative story is told that incorporates at least some scientific principles to give it a sense of believability. It is a mixture of science and imagination. The term "science fiction" is generally credited to writer Hugo Gernsback (1884–1967), who published a magazine devoted to such stories and started book clubs for science fiction fans. His legacy lives on in the Hugo award, a literary award given each year by the World Science Fiction Society.

Jules Verne

The French author Jules Verne (1828–1905) was one of the earliest science fiction writers to incorporate space travel in his stories. In 1865 he wrote *De la terre a la lune* (*From Earth to the Moon*), a tale of an ambitious gun club in America. The men build a massive cannon in Florida and shoot a metal sphere toward the Moon. Inside are three astronauts who plan to explore the lunar surface. Their target misses the mark, and they wind up orbiting the Moon instead. It was the first space travel story based on physics, rather than pure fantasy.

Five years later Verne published the sequel *Autour de la lune* (*'Round the Moon*). The astronauts use small onboard rockets to propel their sphere safely back to Earth, where it splashes down and floats in the ocean. A nearby ship rescues the men and takes them home to a heroes' welcome. The similarities are striking between these stories and the actual events of the Apollo flights one hundred years later.

Edward Everett Hale

In 1869 American writer Edward Everett Hale published a remarkable science fiction tale in the *Atlantic Monthly*. His tale, "The Brick Moon," describes the work of some clever American inventors who decide to build a large beacon to sit in the sky and glow as a constant reference point for ships at sea. To accomplish this goal the men build a large brick sphere set atop a hill. A track leads down the hill to two giant spinning wheels that are set in a gorge and turned by water from a rushing river. The wheels are to catapult the brick moon into space.

One day the brick moon accidentally rolls away from its restraints and is flung into space with some of the workers and their families aboard. For months their friends on the ground search the night skies with telescopes until finally they spot the satellite in Earth orbit. They are amazed to see the moon inhabitants living happily on the satellite surface. On occasion the inhabitants send signals back to Earth by forming a long line and simultaneously making large and small jumps into the air to spell out messages in Morse code.

The story is prophetic in one interesting respect. During brick moon construction the inventors are plagued by constant design changes, funding problems, and criticism from the public and the press. These difficulties would become common ones for the space programs that later developed.

H. G. Wells

Around the turn of the twentieth century the English author Herbert George Wells (1866–1946) wrote popular space travel stories including *The War of the Worlds* (1898) and *The First Men on the Moon* (1901).

The War of the Worlds featured Martian invaders landing spacecraft near London and terrorizing the population with destructive machines and poisonous gas. In the end the humans prevail when a common germ kills the Martians. During the 1930s the story was made into a radio play and rewritten for an American audience. On October 30, 1938 (the night before Halloween), the play was broadcast as a mock newscast. Unfortunately some listeners thought the "news" was real, and there were scattered incidents of panic.

The First Men on the Moon also included unfriendly aliens, this time on the Moon. Some daring explorers from Earth travel to the Moon and are captured by ant-like creatures called Selenites. The name is derived from Selene, the mythical Greek goddess of the Moon. The story features little actual science and is generally considered more of a romantic adventure tale set in space.

George Méliès

In 1902 the French director Georges Méliès created the first known science fiction motion picture. *Le voyage dans la lune* (*Trip to the Moon*) is an eleven-minute silent film very loosely based on the Jules Verne stories about Moon travel.

This time five brave Frenchmen are catapulted to the Moon where their rocket actually impacts the giant right eye of the "man in the Moon." The astronauts begin exploring, but are captured by unfriendly Selenites. The explorers manage to escape back to their spacecraft and push it off the edge of the Moon to fall back to Earth. They splash down in the sea and return to France as heroes.

THE WRIGHT STUFF

On December 17, 1903, brothers Orville and Wilbur Wright made history at Kitty Hawk, North Carolina, with the first sustained flights of a powered aircraft. Each brother took two flights that day. The longest flight covered about 850 feet and lasted just under one minute. The modern age of aviation had begun.

In 1905 the Fédération Aéronautique Internationale (FAI) was formed in Europe by representatives from Belgium, France, Germany, Great Britain, Italy, Spain, Switzerland, and the United States. The FAI became the official organization for cataloguing and verifying aeronautical feats around the world. Its purpose is "to advance the science and sport of aeronautics."

Aviation got off to a slow start in the United States. The earliest planes were notoriously dangerous, and several aspiring adventurers were killed flying them. Neither the American public nor the federal government was convinced that airplanes were safe and effective. The mood in Europe was much different. Engineers in France, England, and Germany produced their own versions of reliable and versatile aircraft.

FLYING TAKES OFF

When World War I began in Europe in 1914, the power of aerial warfare became apparent immediately. The United States realized that it was behind its European counterparts in aviation expertise, and in an effort to correct the situation the U.S. government formed the Advisory Committee for Aeronautics (ACA) in 1915. Later the word "National" was tacked on, and the agency became known as the NACA. The NACA began an ambitious campaign of research and development into aircraft design and flight theory.

By the end of the war in 1918 the United States had made great progress in the field of aviation, including the development of commercial airlines and postal air services. In addition, public attitudes about flying were beginning to change. The daring feats of such World War I pilots as the American ace Eddie Rickenbacker (1890–1973) and the German legend Manfred von Richthofen (The Red Baron; 1882–1918), had brought an air of excitement to flying.

In 1919 New York City hotel owner Raymond Orteig offered a $25,000 prize to the first aviator that could fly nonstop from New York to Paris or from Paris to New York. The Orteig Prize inspired one of the greatest flying feats of the century. A young man named Charles A. Lindbergh (1902–74) convinced businessmen in St. Louis, Missouri, to finance his attempt to win the prize. "Lucky Lindy" had earned his flying reputation as a successful airmail pilot and barnstormer (one who performs flying stunts at air shows). On the morning of May 20, 1927, he took off from Long Island, New York, in his monoplane the *Spirit of St. Louis*. Thirty-three and a half hours later he landed on a Paris airstrip amid throngs of cheering spectators. Lindbergh was an instant hero, and flying was suddenly of vital interest to Americans, who thrilled to the adventures of such pilots as Amelia Earhart, Wiley Post, Howard Hughes, and Douglas (Wrong-Way) Corrigan.

Meanwhile the NACA continued its work in aviation science. Orville Wright joined the organization and remained a member until his death in 1948. During these decades the NACA drove many developments within the field of aeronautics, except for one—rockets.

PIONEERS OF ROCKET SCIENCE

There are three men in history considered the founders of modern rocket science: Konstantin Tsiolkovsky of Russia, Hermann Oberth of Germany, and Robert Goddard of the United States. All three were working on rocket science during the early years of the twentieth century. Although they were scattered around the world, they reached similar conclusions at about the same time.

Konstantin Tsiolkovsky

Konstantin Tsiolkovsky (1857–1935) was inspired by his love of science and the stories of Jules Verne. The Russian schoolteacher tried his hand at science fiction, before and after becoming a scientist. Tsiolkovsky studied the theoretical questions of rocket flight—gravity effects, escape velocity, and fuel needs. He developed a simple mathematical equation relating the final velocity of a rocket to the initial velocity, the starting and ending mass of the rocket, and the velocity of the rocket exhaust gases. Tsiolkovsky's equation became a fundamental concept of rocket science and is still taught today.

In 1895 Tsiolkovsky wrote *Dreams of Earth and Sky*. The book described how a satellite could be launched into orbit around Earth. Later publications included *Exploration of the Universe with Reaction Machines* and *Research into Interplanetary Space by Means of*

FIGURE 1.1

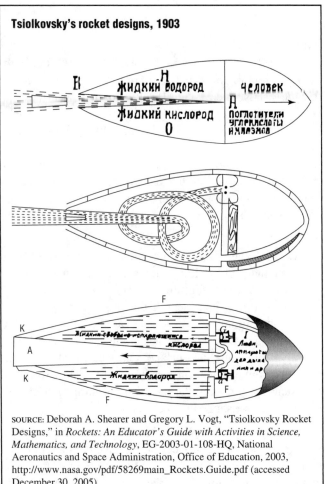

Tsiolkovsky's rocket designs, 1903

SOURCE: Deborah A. Shearer and Gregory L. Vogt, "Tsiolkovsky Rocket Designs," in *Rockets: An Educator's Guide with Activities in Science, Mathematics, and Technology*, EG-2003-01-108-HQ, National Aeronautics and Space Administration, Office of Education, 2003, http://www.nasa.gov/pdf/58269main_Rockets.Guide.pdf (accessed December 30, 2005)

Rocket Power, both published in 1903. Figure 1.1 shows some of Tsiolkovsky's designs for liquid propelled rockets. Two decades later he wrote the *Plan of Space Exploration* (1926) and *The Space Rocket Trains* (1929).

Tsiolkovsky believed that rockets launched into space would have to include multiple stages. That is, instead of having one big cylinder loaded with fuel, the fuel must be divided up amongst smaller rocket stages linked together. As each stage uses up its fuel it could be jettisoned away so the remainder does not have to carry dead weight. Tsiolkovsky reasoned that this was the only way for the mass of a rocket to be reduced as its fuel supply was depleted.

In a 1911 letter Tsiolkovsky predicted, "Mankind will not remain on Earth forever, but in its quest for light and space will penetrate beyond the confines of the atmosphere, at first timidly, and later will conquer for itself all of solar space."

Hermann Oberth

Hermann Oberth (1894–1989) was born in Transylvania, Romania, but later became a German citizen. He also was a fan of Jules Verne. As a teenager Oberth studied mathematics and began developing sophisticated rocket theories. He studied medicine and physics at the University of Munich. During the 1920s he wrote important papers titled *Die Rakete zu den Planetenräumen* ("The Rocket into Interplanetary Space") and *Wege zur Raumschiffart* ("Methods of Achieving Space Flight").

In 1923 Oberth predicted that rockets "can be built so powerfully that they could be capable of carrying a man aloft." He proposed bullet-shaped rockets for manned missions to Mars and an Earth-orbiting space station for refueling rockets. Like his Russian counterpart, Oberth advocated multi-stage rockets fuelled by liquid propellants.

Oberth's writings were hugely popular and influenced movie producer Fritz Lang to make a 1929 movie about space travel called *Die Frau im Mond* (*The Woman in the Moon*). Oberth served as a technical adviser on the film. He also inspired the German rocket club known as *Verein für Raumschiffahrt* (VfR) or Society for Spaceship Travel. The VfR put Oberth's theories into practice, building and launching rockets based on his designs.

As World War II drew near during the 1930s the Nazi government put Oberth and other VfR members to work developing rockets for warfare, rather than for space flight.

Robert Goddard

Robert Goddard (1882–1945) was an American physicist born in Worcester, Massachusetts. In a speech made in 1904 he said, "It is difficult to say what is impossible, for the dream of yesterday is the hope of today and reality of tomorrow"; and Goddard spent the rest of his life making rocket flight a reality. After graduating from college he taught physics at Clark University in his hometown. He also spent time on a relative's farm experimenting with explosive rocket propellants. Unlike his Russian and German counterparts, Goddard's rocket science was more experimental than theoretical. In all, he was granted seventy patents for his inventions. The first two came in 1914 for a liquid-fuel gun rocket and a multi-stage step rocket. He is believed to be the first person to prove experimentally that a rocket can provide thrust in a vacuum.

Much of Goddard's research was funded by the Smithsonian Institution and the Guggenheim Foundation. The U.S. government showed little interest in rocket science except for its possible use in warfare. Late in World War I Goddard presented the military with the concept for a new rocket weapon, later called the bazooka. After the war Goddard worked part time as a weapons consultant to the armed forces.

In 1920 the Smithsonian published Goddard's famous paper, "A Method of Attaining Extreme Altitudes," in which he described how a rocket could be sent to the Moon. The idea was greeted with skepticism from scientists and derision from the media. The *New York Times* published a scornful editorial ridiculing Goddard for this fanciful notion. Goddard was stung by the criticism and spent the rest of his life avoiding publicity. His low-profile approach kept his work from being well known for many years. One person who did take a keen interest in him was the aviator Charles Lindbergh. Lindbergh played a key role in securing funding for Goddard's rocket research from the Guggenheim Foundation (http://www.charleslindbergh.com/rocket/).

On March 16, 1926, Goddard achieved the first known successful flight of a liquid propelled rocket. (See Figure 1.2.) Throughout the next decade he labored quietly in the desert near Roswell, New Mexico, developing increasingly more powerful rockets. In 1935 Goddard launched the first supersonic liquid fuel rocket. (Supersonic means faster than the speed of sound. Sound waves travel at about 700 miles per hour, depending on the air temperature.) A year later Goddard's achievements finally received recognition when the Smithsonian published another of his papers titled "Liquid Propellant Rocket Development."

He continued his work until 1941 when the United States entered World War II. Until his death in 1945 Goddard worked with the military to develop rocket applications for aircraft. Three decades later the *New York Times* finally issued an apology for its 1920 editorial about him. The date was July 17, 1969, and three American astronauts were on their way to the Moon. The newspaper admitted that Goddard had been right after all. The Goddard Space Flight Center near Washington, D.C., and the Goddard crater on the Moon are both named after him.

Wernher von Braun

World War II (1939–45) ushered in the rocket age. The Nazi government of Germany was eager to use rockets against its enemies. The great talents and minds of the VfR were directed to forget about space travel and concentrate on warfare. During the early 1940s Germany developed the most sophisticated rocket program in the world. At its helm was a brilliant young man named Wernher von Braun (1912–77).

Von Braun had been an assistant to Hermann Oberth during the 1930s and an active member of the VfR. He was put in charge of developing a rocket weapon to terrorize the British population. Von Braun's team included Oberth and hundreds of people who worked at a remote island called Pennemünde. They developed the

FIGURE 1.2

Goddard's rocket design, 1926

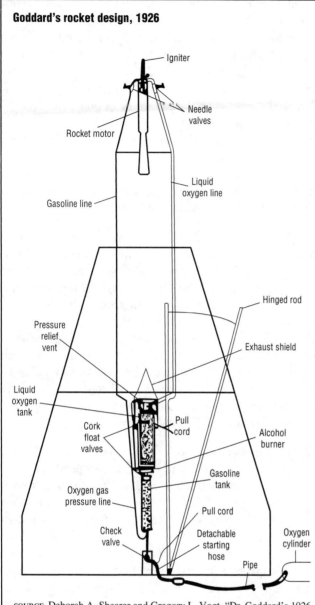

Igniter

Needle valves

Rocket motor

Liquid oxygen line

Gasoline line

Hinged rod

Pressure relief vent

Exhaust shield

Liquid oxygen tank

Cork float valves

Pull cord

Alcohol burner

Gasoline tank

Oxygen gas pressure line

Pull cord

Check valve

Detachable starting hose

Oxygen cylinder

Pipe

SOURCE: Deborah A. Shearer and Gregory L. Vogt, "Dr. Goddard's 1926 Rocket," in *Rockets: An Educator's Guide with Activities in Science, Mathematics, and Technology*, EG-2003-01-108-HQ, National Aeronautics and Space Administration, Office of Education, 2003, http://www.nasa.gov/pdf/58269main_Rockets.Guide.pdf (accessed December 30, 2005)

rocket-powered *Vergeltungswaffens* or weapons of vengeance. They were called V weapons, for short.

There were two series of V weapons. The V-1 carried a ton of explosives and traveled at a top speed of about 400 miles per hour. This was slow enough that British gunners could blow apart the V-1's as they descended through the air. Although thousands of V-1's were launched against England, roughly half of them never impacted the ground.

Far more lethal was the second V weapon called the V-2. This was truly a rocket with a top speed around 2,000 miles per hour. The V-2's traveled far too fast to be shot down and terrified the British public. Approximately 1,000 V-2 rockets rained down on England during World War II, killing 115,000 people.

On September 8, 1944, the first V-2 rocket fell on London. Von Braun reportedly turned to his colleagues and said "the rocket worked perfectly, except for landing on the wrong planet." The tide had already turned against Germany. By early 1945 the country was being invaded by the Soviets from the east and the Allies from the west. Von Braun moved his team near the Germany-Switzerland border to be in position to surrender to American forces.

A negotiated surrender was worked out in which von Braun turned over himself, people on his team, and vital plans, drawings, rocket parts, and documents. In exchange the U.S. Army agreed to transport the team to America and fund their work on an American rocket program. The Army called the agreement Operation Paperclip. They had no way of knowing that this move was going to put Americans on the Moon.

THE X SERIES

Even before World War II ended the United States began developing rocket-powered planes. In 1943 the NACA initiated the research program in conjunction with the Air Force and Navy. Because the planes were experimental, they were given the name X-aircraft. In 1944 a company called Bell Aircraft began work on the X-1. At first it was called the XS-1, with the "S" standing for supersonic. Later the "S" was dropped from the name.

On October 14, 1947, a young Air Force captain named Charles (Chuck) Yeager (1923–) flew the X-1 at the speed of sound, which is known as Mach 1. The X-1 was only the first of many high-performance planes tested in the program. Eventually X planes flew at hypersonic speeds, that is, speeds greater than Mach 5. The X-15 was a rocket-fueled plane tested during the late 1950s and early 1960s. It was taken up to an altitude of approximately 45,000 feet by a carrier plane, a B-52 aircraft, and released. A rocket engine was then fired to propel the X-15 to incredible speeds and heights. An X-15 flew at Mach 6.04, the fastest suborbital speed ever reached, on November 9, 1961. On August 22, 1963, an X-15 soared across the boundary into space to an altitude of sixty-seven miles. This record would remain unbroken for more than four decades.

The X-series were high-speed, high-altitude planes unlike any ever built before. Most of them were tested over desolate desert areas near Muroc, California. Daring young test pilots flew the X-series planes. Unfortunately this was a very dangerous profession. Numerous pilots were killed or seriously injured while testing X-series planes. The pilots who survived became the first men considered for the nation's astronaut program.

A COLD WAR IN SPACE BEGINS

The term Cold War is used to describe U.S. relations with the Soviet Union from the end of World War II to the late 1980s. During this period, chiefly marked by a mutual mistrust and rivalry that led to a buildup of arms, both nations developed extensive nuclear weapons programs. Each thought the other was militarily aggressive, deceitful, and dangerous. Each feared the other wanted to take over the world. This paranoia was in full force when space travel began.

The International Geophysical Year

In 1952 a group of American scientists proposed that the International Council of Scientific Unions (ICSU) should sponsor a worldwide research program to learn more about Earth's polar regions. Eventually the project was expanded to include the entire planet and the space around it. The ICSU decided to hold the project between July 1957 and December 1958 and call it the International Geophysical Year (IGY). Geophysics is a branch of earth science concerned with physical processes and phenomena in the Earth and its vicinity.

The IGY time period was selected to coincide with an expected phase of heightened solar activity. Approximately every eleven years the Sun undergoes a one- to two-year period of extra radioactive and magnetic activity. This is called the solar maximum. The ICSU hoped that rocket technology would progress enough to put satellites in Earth orbit during the next solar maximum and collect data on this phenomenon.

Sixty-seven countries participated in various ways in the IGY project. The American delegation to the ICSU was led by the National Academy of Sciences (NAS). The NAS put together a team of scientists from businesses, universities, and private and military research laboratories to conduct American activities during the IGY.

Ballistic Missiles

Following World War II both the United States and the Soviets began researching the feasibility of attaching warheads to long-range rockets capable of traveling half way around the world. These weapons were eventually called intercontinental ballistic missiles, or ICBMs. They could be equipped with conventional or nuclear warheads. The United States had introduced the nuclear warfare age by dropping atomic bombs on Japan to end World War II in August 1945.

By the early 1950s the U.S. Air Force was actively testing three different ICBMs under the Navaho, Snark, and Atlas programs. This work was highly classified as a matter of national security. The United States and Soviet Union both engaged in massive spying campaigns throughout the Cold War. In 1955 American spies brought word that the Soviets were close to completing ICBMs capable of reaching U.S. cities.

The Soviet rocket work was spearheaded by Sergei Korolev (1906–66). He oversaw development of the R-7, the world's first ICBM, and is considered the father of the Soviet space program.

Sputnik 1

The United States worked throughout the mid-1950s to construct a successful science satellite for the IGY. This work proceeded separately from ICBM development. However, at the time only the military had the expertise and resources to build rockets capable of leaving Earth's atmosphere. The Navy was charged with developing a rocket capable of carrying a package of scientific instruments into Earth orbit. In 1957 testing was still ongoing and proceeding poorly when the United States got a shock.

On the evening of October 4, 1957, the Soviet Union news service announced that the nation had successfully launched the first-ever artificial satellite into Earth orbit. It was called Sputnik, which means "companion" in English. The word also translates as "satellite," because a satellite is Earth's companion in an astronomical sense. The satellite weighed 184 pounds and was about the size of a basketball. It had been launched atop an R-7 Semiorka rocket. The satellite circled the Earth every ninety-eight minutes.

A Secret Surprise

The launch announcement of Sputnik 1 was a huge disappointment and surprise to American scientists. They knew their Soviet counterparts were working on a science satellite for the IGY but had no idea the Soviets had progressed so quickly. The American scientists had openly shared information about their research during ICSU meetings. On the other hand, the Soviet government forbade its scientists from disclosing any details about their work. Sputnik had been developed and launched in near total secrecy. According to an account in Time, Lloyd Berkner, the president of the ICSU, learned about the launch while at a dinner party at the Soviet embassy in Washington, D.C., when a reporter from the New York Times whispered the news to him (http://www.time.com/time/80days/571004.html).

The American public was even more shocked by the announcement. Millions went outside in the darkness to look for the satellite in the night sky. Witnesses said it was a tiny twinkling pinpoint of light that moved steadily across the horizon. The satellite continuously broadcast radio signals that were picked up by ham radio operators

FIGURE 1.3

Explorer I in space

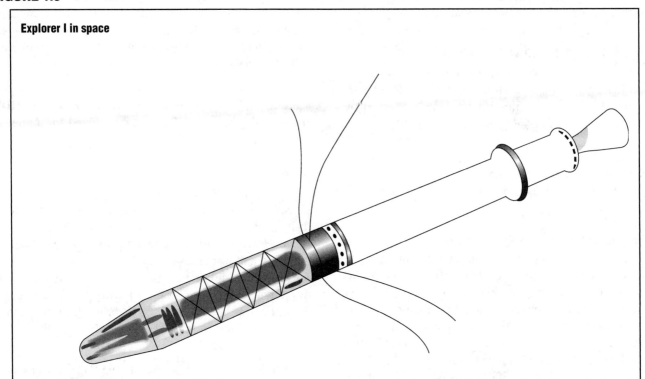

SOURCE: Gregory L. Vogt, "Artist's Concept of Explorer I in Space," in *Space-Based Astronomy: An Educator's Guide with Activities in Science, Mathematics, and Technology*, EG-2001-01-122-HQ, National Aeronautics and Space Administration, Office of Space Science, 2003, http://www.nasa.gov/pdf/58277main_Space.Based.Astronomy.pdf (accessed December 30, 2005)

all over the world. Ham radio is communication using short-wave radio signals on small amateur stations.

The *Sputnik 1* signals were another unpleasant surprise for American scientists. It had been universally agreed that IGY satellites would broadcast radio signals at a frequency of 108 megahertz. The United States had already built a satellite tracking system designed for that frequency. *Sputnik 1* transmitted at much lower frequencies, ensuring that U.S. scientists would not be able to pick up its data.

Sputnik 2—A Dog in Space

The success of *Sputnik 1* caught America off-guard and unprepared. The public realized for the first time that the Soviets probably had the capability to launch long-range nuclear missiles against the United States. Only a month later there was even further dismay when the Soviets launched a second Sputnik satellite.

Sputnik 2 was much larger than its predecessor and carried a live dog, a huskie-mix named Laika, into orbit. The American press nicknamed her "Muttnick." It was a one-way trip for her as the Soviet scientists had not yet worked out how to bring the spacecraft safely back to Earth. At the time the Soviet news agency bragged that Laika survived for a week aboard the spacecraft. Decades later scientists admitted that Laika died only hours after

launch when she panicked and overheated in her tiny cabin.

America Reacts

The American public was scared by the size of *Sputnik 2*. It weighed more than 1,000 pounds. It was common knowledge that the United States did not have a rocket capable of carrying that much weight into space. There was an uproar in the media, and politicians demanded to know how the Soviet Union had gotten so far ahead of the United States in space technology. President Dwight Eisenhower charged the U.S. military to do whatever it took to put a satellite in space.

The U.S. Navy's efforts to build a satellite had proved unsuccessful. The military turned to Wernher von Braun and his team of rocket scientists working for the U.S. Army. On January 31, 1958, the first American satellite soared into orbit. It was named *Explorer I* and rode atop a Jupiter-C rocket developed by the von Braun team at Huntsville, Alabama. (See Figure 1.3.)

A few months later the Soviets answered with *Sputnik 3*, a miniature physics laboratory sent into orbit to collect scientific data.

In October 1958 the United States formed a new federal agency to oversee the nation's space endeavors. It was named the National Aeronautics and Space Administration

(NASA). Although it was a civilian agency charged with operating peaceful missions in space, NASA would rely heavily on military resources to achieve its goals.

FIRST MAN IN SPACE

On April 12, 1961, cosmonaut Yuri Gagarin (1934–68) became the first human ever to travel beyond Earth's atmosphere, enter the frontier of space, and return safely to Earth. Gagarin was born in a village near Gzhatsk (now Gagarin) in central Russia. He grew up in a peasant family, dreaming of becoming a pilot. Before being recruited to be a cosmonaut, Gagarin was serving as a lieutenant in the Soviet air force.

His flight took him roughly 200 miles above Earth and he spent 108 minutes circling the planet, completing one entire orbit and part of another one. His cramped spacecraft was equipped with a radio for communicating with ground control. Looking down at the planet beneath him he reported, "The Earth is blue. How wonderful. It is amazing."

The weightlessness bestowed by space travel had always been a worry for scientists. There is a common misconception among the public that there is no gravity in space. This is not true. Actually the force of gravity remains very strong for great distances around Earth. Objects and people that leave Earth's atmosphere experience weightlessness, because they are in free fall toward Earth throughout their trip.

At the time of Gagarin's flight scientists were not sure how the human body would react to weightlessness. His spacecraft included a computerized automatic pilot, in case Gagarin lost consciousness or was unable to move. This fear proved to be unfounded. The mission showed that humans can not only withstand weightlessness but can function quite well in it.

Gagarin returned to Earth safely. He ejected from his spacecraft somewhere over Russia and parachuted to the ground. He became a national hero and an international sensation. His picture was on the front page of every major newspaper in the world.

The scientific teams in the United States were impressed with Gagarin's accomplishment but also envious of it. A NASA spokesman congratulated the Soviets for their achievement and summed up the U.S. space program with these glum words: "So close, but yet so far." In Huntsville, Alabama, rocket scientist Wernher von Braun was more blunt, saying, "To catch up, the U.S.A. must run like hell."

RACE TO THE MOON

It was a month later that the first American entered space. On May 5, 1961, Alan Shepard (1923–98) soared to an altitude of 116 miles in a spaceship named *Freedom 7*. He spent fifteen minutes and twenty-eight seconds in a suborbital flight. Suborbital means less than one orbit. In other words, a suborbital flight does not complete an entire circle around Earth. Shepard's flight was much shorter in distance and time than Gagarin's flight had been. A few months later the second cosmonaut in space, Gherman Titov (1935–2000), completed seventeen and a half orbits around Earth. NASA knew it would be a year or more before it could accomplish a similar feat.

The United States was tired of coming in second place. Because there was no way to beat the Soviets at the orbital space race, President Kennedy decided to start a new race—one where both sides would start even. His advisers recommended that the United States put a manned spacecraft in orbit around the Moon or even land a man on the Moon. Either one would require development of a huge new rocket to supply the lifting power needed to boost a spaceship out of Earth orbit. Neither the Soviets nor the Americans had such a rocket.

On May 25, 1961, President John F. Kennedy revealed his choice to the world in a speech called the "Special Message to the Congress on Urgent National Needs." It is commonly known as the Moon challenge speech. His words ignited the biggest race in human history: "First, I believe that this nation should commit itself to achieving the goal, before this decade is out, of landing a man on the moon and returning him safely to the earth. No single space project in this period will be more impressive to mankind, or more important for the long-range exploration of space; and none will be so difficult or expensive to accomplish."

AIMING FOR DRY SEAS

Suddenly all eyes were on the Moon. Earth's closest neighbor had been a subject of fascination since the first humans gazed up at the night sky.

Most of the features on the Moon were named during the 1600s by an Italian astronomer named Giovanni Riccioli (1598–1671). Riccioli was a Jesuit priest, a member of the Roman Catholic order the Society of Jesus, which is devoted to missionary and educational work. At the request of the church Riccioli devoted his life to astronomy and telescopic studies. At the time the writings of Kepler and Copernicus were popular and controversial. In keeping with church doctrine, Riccioli disputed Copernicus's claim that the Earth was not the center of the universe.

Despite this gross error Riccioli's work proved very useful to later scientists. He published a detailed lunar map that he developed with fellow Jesuit and Italian physicist Francesco Grimaldi (1613–63). This map featured Latin names for lunar features. Elevations and depressions were named after famous astronomers and philosophers. Large, dark, flat areas that looked like bodies of water were named Oceans or Seas.

Four hundred years later humans on opposite sides of Earth took aim at these features. During the early and mid-1960s NASA and the Soviets sent dozens of photographic probes to take pictures of the Moon. Some probes proved successful, and some did not. Four NASA probes crashed into the Moon, but they had beamed back valuable photographs before impacting the lunar surface. In 1966 the Soviet probe *Luna 9* softly set down in the Oceanus Procellarum (Ocean of Storms), the largest of the lunar "seas." Four months later NASA's *Surveyor 1* probe landed nearby.

Both countries needed lunar data to support their efforts to send humans to the Moon. During this time Soviet officials did not even acknowledge that they had a manned lunar program. Those in the U.S. program suspected that they did but could not be sure. It was not until years later, when Sergei Leskov recounted the story in "How We Didn't Get to the Moon" in the Russian publication *Izvestiya* (August 18, 1989) that the United States learned how badly the Soviets had tried to beat them to the Moon.

HARD WORK

The U.S. effort to put men on the Moon was named the Apollo program. It actually included three phases:

- Mercury—Suborbital and orbital missions of short duration.

- Gemini—Longer duration orbital missions including extravehicular activity (space walking) and docking of spacecraft in space.

- Apollo—Manned lunar landings in which a module containing two astronauts softly lands on the Moon. A third astronaut remains in lunar orbit while the other two explore the Moon's surface.

Alan Shepard's historic flight of 1961 was considered the first Mercury mission. Over the next two years five more successful Mercury flights were conducted. In 1965 a series of ten manned Gemini missions began. They were completed near the end of 1966.

Soon after it started, it became apparent that the Moon program was going to be expensive. On September 12, 1962, President Kennedy reinforced his commitment to the project during a speech at Rice University in Houston, Texas. Kennedy said, "We choose to go to the moon. We choose to go to the moon in this decade and do the other things, not because they are easy, but because they are hard, because that goal will serve to organize and measure the best of our energies and skills, because that challenge is one that we are willing to accept, one we are unwilling to postpone, and one which we intend to win."

ROCKETS ARE KEY

One key goal in the United States and the Soviet Union was development of a large and powerful rocket—a so-called super booster. NASA called its superbooster a Saturn rocket. The Soviets named their rocket the N-1.

Development of the Saturn rocket series began in 1961 under the direction of Wernher von Braun. He had actually been pitching the idea to the military for several years. Before the Apollo program NASA utilized relatively small rockets capable of lifting a few hundred to a few thousand pounds into Earth orbit.

The Scout rocket was used to launch small satellites and probes weighing up to 300 pounds. It was devised by combining aspects of rockets used by the armed forces (Polaris and Vanguard rockets of the Navy and the Sergeant rockets of the Army). The Thor, Atlas, and Titan series evolved from Air Force rockets first developed as ICBMs. During the mid-1960s the military replaced most of its Atlas rockets with Minuteman missiles. Modified Atlas rockets were used to launch satellites and for the Mercury program. The Titan II was used during the Gemini program.

The Saturn series evolved from von Braun's Jupiter rockets. Legend has it that the Saturn got its name because it was one step beyond the Jupiter rocket, just as Saturn is the next step beyond Jupiter in the solar system. The Saturn 5, with a height of 364 feet (111 meters) and a diameter of thirty-three feet (ten meters), is the largest rocket ever built. (See Figure 1.4.) It had to be to push the 100-ton Apollo spacecraft toward the Moon.

APOLLO: TRAGEDY AND TRIUMPH

The Soviet space program continued to flourish. In September 1968 an unmanned probe called *Zond 5* became the first spacecraft to travel around the Moon and return to Earth. The pressure was on NASA to speed up the Apollo missions.

On January 27, 1967, three American astronauts—Virgil "Gus" Grissom, Edward White, and Roger B. Chaffee—were killed when a flash fire raced through their capsule during a routine practice drill. They were the first human casualties of the space program. To honor their memory their tragic mission was named *Apollo 1*. The tragedy stunned the nation. Although some politicians called for the program to end, Apollo continued.

On October 11, 1968, the next Apollo mission was launched. *Apollo 7* successfully conducted a flight test and returned to Earth. It was followed in rapid succession by the more ambitious missions of *Apollo 8, Apollo 9,*

FIGURE 1.4

Saturn 5 rocket being transported to launch pad

SOURCE: Deborah A. Shearer and Gregory L. Vogt, "Saturn 5 Rocket Being Transported to the Launch Pad," in *Rockets: An Educator's Guide with Activities in Science, Mathematics, and Technology*, EG-2003-01-108-HQ, National Aeronautics and Space Administration, Office of Education, 2003, http://www .nasa.gov/pdf/58269main_Rockets.Guide.pdf (accessed December 30, 2005)

and *Apollo 10*, each of which tested a lunar or command module in lunar or Earth orbit. The mission to set humans on the Moon was named *Apollo 11* and scheduled for July 1969.

By this time the Soviets had desperately tried to get their own manned lunar program going. However, the N-1 rocket kept failing its launch tests. The Soviets realized that it would not be ready before the *Apollo 11* launch. Still hoping to steal some of the thunder from the Americans, the Soviets launched a robotic probe named *Luna 15* to the Moon. It was designed to gather samples from the lunar surface and return to Earth before the *Apollo 11* expedition. Launched on July 13, 1969, *Luna 15* completed fifty-two Moon orbits before it crashed into the lunar surface on July 21, 1969, and was lost.

Meanwhile, on July 20, 1969, *Apollo 11* set down safely on the Moon near the Sea of Tranquility. Late that evening astronaut Neil Armstrong stepped out of the spacecraft to become the first human ever to stand upon the Moon. Approximately half a billion people on Earth watched the historic event on television. Four days later the *Apollo 11* crew returned to Earth to a hero's welcome. America had won the space race.

There were six more Apollo missions to the Moon before the program ended in 1972.

THE RIGHT STUFF

NASA's space program introduced a new kind of hero to American culture—the astronaut. When the

Mercury program began NASA selected seven men to be astronauts. They were called the "Mercury Seven." The men were all successful military test pilots known for their bravery and professional piloting skills:

- M. Scott Carpenter (1925–)
- L. Gordon "Gordo" Cooper, Jr. (1927–2004)
- John H. Glenn, Jr. (1921–)
- Virgil "Gus" Grissom (1926–67)
- Walter Schirra, Jr. (1923–)
- Alan Shepard, Jr.
- Donald "Deke" Slayton (1924–93)

The men had to pass strenuous batteries of physical, mental, and medical tests to become astronauts and begin their training to go into space. To the American public the Mercury Seven captured the bold and daring spirit of famous flyers like the Red Baron and Charles Lindbergh. They were instant superstars and began receiving thousands of fan letters. Once NASA realized the great popularity of the astronauts they used them as good-will ambassadors for the agency. Astronauts traveled throughout the country speaking to civic groups and clubs to elicit public support for the space program.

NASA scientists originally envisioned astronauts as mere guinea pigs for space experiments. They were intended to be passive passengers covered with medical sensors and sealed inside space capsules completely controlled by operators on the ground through onboard computers. The astronauts rebelled at this notion and insisted on many changes, including installation of windows and manual piloting controls on the space capsules. When the Gemini program began, NASA selected nine more astronaut candidates and soon dozens after that. By the end of the Apollo program thirty-four American astronauts had traveled into space.

In September 1979 the story of the original Mercury Seven was profiled in a book titled *The Right Stuff* by Tom Wolfe. In the book Wolfe describes the tremendous pressures put upon the first astronauts during the space program, their dedication to serving their country, and how they reacted to fame and glory. In 1983 the book was made into a popular movie of the same name.

DÉTENTE IN SPACE

During the early years of space flight, American relations with the Soviet Union were at their worst. Only months after the Soviets put their first cosmonauts in space the Soviet Premier Nikita Khrushchev made a veiled threat: "We placed Gagarin and Titov in space, and we can replace them with other loads that can be directed to any place on Earth." The meaning was clear to the American public. The Soviet Union's powerful rockets could carry nuclear warheads just as easily as they carried humans.

In October 1962 U.S. spy planes captured photographs of nuclear missile installations being built by the Soviets on the island of Cuba, only ninety miles from the coast of Florida. President Kennedy and his advisers considered their options to stop this threat: lodging diplomatic protests, attacking and destroying the facilities, or blockading the seas around Cuba to block Soviet ships from reaching it. Diplomacy had proved ineffective with the Soviets, and a military attack could launch an all-out nuclear war. In the end Kennedy chose the blockade.

The Soviets protested angrily and shot down an American spy plane as it flew over Cuba, killing the pilot. The situation became very tense, and nuclear war seemed imminent. Finally, the two sides agreed that the Soviets would dismantle the bases in Cuba, and the United States would remove Jupiter missiles from Turkey. The Cuban Missile Crisis, as it was later called, is now considered one of the most dangerous events of the entire Cold War.

Détente is a French word that means a relaxation of strained relations. The United States and the Soviet Union occasionally enjoyed periods of détente during the Cold War, particularly in their space activities. A few months prior to the Cuban Missile Crisis the two nations agreed to cooperate in launching some meteorological and other science satellites and share the data obtained. In 1965 a joint project was undertaken in which U.S. and Soviet scientists shared information they had learned about space biology and medicine.

In 1967 the two countries negotiated a United Nations Treaty regarding the application of international law to outer space. It was called the *Treaty on Principles Governing the Activities of States in the Exploration and Use of Outer Space, including the Moon and Other Celestial Bodies*. It is more commonly known as the Outer Space Treaty.

The treaty provides a basic framework for activities allowed and not allowed in space and during space travel. The main principles are:

- Nations cannot place nuclear weapons or other weapons of mass destruction in Earth orbit or elsewhere in space.
- Outer space is open to all humankind and all nations for exploration and use.
- Outer space cannot be appropriated or claimed for ownership by any nation.
- Celestial bodies can only be used for peaceful purposes.

- Nations cannot contaminate outer space or celestial bodies.

- Astronauts are "envoys of mankind."

- Nations are responsible for all their national space activities whether conducted by governmental agencies or non-governmental organizations.

- Nations are liable for any damage caused by objects they put into space.

On January 27, 1967, the Outer Space Treaty was signed by the United States, the Soviet Union, and the United Kingdom. Over the next four decades it would be signed by more than one hundred nations.

In 1969 NASA proposed development of U.S. and Soviet spacecraft that could dock with each other in space for future missions of mutual interest. In July 1975 the docking procedure proved to be successful during the Apollo-Soyuz Test Project. The mission was considered largely symbolic, and many people considered it wasted money that could have been spent on space exploration.

Near the end of the Apollo program the two countries agreed to a number of cooperative projects including sharing of lunar samples, weather satellite data, and space medical data.

PARKED IN LOW EARTH ORBIT

In the minds of most Americans the space race was over the day *Apollo 11* set down on the Moon. Although NASA carried out six more Apollo missions, public interest and political support for them faded quickly. Neither the U.S. nor Soviet government was interested in racing to somewhere else in space. Both governments decided to concentrate on putting manned scientific space stations in low Earth orbit (LEO).

LEO is approximately 125–1,200 miles above Earth's surface. Below this altitude air drag from Earth's atmosphere is still dense enough to pull spacecraft downward. Just above LEO space lies a thick region of radiation known as the inner Van Allen radiation belt. This region poses a hazard to human life and to sensitive electronic equipment.

LEO was and is the orbit of choice for most satellites and for all crewed missions. Spacecraft in LEO travel at about 17,000 miles per hour and circle the Earth once every ninety minutes or so.

Soviet Programs

Between 1971 and 1986 the Soviets put eight space stations into LEO or just below it. These included stations called *Salyut 1* through *Salyut 7* and the more ambitious *Mir* station. Dozens of cosmonauts visited and inhabited the stations, often for many months. The Soviets repeatedly set and broke human space duration

records at their stations. In 1995 cosmonaut Valeri Polyakov (1942–) completed a 438-day mission aboard the *Mir* station. Even in 2004 this stood as the longest period of time spent in space by any human. Eventually all of the Soviet stations fell out of orbit and were destroyed by reentry to Earth's atmosphere. Although none were meant to be "permanent," the *Mir* station did stay in orbit for fifteen years.

U.S. Programs

In 1973 NASA launched its own series of space stations called *Skylab 1* through *Skylab 4*. They orbited within LEO at an altitude of 268–270 miles above Earth. NASA wanted to build a very large space station in which to conduct scientific investigations in LEO. Without political support the agency had to put this plan on hold. Instead NASA concentrated on a new type of reusable space plane called a space shuttle. The space shuttle was to be the workhorse of the American space program, ferrying astronauts and supplies back and forth to the space station, once it was built.

NASA got the funding it needed to develop the shuttle by promising to build a vehicle that could carry military, weather/science, and commercial satellites into LEO. Prior to the shuttle program all satellites were launched aboard expendable rockets that could not be reused. The reusability of the shuttle was one of its best selling points. Also, each shuttle could carry a crew of five to seven people that could conduct scientific experiments in LEO and deploy and repair satellites as needed.

Despite a number of design challenges the first space shuttle was ready for flight by 1981. On April 12, 1981, the first test mission was conducted. Before the end of the year a space shuttle carried an orbiting solar observatory into LEO. Two dozen more missions were carried out before disaster struck in 1986. By this time there were four space shuttles in NASA's fleet. Missions were rotated between the vehicles in order to perform needed maintenance and repairs.

On January 28, 1986, the space shuttle *Challenger* exploded seventy-three seconds after liftoff. All seven crewmembers aboard were killed. The shuttle fleet was grounded for more than two years, while NASA restructured the program and redesigned key elements of the spacecraft. In October 1988 space shuttle flights resumed once again.

By this time the Soviet Union was politically disintegrating. Within three years America's former archenemy had splintered into dozens of individual republics, and the Cold War was over. The republic of Russia took over the space program begun by the Soviet Union. The Soviet space program came under the operation of a new government agency named Rosaviakosmos.

International Plans

NASA and Rosaviakosmos entered a new era of cooperative space ventures. American astronauts visited the space station *Mir* and Russian cosmonauts traveled aboard U.S. space shuttle missions. In 1993 the United States invited Russia to join in building an *International Space Station* (*ISS*) to be put in LEO. The Russians agreed. The *ISS* program eventually included Canada, Japan, and eleven European nations as full partners and Brazil as a contributing partner.

In 1998 *ISS* construction began. The station was designed for continuous human inhabitation and detailed scientific investigations. The Russians and Americans took turns adding components to the station and crewing it with astronauts and cosmonauts. All transport of heavy matériel was delegated to the U.S. space shuttle fleet. The Russians did not have a spacecraft capable of carrying heavy weights into LEO. Throughout 1999 and the next three years nearly all shuttle missions were devoted to *ISS* construction.

During the first mission of 2003 another space shuttle was lost in an accident. On February 1, 2003, the shuttle *Columbia* disintegrated during reentry over the western United States. Again, seven crewmembers were killed. The shuttle fleet was grounded again, and *ISS* construction came to an immediate halt. As of February 2006 the partially constructed station is sitting in LEO with only two crewmembers aboard. Since the *Columbia* disaster Russian spacecraft have ferried supplies to the station and handled crew changes.

In July 2005 the space shuttle program resumed operations with a successful return-to-flight test mission. However, its future and that of the *ISS* are uncertain due to a shift in American space goals announced in February 2004 by President George W. Bush. After sticking with LEO missions for more than thirty years, the United States has a new goal for its human explorers: to return to the Moon and travel to Mars and beyond. It remains to be seen whether the American public will support this expensive program. However, NASA continues to plan for the future.

SPACE-AGE SCIENCE FICTION

The advent of the space age introduced a wealth of information to science fiction authors. They were able to produce works much more sophisticated than those of the past.

One of the most innovative of these authors was Gene Roddenberry (1921–91). During the mid-1960s he created a television show called *Star Trek*. This was a futuristic tale about space exploration set in the twenty-third century. A mixed crew of humans and aliens travel around the galaxy in the starship *Enterprise*. The tele-vision show was not popular during its original run but developed a loyal fan base over the next few decades and spawned a number of movies.

In 1974 more than 100,000 *Star Trek* fans wrote to the U.S. government requesting that one of the newly developed space shuttles be named *Enterprise*. NASA gave the name to the prototype shuttle model used for flight testing.

Another notable science fiction work of the 1960s was the 1968 film *2001: A Space Odyssey*, based on a story by Arthur C. Clarke. Astronauts exploring the Moon find a mysterious artifact. Believing that it came from Jupiter they set off for that planet on an amazing spacecraft. The ship is equipped with a supercomputer named HAL that malfunctions and turns against the human crew. The film features little dialogue, but became a hit for its very imaginative plot and spectacular views of futuristic space travel.

In 1977 the science fiction film *Star Wars* debuted and became one of the most popular movies of all time. Set "a long time ago in a galaxy far, far away," the film tells the story of an adventurous young man who leaves his home world to join a band of rebels fighting against a tyrannical empire. The movie was renowned for its story, characters, adventure, and special effects. The *Star Wars* franchise went on to include five more highly successful films and a book series.

Hollywood movies featuring hostile space aliens invading Earth were a staple of 1950s pop culture. Such films captured the paranoia and fear that Americans felt about the communist threat from the Soviet Union. Beginning in the 1970s a kinder, gentler viewpoint of aliens emerged in movies such as *Close Encounters of the Third Kind* (1977), *ET: The Extraterrestrial* (1982), *Cocoon* (1985), and *Contact* (1997). However, horrific and murderous aliens remain a staple of science fiction films, as evidenced in the popularity of *Alien* (1979) and its sequels, *Independence Day* (1996), and *War of the Worlds* (2005).

ROBOTIC SPACE EXPLORERS

Space programs centered on human explorers are very expensive. It is far cheaper to build and send mechanized (robotic) spacecraft to do the exploring. During the 1960s the Apollo program dominated the spotlight, but it was not the only space exploration project in operation.

Beginning in 1962 NASA launched robotic probes that flew by Mercury, Venus, or Mars and beamed back photographs of them. During the 1970s more sophisticated robotic spacecraft landed on Mars or were sent to fly by the outer planets (Jupiter, Saturn, Uranus, Neptune, and Pluto). These missions were given heroic names, including Mariner, Viking, and Voyager.

In 1990 a robotic spacecraft called *Magellan* set down on the surface of Venus. It was named after Ferdinand Magellan (c. 1480–1521), the Portuguese explorer who led the first sailing expedition to circle the world. In 1995 a spacecraft named after Galileo Galilei began orbiting Jupiter.

Interplanetary exploration is tough, even for machines. During the 1990s NASA lost three robotic spacecraft on their way to Mars. Another one safely made it to the Martian surface. In the early 2000s NASA sent two more missions to Mars. The first was called *Mars Odyssey* and included a probe that went into orbit around the planet in 2001. It was joined two years later by a European Space Agency (ESA) probe called the *Mars Express*. Unfortunately, a lander from this mission was lost on its way to the surface. In 2004 NASA's *Mars Exploration* spacecraft successfully put down two rovers on Mars named *Spirit* and *Opportunity*. As of February 2006 these two rovers continued to explore the surface of Mars while the three surviving orbiters circled overhead.

In 2004 NASA's *Cassini* spacecraft went into orbit around Saturn after a seven-year journey from Earth. The orbiter released the ESA-provided *Huygens* probe, which provided the first-ever close-up photographs of Titan, Saturn's largest moon. In 2005 two new robotic missions were launched toward other planets: NASA's *Mars Reconnaissance Orbiter* and the ESA's *Venus Express*. They were expected to reach their destinations in 2006.

Not all space exploration requires long-distance travel. Advances in computers and telescopes have allowed scientists to do a lot of exploring with robotic spacecraft stationed nearby Earth. Dozens of these high-technology machines take photographs, measure radiation waves, and collect data on galactic and solar phenomena.

The latest generation of robotic explorers are designed to snatch samples in outer space and return them to Earth. The first such mission to return was conducted by a NASA spacecraft named *Genesis*. In 2004 it crash-landed in the Utah desert following an equipment malfunction during reentry. Fortunately, some of its valuable cargo was saved—samples of the solar wind (charged particles emitted from the sun) collected a million miles from Earth. Also in 2004 the NASA spacecraft *Stardust* sailed nearby the comet Wild 2 as it journeyed around the sun. *Stardust* collected dust particles believed to be 4.5 billion years old and returned them safely to Earth in January 2006.

In November 2005 Japan's *Hayabusa* spacecraft landed on the asteroid Itokawa between Earth and Mars to collect dust samples. Equipment and communication problems have plagued the mission, and scientists are hopeful, but not certain, that the samples were collected and will return to Earth in 2010.

APPLICATION SATELLITES

Application satellites are spacecraft put into Earth orbit to serve as tools of earth science or for navigation, communication, or other commercial purposes. Although they are not really space explorers, they would not be possible without the technology of the space age.

On April 1, 1960, NASA launched the first successful meteorological satellite named *TIROS 1* (Television Infrared Observation Satellite). The spacecraft was equipped with television cameras in order to film cloud cover around the Earth. This is an example of an active satellite (one that collects data or performs some other activity and transmits signals back to Earth).

Over the next few decades weather satellites grew increasingly more sophisticated in their capabilities. The primary systems used following *TIROS* were named *NIMBUS*, *TOS* (TIROS Operational Satellite), *ITOS* (Improved TOS), *SMS* (Synchronous Meteorological Satellite), NOAA (National Oceanic and Atmospheric Administration), and *GOES* (Geostationary Operational Environmental Satellite).

Other application satellites perform various duties for earth scientists, such as mapping oceans and land masses or measuring the heat and moisture content of Earth's surface. Famous examples include *Landsat 7*, *Seasat*, and the *HCMM* (Heat Capacity Mapping Mission).

On August 12, 1960, NASA launched its first communications satellite named *ECHO 1*. It was a large metallic sphere that reflected radio signals. This is an example of what is called a passive satellite. *ECHO 1* did not collect data or emit transmissions, it simply reflected radio signals sent from one location on Earth to another location on Earth. Other famous communications satellite series include *Relay*, *Telstar*, *ATS* (Applications Technology Satellite), *CTS* (Communications Technology Satellite), *Comsat*, *Satcom*, *Marisat*, *Galaxy*, and *Syncom*.

Most communications satellites have commercial purposes and are financed by large corporations. However, there have been a number of small communications satellites put into space to service ham radio enthusiasts. Ham radio is communication between amateur radio station operators. These operators use two-way radio stations to communicate with each other via voice, computer, television transmission, and/or Morse code. Ham radio evolved from the telegraph system, which was common before the telephone came into widespread use. In 1960 a group of ham radio operators in California got together and built a ten-pound transmitter satellite. They called it *OSCAR* (*Satellite Carrying Amateur*

Radio). Amazingly they convinced the U.S. Air Force to launch *OSCAR* into space piggybacked on the military's *Discoverer XXXVI* satellite.

On December 12, 1961, *OSCAR* rode into orbit with the military satellite boosted by a powerful Thor Agena B rocket. *OSCAR* operated for three weeks broadcasting the message "hi" over and over in Morse code. Hundreds of ham radio operators around the world reported picking up the signal. The tiny satellite was destroyed when it re-entered Earth's atmosphere on January 31, 1962, after 312 orbits. The OSCAR project was so successful it spawned formation of the Amateur Satellite Corporation (AMSAT), a nonprofit volunteer organization that designs, builds, and operates amateur satellites. Dozens of AMSAT satellites have been put into orbit over the decades.

NASA maintains a whole series of communication satellites in Earth orbit that allow the agency to communicate with astronauts and relay data to robotic spacecraft during their missions. They are called Tracking and Data Relay Satellites.

Many communication satellites are placed in Earth orbit 22,241 miles from the planet's surface. At this distance they are anchored in place by Earth's gravity and in synch with the Earth's revolution rate. In other words, they move around the Earth at the same speed that the Earth is revolving around its axis. This is called a geosynchronous orbit. Satellites placed in geosynchronous orbit even with Earth's equator appear from Earth to hover in space at the exact same location all the time. This is called a geostationary orbit.

Navigation is the act of determining one's position relative to other locations. Before the invention of satellites navigational signals were transmitted by land-based systems using antennas. (See Figure 1.5.) These antennas sent low-frequency radio signals that traveled along the surface of Earth or reflected off the ionosphere to reach their target receptor. The ionosphere is a layer of atmosphere that begins about thirty miles above the Earth's surface. Atmospheric gases undergo electrical and chemical changes within the ionosphere. This is what gives it reflective properties.

During the 1970s the U.S. military developed a space-based navigational system called the Global Positioning System (GPS). This system relies on satellites in Earth orbit to handle signal transmissions as shown in Figure 1.6. During the 1980s GPS was made available for international civil use. Over the next two decades it became one of the most popular navigational tools in the world.

SPACESHIPONE

SpaceShipOne was the first privately built and financed craft to fly a human into space. For more than

FIGURE 1.5

Terrestrial navigation system

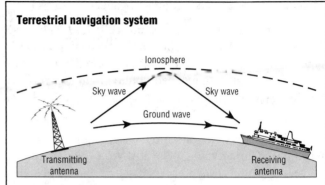

SOURCE: Joe Kunches, "Figure 1. The Paths Taken by a Radio Wave Transmitted by a Terrestrial Navigation System," in *Space Environment Topics: Navigation*, U.S. Department of Commerce, National Oceanic and Atmospheric Administration, Space Environment Laboratory, 1995, http://www.sec.noaa.gov/info/Navigation.pdf (accessed December 28, 2005)

FIGURE 1.6

Space-based navigation system

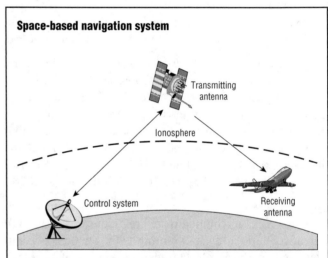

SOURCE: Joe Kunches, "Figure 2. Ground Stations Monitor the Information from the Satellites and Upload Changes as Needed to Ensure that the Navigation Information is Accurate," in *Space Environment Topics: Navigation*, U.S. Department of Commerce, National Oceanic and Atmospheric Administration, Space Environment Laboratory, 1995, http://www.sec.noaa.gov/info/Navigation.pdf (accessed December 28, 2005)

three decades the only way for humans to access space was through government-operated space programs. This all changed on June 21, 2004, when *SpaceShipOne* carried pilot Mike Melvill (1941–) to an altitude of 62.2 miles (100.1 kilometers). One hundred kilometers is the boundary of space as defined by the Fédération Aéronautique Internationale (FAI).

SpaceShipOne was designed and built by Scaled Composites, a California-based firm. The funding was provided by American billionaire Paul G. Allen (1953–), cofounder of the Microsoft Corporation. Allen financed the project as a way to have some meaningful impact on

FIGURE 1.7

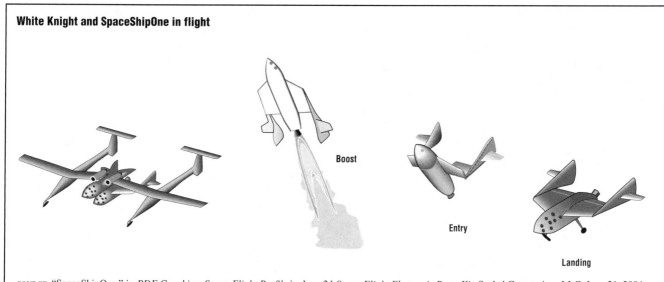

White Knight and SpaceShipOne in flight

Boost

Entry

Landing

SOURCE: "SpaceShipOne," in *PDF Graphics: Space Flight Profile* in *June 21 Space Flight Electronic Press Kit*, Scaled Composites, LLC, June 21, 2004, http://scaled.com/projects/tierone/june21presskit.htm (accessed January 4, 2005) © 2004 Mojave Aerospace Ventures LLC; SpaceShipOne is a Paul G. Allen Project.

space exploration. During development of *SpaceShip-One*, Allen became aware of the Ansari X Prize—a $10 million prize offered by private investors to the developers of the first nongovernmental reusable space plane. The Ansari X Prize was the brainchild of Peter Diamandis (1961–), an aerospace engineer and entrepreneur in space tourism. His inspiration was the Orteig prize, which was won in 1927 by aviator Charles Lindbergh, when he flew nonstop across the Atlantic Ocean between New York and Paris. Lindbergh's flight incited interest and investment in aviation. Diamandis believed his prize would launch another new industry, private space travel.

In 1994 Diamandis started the X Prize Foundation to raise money for the prize. His first investors were businesspeople in St. Louis, Missouri, the same city that played a key role in Lindbergh's flight many years before. Over the next decade Diamandis received support from a number of backers, including science fiction author Arthur C. Clarke, Apollo astronaut Buzz Aldrin, and Erik Lindbergh, the grandson of Charles Lindbergh. However, the foundation still struggled to raise the funds it needed.

In 2004 the foundation received a financial boost. Space enthusiasts Anousheh Ansari and her brother-in-law Amir Ansari made a multi-million-dollar contribution to the prize fund. The Ansaris were born in Iran but had immigrated to the United States and formed a successful telecommunications business. The X prize was renamed the Ansari X Prize in their honor.

By 2004 dozens of teams were developing rockets and spacecraft to vie for the prize. The rules required that the spacecraft carry three people (or at least one person plus the equivalent weight of two people) to an altitude of at least 100 kilometers. The feat had to be accomplished twice within a two-week period using the same spacecraft.

Scaled Composites used a two-part flight sequence to boost *SpaceShipOne* into space. A manned twin-turbojet plane called *White Knight* lifted off from a runway carrying the smaller manned space plane attached to its belly. (See Figure 1.7.) After reaching an altitude of approximately 47,000 feet, the space plane was released. Immediately its rockets were fired to propel it vertically into space. It then turned and reentered the atmosphere and glided to a landing at the same air strip from which it took off.

On September 29, 2004, *SpaceShipOne* achieved an altitude of 63.9 miles (102.8 km) with pilot Mike Melvill at the controls. The successful flight garnered international media coverage and increased interest in the competition. On October 4, 2004, a large crowd gathered at an airstrip in Mojave, California, to watch *SpaceShipOne* attempt to make history. They were not disappointed. Brian Binnie (1953–) took the space plane to an altitude of 69.6 miles (112 km) to win the Ansari X Prize. The news was broadcast around the world, and President George W. Bush telephoned the team to offer his congratulations.

The people behind the Ansari X Prize and *SpaceShipOne* purposely used parallels to events in aeronautical history to build their legacy. The concept of the prize drew upon the symbolism and romanticism attached to

Lindbergh's heroic flight. Important announcements and flights were conducted on dates of significance to space enthusiasts. The first test flight of *SpaceShipOne* to break the sound barrier occurred on December 17, 2003—the one hundredth anniversary of the first powered flight of the Wright Brothers. The Ansari's multi-million-dollar contribution to the X Prize fund was announced on May 5, 2004, the forty-third anniversary of Alan Shepard's flight into space. The date October 4 was chosen as the day for *SpaceShipOne*'s prize-winning flight attempt, because it was the date in 1957 when the Soviet Union launched *Sputnik 1*, the first artificial satellite to go into space.

According to Diamandis, the X in X prize stood for the Roman numeral ten (as in the $10 million prize) and for the X in "experimental" (as in the famous X series of experimental planes flown during the 1950s and 1960s). The X series flights were extremely important to the development of space travel. The *White Knight* carrier plane was named after two of the test pilots who flew the X-15. *SpaceShipOne*'s prize-winning flight broke the altitude record set by an X-15 in 1963, a record that had stood for more than four decades.

However, the achievements of *SpaceShipOne* will likely be remembered for their effects on the future of aeronautics, not their ties to the past. The first nongovernmental manned space flight offers tantalizing prospects for private individuals to travel into space. It may represent the birth of a new industry and a means for many people to experience the adventure of space flight. (For further information on *SpaceShipOne*, see Chapter 3.)

SPACE COMMERCE

The Space Age introduced new areas of commerce for the world's entrepreneurs. Companies engaged in aviation, aeronautics, and aerospace activities have been the most direct beneficiaries. However, other industries have seized upon space-based opportunities, primarily the businesses of commercial satellite services and space tourism.

Commercial Satellite Services

In 1962 the U.S. Congress passed the Communications Satellite Act (P.L. 87-624), opening the door for commercial use of satellites in space. For decades these satellites could only be launched by national space agencies (such as NASA). In 1980 Arianespace (a subsidiary of the European Space Agency) became the world's first commercial space transportation company. Arianespace began offering satellite launches using Ariane rockets at its spaceport in French Guiana (a small country along the northern coast of South America). Its first client was an American telecommunications company.

In 1984 the Commercial Space Launch Act (P.L. 98-575) was passed in the United States. The act granted power to the U.S. private sector to develop and provide satellite launching, reentry, and associated services and noted that this "would enable the United States to retain its competitive position internationally, contributing to the national interest and economic well-being of the United States."

Boeing Corporation is a large U.S. aerospace company and a prime NASA contractor. In 1995 Boeing formed a satellite launching business with Russian, Norwegian, and Ukrainian partners. Sea Launch Company, LLC, is headquartered in Long Beach, California, and operates a rocket launch platform on a modified oil-drilling platform in the South Pacific Ocean. As of 2005 Boeing owned a 40% share in the company. Its partners include RSC-Energia of Russia (25% share); Kvaerner ASA of Norway (20% share); and SDO Yuzhnoye/PO Yuzhmash of Ukraine (15% share). More than a dozen launches of commercial satellites have taken place from the Sea Launch facility since the first launch in 1999.

The Commercial Space Act of 1998 (P.L. 105-303) encouraged NASA to set policies to encourage and facilitate the participation of the private sector in the operation, use, and servicing of the ISS. This act received little attention until 2004, when President Bush announced his plan to retire the space shuttle fleet by 2010. The United States expects it will need commercial services to take over many of the tasks historically performed by the shuttle for the ISS program. This should open up many new opportunities in space for enterprising companies.

Space Tourism

Prior to the 2000s space tourism was limited to occasional space station visits taken by a handful of individuals for multi-million-dollar fees. These trips were granted by the Soviet (and later Russian) Space Agency to raise badly needed funds. Private space tourism companies formed and accepted deposits for future spaceflights on not-yet-developed spacecraft, but these ventures sounded like science fiction to most people. This all changed in 2004 with the successful suborbital excursions of *SpaceShipOne*. Suddenly space visits by private individuals became a viable possibility. New space tourism businesses formed, and the U.S. government rushed to set regulations for private space transportation, an entirely new industry.

COMMERCIAL PASSENGERS ON RUSSIAN MISSIONS. During the late 1980s the Soviet space program was in dire need of money. The Soviet Union was splintering into individual nations, and funds for space travel were in short supply. In 1990 the agency received $28 million from a Japanese media company to take one of their journalists aboard the space station *Mir*. A year later a

TABLE 1.1

Number of successful space launches worldwide, 1998–2005

	1998	1999	2000	2001	2002	2003	2004	2005	Total	Percent of launches
India	0	1	0	2	1	2	1	1	**8**	2%
Japan	1	0	0	1	3	2	0	2	**9**	2%
Sea	0	1	2	2	1	3	3	4	**16**	3%
China	6	4	5	1	4	6	8	5	**39**	8%
ESA*	11	10	11	7	11	4	3	5	**62**	12%
U.S.	34	28	28	21	17	23	16	12	**179**	35%
Russia	24	26	35	23	23	21	22	23	**197**	39%
Total	**76**	**70**	**81**	**57**	**60**	**61**	**53**	**52**	**510**	

Other launches: North Korea (1998) and Israel (2002 and 2004)
*ESA=European Space Agency

SOURCE: Adapted from "Table of World Space Launches," in *Office of Spaceflight Space Statistics*, National Aeronautics and Space Administration, January 2006, http://www.hq.nasa.gov/osf/spacestat.html (accessed December 30, 2005)

London bank paid an undisclosed amount of money to allow a British woman to spend a "space vacation" aboard *Mir*.

In 2001 the Russian space agency Rosaviakosmos charged an American businessman $15 million for a "space vacation" aboard the *ISS*. The next year Rosaviakosmos raised the price. A South African space tourist paid $20 million to visit the *ISS*. In September 2005 an American scientist and businessman spent a week on the *ISS* for a $20 million fee.

Russia's *ISS* partners (including NASA) have not shown any interest in space tourism. The United States initially refused to let tourists aboard the *ISS* but relented after heated negotiations with Rosaviakosmos.

The Russian agency has stated publicly that it hopes to develop space tourism as a thriving business. It sells tourist packages that allow people to undergo simulated cosmonaut training at its facilities in Star City.

THE FUTURE OF PRIVATE SPACE TRAVEL. All three tourist trips to the *ISS* were brokered by an American company called Space Adventures. The company, which is based in Virginia, was founded in 1998 by aerospace engineer and space entrepreneur Peter Diamandis. It also has plans to market suborbital flights aboard a new space plane being developed by a Russian contractor. The plane will take tourists to an altitude just over sixty-two miles from Earth. According to the company, passenger flights are expected to begin before 2008.

Other companies known to be developing commercial spacelines include Virgin Galactic, Space X, Rocketplane, and Armadillo Aerospace.

The Commercial Space Launch Amendments Act of 2004 instructed the Federal Aviation Administration (FAA) to begin formulating rules to govern the transport of passengers into space aboard commercial spacecraft. In December 2005 the FAA issued more than 100 pages of proposed rules dealing with issues such as crew training, pilot certification, and requirements for informed consent about the risks of space flight. Final rules were expected in mid-2006.

Those people who do not make it into space during their lifetime also have another option. Several companies around the world already offer services where the cremated ashes of a loved one can be carried into space. The service costs anywhere from $5,000 to $15,000.

SPACE LAUNCHES

NASA tracks the number of spacecraft launches conducted worldwide each year. Table 1.1 shows the figures for 1998 through 2005. The United States and Russia accounted for 74% of all launches. However, both countries launch satellites for other nations.

Only a handful of launches take place each year to support human spaceflight programs operated by the United States, Russia, and China. The vast majority of launches take place to put unmanned commercial, science, and military satellites into Earth orbit. The science missions are largely devoted to earth science, studying Earth's weather, climate patterns, atmospheric conditions, etc. Military satellites perform reconnaissance (spying) from space or support the communication and navigation needs of armed forces around the world.

SPACE CASUALTIES

Exploration has always been dangerous. Many ancient explorers died during their journeys across deserts, seas, mountains, and jungles. Space exploration has its own casualties.

During the earliest days of space travel dozens of animals were sacrificed for the space program. The United States sent a variety of small animals and primates up in rockets to test the safety of space flight for humans.

Very few survived the flight or the examination afterward. Some of the so-called astro-monkeys and astro-chimps that lost their lives were named Able, Albert, Bonny, Goliath, Gordo, and Scatback. The Soviets preferred dogs to test their spacecraft. Dogs named Bars, Laika, Lisichka, Mushka, and Pchelka died as a result.

Space programs in both countries have suffered human losses throughout the years as well:

- January 27, 1967—*Apollo 1* crew died during a flash fire aboard a capsule on the launch pad undergoing routine test. The casualties were Virgil (Gus) Grissom, Edward White, and Roger Chaffee.

- April 24, 1969—*Soyuz 1* cosmonaut Vladimir Komarov died during descent to Earth when his parachutes failed to function properly.

- June 29, 1971—*Soyuz 11* crew died during descent to Earth when their spacecraft lost its atmosphere due to a leaky valve. The casualties were Georgiy Dobrovolskiy, Vladislav Volkov, and Viktor Patsayev.

- January 28, 1986—The space shuttle *Challenger* crew died shortly after launch due to an explosion caused by leaking hot gases. The casualties were Francis "Dick" Scobee, Mike Smith, Judy Resnik, Ron McNair, Ellison Onizuka, Gregory Jarvis, and Christa McAuliffe.

- February 1, 2003—The space shuttle *Columbia* crew died during Earth reentry when a damaged wing allowed hot gases to enter the spacecraft, tearing the shuttle apart. The casualties were Rick Husband, William McCool, David Brown, Kalpana Chawla, Michael Anderson, Laurel Clark, and Ilan Roman.

In January 2004 NASA administrator Sean O'Keefe announced that the last Thursday in January will become a day of remembrance for lives lost in the American space program. Each year on this day NASA employees will observe a moment of silence, and flags will be flown at half-staff to honor the dead.

Like all journeys of discovery space exploration is a bold and perilous undertaking. Major sacrifices have been made to move humankind closer to the stars. In 1962 President Kennedy aptly described the combination of fear, hope, and yearning that characterizes every journey into space: "As we set sail we ask God's blessing on the most hazardous and dangerous and greatest adventure on which man has ever embarked."

CHAPTER 2
SPACE ORGANIZATIONS, PART 1: NASA

It is the policy of the United States that activities in space should be devoted to peaceful purposes for the benefit of all mankind.

—National Aeronautics and Space Agency Act of 1958

Once it became obvious that space exploration was an achievable reality, it became a national priority for rich and powerful countries. Following World War II there were only two superpowers in the world—the United States and the Union of Soviet Socialist Republics, and they considered each other enemies.

Both superpowers had military, scientific, and political reasons to pursue space travel. Outer space was a potential battlefield and provided an opportunity to spy on enemies on the other side of the world. Scientists, however, valued space travel for another reason. They wanted to gather data from space to help them unravel the mysteries of the universe. From a political standpoint, a successful space program was a source of national pride, a symbol of national superiority. This motivation above all others drove the earliest decades of space exploration.

The Soviet Union's space program was under the control of the military. In contrast, the United States split its space program into two parts. The U.S. military was given control over space projects related to national defense. A new civilian agency called the National Aeronautics and Space Administration (NASA) was formed in 1958 to oversee peaceful space programs.

Throughout its history NASA has been associated with spectacular feats and horrific disasters in space exploration. It has received great praise for its successes and harsh criticism for its failures. Space travel is an expensive enterprise. As a government agency NASA is bound by federal budget constraints. This budget rises and falls according to the political climate. American presidents set space goals, but the U.S. Congress sets NASA's budget.

In 1961 President John F. Kennedy charged NASA with the monumental task of putting a man on the Moon before the end of the decade. Congress allocated billions of dollars to NASA and this goal was accomplished. Later presidents also set grand goals for the agency, but none of these were realized. Every major endeavor went over budget and fell behind schedule. The public seemed to lose interest in space travel. Congress lacked the political motivation to increase NASA's funding. In 1965 NASA's budget comprised nearly 4% of the federal budget. By 1974 this percentage was less than 1%. It has remained near this level for more than thirty years. (See Figure 2.1.)

Since the 1980s NASA's reputation has suffered. Between 1986 and 2003 the agency experienced a string of failures. Two spacecraft sent to Mars were lost. A space telescope was launched into space with a faulty mirror. Worst of all, two catastrophic disasters killed fourteen astronauts. Critics complained that NASA had become overconfident, too bureaucratic, and had lost its technological edge.

In 2004 and 2005 NASA got a huge boost in prestige with the success of its robotic missions to Mars and Saturn. This was accompanied by a declaration from President George W. Bush that NASA should set bold new goals to send crewed missions to the Moon and Mars. It remains to be seen whether Congress will fund these enterprises and whether NASA will be able to overcome the many obstacles in its path to space.

A NEW AGENCY IS BORN

NASA was founded on October 1, 1958, following enactment of the National Aeronautics and Space Act of 1958 (P.L. 85-568). The stated purpose of the act was "to provide for research into problems of flight within and outside the earth's atmosphere, and for other purposes."

FIGURE 2.1

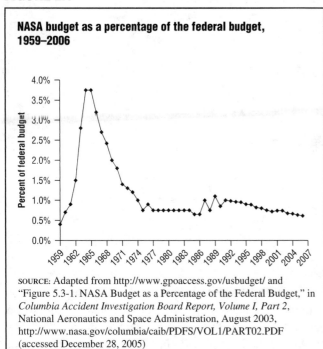

NASA budget as a percentage of the federal budget, 1959–2006

SOURCE: Adapted from http://www.gpoaccess.gov/usbudget/ and "Figure 5.3-1. NASA Budget as a Percentage of the Federal Budget," in *Columbia Accident Investigation Board Report, Volume I, Part 2,* National Aeronautics and Space Administration, August 2003, http://www.nasa.gov/columbia/caib/PDFS/VOL1/PART02.PDF (accessed December 28, 2005)

The act specifically mandated that NASA would be a civilian agency with control over all nonmilitary aeronautical and space activities within the United States. The research and development of weapons and national defense systems remained under the control of the U.S. Department of Defense (DOD). However, the act called for sharing of information between the two agencies. Cooperation by NASA in space ventures with other countries was allowed if the purpose was "peaceful application of the results."

The National Aeronautics and Space Act of 1958 outlined eight objectives for NASA:

- To expand human knowledge about atmospheric and space phenomena

- To improve all aspects of aeronautical and space vehicles

- To develop and operate vehicles capable of carrying supplies, equipment, scientific instruments, and living organisms into space

- To conduct long-range studies into the potential benefits, opportunities, and problems associated with astronautical and space activities

- To preserve the role of the United States as a leader in aeronautical and space science and technology and its application

- To share discoveries of military value with agencies involved in national defense

- To cooperate with other nations in peaceful ventures

- To cooperate with other U.S. agencies in utilizing national scientific and engineering resources in the most effective and efficient means possible

PEACEFUL VERSUS MILITARY PURPOSES

The new agency consolidated the resources of several government organizations, chiefly the National Advisory Committee for Aeronautics (NACA). The NACA was originally named the Advisory Committee for Aeronautics when it was formed in 1915 to "supervise and direct the scientific study of the problems of flight, with a view of their practical solution." At the time World War I was raging in Europe. German zeppelins had dropped bombs on Britain earlier that year, ushering in a new means of waging war.

Although the NACA's formation was driven by war, the agency also conducted aircraft research and set policy and regulations for commercial and civil aviation. In 1926 the Air Commerce Act was passed, freeing the NACA of regulatory responsibilities. The agency turned its full attention to aeronautical research and development at its Langley Aeronautical Laboratory in Virginia and later Ames Aeronautical Laboratory at Moffett Field, California, and a testing facility at Wallops Island, Virginia. NACA research and development benefited both military and civilian aviation.

On October 14, 1947, Air Force Captain Charles (Chuck) Yeager (1923–) made the first supersonic flight in an X-1 aircraft, a rocket-powered research plane developed by the Air Force and the NACA. NACA played an integral role in developing and testing the X-series of experimental aircraft. In 1952 the NACA began researching the challenges of spaceflight. Two years later the agency recommended that the U.S. Air Force develop a manned research vehicle to travel beyond earth's atmosphere.

In October 1957 the Soviet Union launched *Sputnik 1,* the world's first artificial satellite. A few months later a private space organization called the American Rocket Society urged President Dwight D. Eisenhower to establish a new agency to assume responsibility for all U.S. nonmilitary space projects. The agency was to pursue "broad cultural, scientific, and commercial objectives" and be independent of the Department of Defense. The director of the NACA provided a counter-recommendation that the new agency operate under joint control of the NACA, the Department of Defense, the National Academy of Sciences, and the National Science Foundation in support of military and nonmilitary projects.

In January 1958 President Eisenhower wrote Soviet Premier Nikolai Bulganin and proposed that the two countries "agree that outer space should be used only for peaceful purposes." The Soviet Premier refused to agree to the proposal unless the United States ceased all nuclear weapons testing and disbanded all its military bases on foreign soil. These conditions were unacceptable to the United States.

In July 1958 President Eisenhower signed the National Aeronautics and Space Act of 1958, turning over all nonmilitary space projects to NASA. The new agency absorbed the personnel and facilities of the NACA, which ceased to exist.

NASA began with approximately 8,000 employees and an annual budget of $100 million (about $700.7 million in 2006 dollars, according to the Federal Reserve Bank of Minneapolis). About half of NASA's first employees were civilian personnel working on space projects at the Army's Redstone Arsenal in Huntsville, Alabama. They included a rocketry team headed by Wernher von Braun (1912–77). Von Braun was a German rocket scientist who moved to the United States following World War II, and he played a major role in building America's rocket program at NASA.

T. Keith Glennan (1905–95) served as the agency's first administrator. Under his direction NASA took control of the DOD's Jet Propulsion Laboratory in California and parts of the Naval Research Laboratory in Washington, D.C. NASA also took over several satellite and lunar probe programs being operated by the Air Force and the DOD's Advanced Research Projects Agency. The military retained control over reconnaissance satellites, ballistic missiles, and a handful of other DOD space projects that were then in the research stage.

NASA historians say that President Eisenhower believed that the civilian space program should be "small in scale and limited in its objectives" (http://history.nasa.gov/SP-4213/ch2.htm).

NASA SHOOTS FOR THE MOON

NASA did not stay small for long. The agency had grand plans. In February 1960 NASA presented to Congress a ten-year plan for the nation's space program. It included an array of scientific satellites; robotic probes to the Moon, Mars, and Venus; development of new and powerful rockets; and manned spaceflights to orbit the Earth and the Moon. NASA estimated the program would cost around $12 billion.

Congress was politically motivated to support the program. The Soviet Union had already landed a probe on the Moon as part of its Luna Project. At the time NASA was continuing the Pioneer Project begun by the NACA to obtain data from probes sent to the Moon. The first three Pioneer rockets had launched during 1958 but failed to escape Earth's gravity. In March 1959 *Pioneer 4* was the first U.S. spacecraft to escape Earth's gravity. It passed within 37,300 miles of the Moon. However, the Soviet's *Luna 1* probe had already passed much closer to the Moon. The United States was behind in the space race.

After John F. Kennedy was elected president in November 1960, he charged Vice President Lyndon Johnson with finding a way for the United States to beat the Soviets to a major space goal. NASA pushed for a manned lunar landing, and Johnson agreed. On May 25, 1961, President Kennedy asked Congress to provide financial support to NASA to put a man on the Moon before the end of the decade.

NASA Deputy Administrator Hugh Dryden (1898–1965) named the Moon effort the Apollo program. It was named after the mythical Greek god Apollo who drove the chariot of the Sun across the sky.

Catching Up

NASA had a lot of work to do just to catch up with the Soviets in space. In April 1961 they had put the first human in Earth orbit. Cosmonaut Yuri Gagarin (1934–68) circled the Earth one time in a flight that lasted 108 minutes. It would be nearly a year before NASA could even come close to this achievement, with the flight of John Glenn (1921–) in *Friendship 7* on February 20, 1962.

When it was first created in 1958, NASA was concerned with finishing ongoing NACA projects. These included a weather satellite, a military spy satellite, and the Pioneer lunar space probes. These probes were intended to go into lunar orbit or impact the Moon's surface while sending back photographs and scientific data. *Pioneer 4* provided NASA with valuable new radiation data needed for the ongoing Mercury project.

The Mercury Project

The Mercury project actually began in 1958, only a week after NASA was created. The official announcement was made on December 17, 1958—the fifty-fifth anniversary of the Wright Brothers' flight. The project was named after the mythical Roman god Mercury, the winged messenger.

The Mercury project had three specific objectives:

- Put a manned spacecraft into Earth orbit
- Investigate the effects of space travel on humans
- Recover the spacecraft and humans safely

On May 5, 1961, NASA astronaut Alan B. Shepard, Jr. (1923–98) became the first American in space when he took a fifteen-minute suborbital flight. Shepard's flight was far shorter than Gagarin's had been and included only five minutes of weightlessness. NASA desperately needed more data on the effects of weightlessness on humans.

TABLE 2.1

Project Mercury manned flights

Date of launch	Mercury flight no.	Spacecraft name	Flight type	Highest altitude	Time in space	Astronaut
5/6/1961	3	Freedom 7	Sub orbital	116 miles	15 min 28 sec	Alan Shepard
7/21/1961	4	Liberty Bell 7	Sub orbital	118 miles	15 min 37 sec	Gus Grissom
2/20/1962	6	Friendship 7	3 orbits	162 miles	4 hr 55 min	John Glenn
5/24/1962	7	Aurora 7	3 orbits	167 miles	4 hr 56 min	Scott Carpenter
10/3/1962	8	Sigma 7	6 orbits	176 miles	9 hr 13 min	Walter Schirra
5/15/1963	9	Faith 7	22.5 orbits	166 miles	1 day 10 hr 19 min	Gordon Cooper

SOURCE: Created by Kim Weldon for Thomson Gale, 2004

This was considered a key element to manned flights to the Moon.

Between 1961 and 1963 six Mercury astronauts made six successful spaceflights and spent a total of 53.9 hours in space. (See Table 2.1.) Over this time period NASA's budget increased from $964 million in 1961 to $3.7 billion in 1963. By 1964 it had risen to over $5 billion and would remain at this level for three more years.

Is It Worth It?

The United States paid a high price for NASA's Moon program. It was conducted during one of the most turbulent times in American history. The decade of the 1960s was characterized by social unrest, protest, and national tragedies.

On November 22, 1963, President Kennedy was assassinated in Dallas, Texas. Vice President Lyndon Johnson assumed the presidency. Johnson had always supported the space program and had been instrumental in passing the bill that created NASA. He assured NASA that the Apollo program would continue as planned. On November 29, 1963, Johnson announced that portions of the Air Force missile testing range on Merritt Island, Florida, would be designated the John F. Kennedy Space Center.

Robert Gilruth (1913–2000) was then NASA manager in charge of the Manned Spaceflight Center in Houston, Texas. He promised the country that Apollo would be successful because it employed "the kind of people who will not permit it to fail."

In 1964 social scientist Amitai Etzioni published *The Moon-Doggle*, a book that was extremely critical of NASA. The title was a play on the word "boondoggle" which means a wasteful and impractical project. Etzioni criticized the agency for spending so much money on manned spaceflights when unmanned satellites could achieve more for less money. He also questioned the scientific value (and costs) of sending astronauts to the Moon. Etzioni was not alone in feeling this way. American society was increasingly concerned with pressing social and national issues including the escalating war in Vietnam and civil rights.

The Gemini Project

NASA scientists realized during the Mercury missions that they needed an intermediate step before the Apollo flights. They had to be sure that humans could survive and function in space for up to fourteen days. This was the amount of time estimated then for a round-trip to the Moon. The program designed to test human endurance in space was named the Gemini project after the constellation represented by the twin stars Castor and Pollux. The name was chosen because the Gemini space capsule was designed to hold two astronauts, rather than one.

A major goal of the Gemini project was to successfully rendezvous orbiting vehicles into one unit and maneuver that unit with a propulsion system. This was a feat that would be necessary to achieve the Moon landings. The last Gemini goal was to perfect atmospheric reentry of the spacecraft and perform a ground landing, rather than a landing at sea. All of the goals except a ground landing were achieved.

During 1965 and 1966 NASA completed ten Gemini missions with twenty astronauts that spent a total of more than forty days in space. (See Table 2.2.) The *Gemini IV* mission featured the first extravehicular activity (EVA) by an American. Astronaut Edward White (1930–67) spent twenty-two minutes outside of his spacecraft during a "space walk." The longest duration Gemini flight (*Gemini VII*) took place in December 1965, lasting fourteen days.

Moon Resources

By 1967 NASA scientists and engineers had been studying the details of a Moon landing for more than six years. NASA's budget at the time was $5 billion a year (approximately $30.3 billion in 2006 dollars), with about 90% of that money going to outside contractors and university research programs. In 1967 more than 400,000 people at installations around the country worked in

TABLE 2.2

Gemini program manned flights

Dates	Gemini flight no.	Astronauts	Achievements
March 23, 1965	III	Virgil Grissom John Young	3 orbits. Only Gemini spacecraft to be named (Molly Brown)
June 3–7, 1965	IV	James McDivitt Edward White	First American EVA—a 22 minute spacewalk by White
August 21–29, 1965	V	Gordon Cooper Charles Conrad	120 orbits. First use of fuel cells for electrical power.
December 4–18, 1965	VII	Frank Borman James Lovell	Longest mission at 14 days
December 15–16, 1965	VI-A	Walter Schirra Thomas Stafford	First space rendezvous (with Gemini VII)
March 16, 1966	VIII	Neil Armstrong David Scott	First space docking (with unmanned craft)
June 3–6, 1966	IX-A	Thomas Stafford Eugene Cernan	2 hours of EVA
July 18–21, 1966	X	John Young Michael Collins	Rendezvous with Gemini VIII
September 12–15, 1966	XI	Charles Conrad Richard Gordon	Record altitude (739.2 miles)
November 11–15, 1966	XII	James Lovell Edwin Aldrin	Record EVA by Aldrin (5 hours 30 minutes)

SOURCE: Created by Kim Weldon for Thomson Gale, 2004

support of the Apollo program. NASA's employees numbered about 36,000.

Rangers and Surveyors

A series of nine Ranger probes had been launched between 1961 and 1965. They were designed to flight test lunar spacecraft, take photographs of the Moon, and collect data on radiation, magnetic fields, and solar plasma (charged gases emitted from the sun).

The first two probes in the series failed to escape Earth orbit. *Ranger 3* was supposed to impact the Moon, but missed it by 23,000 miles. On April 26, 1962, *Ranger 4* crashed into the far side of the Moon. It was the first American object to reach another celestial body. Unfortunately its central computer had failed during the flight, and no data was transmitted. After two more failed attempts NASA finally achieved success. On July 31, 1964, *Ranger 7* crashed into the Moon after transmitting the first close-up photographs of the lunar surface.

During 1965 *Ranger 8* and *Ranger 9* captured hundreds more vital photographs before their impacts. Nearly 200 photographs taken by *Ranger 9* were broadcast live on television as the probe hurtled toward the lunar surface.

On June 2, 1966, NASA achieved another milestone when the *Surveyor 1* spacecraft made a controlled "soft landing" on the Moon in the Ocean of Storms. The ability to do a soft landing was considered crucial to putting a human safely on the Moon. *Surveyor 1* returned a host of high-quality photographs. However, NASA was still running behind the Soviet space program. The Soviet spacecraft *Luna 9* had soft-landed in the Ocean of Storms four months before *Surveyor 1* got there. *Luna 9* also provided the first television transmission from the lunar surface.

In all, NASA sent seven Surveyor spacecraft to the Moon between 1966 and 1968. Several lost control and crashed, while others achieved soft landings. The 1967 *Surveyor 6* was particularly successful. During its mission NASA controllers were able to lift the spacecraft about ten feet off the ground and set it softly back down again. NASA was ready to put humans aboard a lunar lander.

Flight Techniques

One of the greatest debates of the Apollo program related to flight techniques from the Earth to the Moon. Some engineers advocated a direct flight. In this scenario the Apollo spacecraft would be launched off the Earth and would proceed directly to the Moon, where it would land. This approach had great appeal because it did not require any docking or rendezvous between spacecraft. However, it did require development of new super-sized boosters.

The second approach was called the Earth orbit flight. In this approach the spacecraft would circle the Earth before flying directly to the lunar surface. In this scenario risky docking maneuvers would have to be accomplished in Earth orbit, so the astronauts could return to Earth if something went wrong.

The last approach was called the lunar orbit technique. In this scenario the spacecraft would fly near the Moon and go into orbit around it. Then a small maneuverable landing module would leave the base unit for the trip to the lunar surface and back. This approach was considered the most risky, because it required rendezvous and docking in Moon orbit.

According to an announcement by D. Brainerd Holmes, NASA Director of the Office of Manned Space Flight (November 7, 1962), more than 700 scientists and engineers had spent one million work-hours analyzing the three Apollo flight choices. In the end the lunar orbit technique was chosen.

Apollo Spacecraft

The Apollo spacecraft had three parts:

- Command Module (CM) containing the crew quarters and flight control section
- Service Module (SM) for the propulsion and spacecraft support systems
- Lunar Module (LM) to take two of the crew to and from the lunar surface

FIGURE 2.2

Apollo launch configuration for lunar landing mission

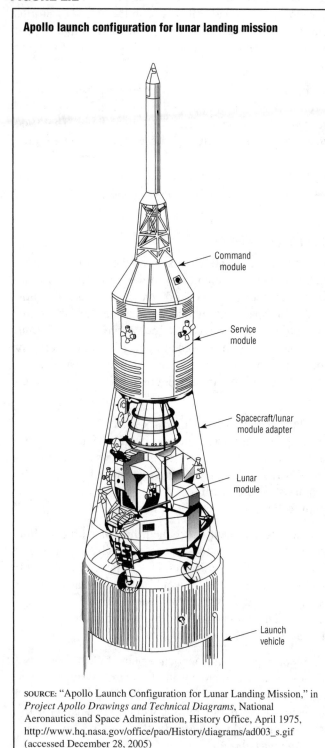

Command module

Service module

Spacecraft/lunar module adapter

Lunar module

Launch vehicle

SOURCE: "Apollo Launch Configuration for Lunar Landing Mission," in *Project Apollo Drawings and Technical Diagrams*, National Aeronautics and Space Administration, History Office, April 1975, http://www.hq.nasa.gov/office/pao/History/diagrams/ad003_s.gif (accessed December 28, 2005)

and service module were called the CSM. When the three modules reached lunar orbit, the lunar module was detached for the journey to and from the Moon's surface.

After the lunar module ascended from the lunar surface it docked with the CSM. Once the two astronauts had moved safely into the CSM, the lunar module was jettisoned away from the spacecraft. Only the CSM made the journey back toward Earth. The service module was jettisoned away just prior to reentry into Earth's atmosphere. The command module with all three astronauts aboard was designed to splash down into the sea.

A Tragic Setback

NASA lost its first astronauts during the Apollo program. In 1966 three unmanned Apollo spacecraft were launched to test the structural integrity of the spacecraft and the flight systems. These were called the Apollo-Saturn missions and were numbered *AS-201* through *AS-203.*

On January 27, 1967, NASA was preparing a spacecraft for mission *AS-204,* the first manned test flight. During a launch pad test of the spacecraft a flash fire broke out and killed all three astronauts in the command module. NASA renamed the mission *Apollo 1* in their honor.

The tragedy temporarily devastated morale at NASA. The agency was not treated kindly by the media. Many newspapers questioned whether a manned lunar mission was worth the risk. Rumors even circulated that the astronauts had been murdered by NASA for criticizing the agency or for other sinister reasons.

The exact cause of the spark that started the fire was never discovered. An extensive investigation conducted by NASA found that a variety of factors contributed to the astronauts' deaths. Some were operational problems—a hatch that was difficult to open, the presence of 100% oxygen in the module, and the use of flammable materials inside the module. The investigation also revealed a number of management and contractor problems. NASA set about redesigning the Apollo modules and reorganized top management staff. The Moon landing that was scheduled for late 1968 was delayed until 1969 due to the *Apollo 1* tragedy.

One Giant Leap

By late 1968 the Apollo program was making tremendous strides. The first manned flight (*Apollo 7*) launched on October 11, 1968. *Apollo 7* included the first live television broadcast from a manned spacecraft. Watching the astronauts on television helped rekindle a feeling of excitement about the space program. The American public grew more excited as one Apollo mission after another was successful. In May 1969 the

Figure 2.2 shows the three modules stacked atop a rocket for launch. The massive Saturn V rocket developed by Wernher von Braun was the launch vehicle selected for the Apollo spacecraft.

The astronauts rode in the command module during launch and reentry. Food, water, and fuel were carried in the service module. While together the command module

Apollo 10 mission featured the first live color television pictures broadcast from outer space.

Two months later *Apollo 11* was launched into space with three astronauts aboard. Their names were Neil Armstrong (1930–), Michael Collins (1930–), and Edwin "Buzz" Aldrin (1930–). At 4:18 PM Eastern Daylight Savings Time (EDT) on July 20, 1969, the lunar module softly landed near the Sea of Tranquility. Astronaut Neil Armstrong reported, "Houston, Tranquility Base here. The Eagle has landed."

At 10:56 PM EDT Armstrong opened the door of the lunar module and climbed down a short ladder. As he put his left foot onto the surface of the Moon he said these words: "That's one small step for man, one giant leap for mankind." It was the first time in history that a human being had set foot on another celestial body.

The event was televised live to a worldwide audience estimated at 528 million people. They watched as Armstrong and Aldrin explored the lunar surface for two hours and thirty-one minutes. The astronauts planted an American flag in the dusty soil and collected forty-eight pounds of Moon rocks. They unveiled a plaque attached to the descent stage (the lower part) of the lunar module. The plaque said, "Here men from planet Earth first set foot upon the Moon. July 1969 AD We came in peace for all mankind." The plaque bore the signatures of all three astronauts and President Richard Nixon.

The two astronauts climbed back into the lunar module. On July 21, 1969, the ascent portion of the module lifted off the lunar surface, leaving the descent stage behind. Armstrong and Aldrin had spent twenty-one hours and thirty-six minutes on the surface of the Moon. They then docked with the command/service module piloted by Collins. Once reunited, the three astronauts headed for Earth, leaving the lunar module ascent stage in orbit around the Moon. On July 24, 1969, their command module safely splash-landed in the Pacific Ocean.

The Can-Do Culture

NASA had achieved something that many people thought could not be done. The agency found itself heaped with praise and congratulations. Putting a man on the Moon was considered an enormous milestone in technological progress. In addition, it had been done before the end of the decade, just as President Kennedy had requested. The achievement fostered a tremendous sense of pride and confidence among NASA personnel. The agency was left with an optimistic conviction that it could do anything, an attitude that came to be known as NASA's "can-do culture."

NASA's critics believe that the agency's can-do culture caused it to make many overly optimistic promises during the following decades. NASA continued to set bold goals for the nation's space program and promised Congress that it could achieve them, just like it had accomplished the Moon landing. The problem was that these goals did not receive nearly as much financial support as the Apollo program received. The Moon landing was possible because NASA was given the necessary resources. Putting a man on the Moon within a decade had taken the talents of hundreds of thousands of people and nearly $24 billion of taxpayers' money. Neither Congress nor the American people were ever inclined again to devote so many resources to a space venture.

Apollo Fizzles Out

NASA launched six more Apollo missions following *Apollo 11*. In November 1969 the *Apollo 12* crew landed near the Ocean of Storms and found the *Surveyor 3* lunar probe sent several years before.

Five months later *Apollo 13* was launched. Two days into the flight an oxygen tank suddenly ruptured aboard the service module. The pressure in the cabin dropped quickly. Fearing the crew would otherwise be lost, NASA devised a way for the astronauts to rely on the limited resources in the lunar module to limp back to Earth. The spacecraft splashed down safely on April 17, 1970.

Once again NASA had achieved a near-miracle. Although on the surface *Apollo 13* appeared like a failure, NASA classified it as a success, because the Agency learned so much about handling emergencies during spaceflight. The experience was later captured in the 1995 movie *Apollo 13*. The movie made famous a phrase uttered by *Apollo 13* commander James Lovell (1928–) following the oxygen tank rupture. Lovell said calmly, "Houston, we have a problem."

The next Apollo launch was postponed, while NASA worked on problems brought to light by the incident with *Apollo 13*. In January 1971 *Apollo 14* successfully reached the Moon for a lunar exploration mission at Fra Mauro. The astronauts took along a new cart, specially designed to hold Moon rocks. Later that year the *Apollo 15* crew took a lunar rover—one of three that NASA had built at a cost of $40 million—that resembled a dune buggy. The astronauts zoomed around the Hadley-Apennine region at a top speed of eight miles per hour. They collected nearly 170 pounds of Moon rocks. The lunar rover was so effective it was used on all the remaining Apollo missions.

In April 1972 *Apollo 16* set down in the Descartes Highlands of the Moon. It was the first mission to explore the highlands and was at the southern-most landing site of any Apollo spacecraft. In December 1972 the *Apollo 17* crew explored highlands and a valley in the Taurus-Littrow area of the Moon. For the first time the mission crew included a scientist, the geologist Harrison "Jack" Schmitt (1935–). The astronauts collected 243 pounds

TABLE 2.3

Apollo program manned missions

Name	Dates	Spacecraft call signs	Crew	Mission time	Note
Apollo 1	January 27, 1967	not used	Virgil I. Grissom (Commander), Edward H. White, Roger B. Chaffee		Spacecraft caught on fire on landing pad during practice drill. All astronauts killed.
Apollo 7	October 11–22, 1968	not used	Walter M. Schirra Jr. (commander), Donn F. Eisele (CM pilot), R. Walter Cunningham (LM pilot)	10 days, 20 hours	CSM piloted flight demonstration in Earth orbit. First live TV from manned spacecraft.
Apollo 8	December 21–27, 1968	not used	Frank Borman (commander), James A. Lovell Jr. (CM pilot), William A. Anders (LM pilot)	6 days, 3 hours	First manned lunar orbital mission. Live TV broadcasts.
Apollo 9	March 03–13, 1969	CM: Gumdrop LM: Spider	James A. McDivitt (commander), David R. Scott (CM pilot), Russell L. Schweickart (LM pilot)	10 days, 1 hour	First manned flight of all lunar hardware in Earth orbit. Schweickart performed 37 minutes EVA. First manned flight of lunar module.
Apollo 10	May 18–26, 1969	CM: Charlie Brown LM: Snoopy	Thomas P. Stafford (commander), John W. Young (CM pilot), Eugene A. Cernan (LM pilot)	8 days, 3 minutes	Practice for Moon landing. First manned CSM/LM operations in cislunar and lunar environment; First live color TV from space.
Apollo 11	July 16–24, 1969	CM: Columbia LM: Eagle	Neil A. Armstrong (commander), Michael Collins (CM pilot), Edwin E. (Buzz) Aldrin Jr. (LM pilot)	8 days, 3 hours, 18 minutes	First manned lunar landing mission and lunar surface EVA.
Apollo 12	November 14–24, 1969	CM: Yankee Clipper LM: Intrepid	Charles Conrad Jr. (commander), Richard F. Gordon Jr. (CM pilot), Alan L. Bean (LM pilot)	10 days, 4 hours, 36 minutes	Lunar landing and lunar exploration.
Apollo 13	April 11–17, 1970	CM: Odyssey LM: Aquarius	James A. Lovell Jr. (commander), John L. Swigert Jr. (CM pilot), Fred W. Haise Jr. (LM pilot)	5 days, 22.9 hours	Mission aborted before spacecraft reached Moon.
Apollo 14	January 31–February 09, 1971	CM: Kitty Hawk LM: Antares	Alan B. Shepard Jr. (commander), Stuart A. Roosa (CM pilot), Edgar D. Mitchell (LM pilot)	9 days	Lunar landing and lunar exploration.
Apollo 15	July 26–August 07, 1971	CM: Endeavor LM: Falcon	David R. Scott (commander), Alfred M. Worden (CM pilot), James B. Irwin (LM pilot)	12 days, 17 hours, 12 minutes	Lunar landing and lunar exploration.
Apollo 16	April 16–27, 1972	CM: Casper LM: Orion	John W. Young (commander), Thomas K. Mattingly II (CM pilot), Charles M. Duke Jr. (LM pilot)	11 days, 1 hour, 51 minutes	Lunar landing and lunar exploration.
Apollo 17	December 07–19, 1972	CM: America LM: Challenger	Eugene A. Cernan (commander), Ronald E. Evans (CM pilot), Harrison H. Schmitt (LM pilot)	12 days, 13 hours, 52 minutes	Last lunar landing mission.

SOURCE: Created by Kim Weldon for Thomson Gale, 2004

of Moon rocks, the most of any Apollo mission. On December 19, 1972, *Apollo 17* splashed down safely in the Pacific Ocean. With the successful completion of the *Apollo 17* mission, the Apollo program was over.

Table 2.3 summarizes information about all of the Apollo missions. In all, NASA put twelve astronauts on the Moon: Neil Armstrong, Buzz Aldrin, Charles "Pete" Conrad Jr. (1930–99), Alan Bean (1932–), Alan B. Shepard Jr., Edgar D. Mitchell (1930–), David R. Scott (1932–), James B. Irwin (1930–91), John W. Young (1930–), Charles M. Duke Jr. (1935–), Eugene A. Cernan (1934–), and Harrison H. Schmitt (1935–). They collected 840 pounds of rocks, soil, and other geological samples from the Moon.

The missions that followed *Apollo 11* never captured the public's imagination the same way that the first Moon landing did. The feeling was that America had already achieved its goal of beating the Soviets to the Moon, and continued lunar exploration held little appeal for many people. Furthermore, the country was engaged in a very costly and demoralizing war in Vietnam. In 1970 NASA's

budget was cut to $3.7 billion per year, down from the $5 billion per year regularly provided in the mid-1960s. NASA had to cancel its planned remaining Apollo missions. *Apollo 18*, *Apollo 19*, and *Apollo 20* never took place.

SPACE SCIENCE SUFFERS

Putting a man on the Moon was conducted mostly for political purposes. It bolstered national pride and prestige. It was largely a symbolic endeavor. Many scientists thought the Apollo program achieved far less in scientific terms than unmanned probes could have accomplished. One reason the program was so expensive was that so many resources had to be devoted to keeping fragile humans alive and well in the harsh environment of space. Critics said this money could have been invested in robotics research and development to produce a fleet of unmanned probes and sample collectors to explore the Moon and far beyond.

The debate over human exploration versus robotic exploration began in the 1950s and still goes on in the

2000s. NASA's Ranger and Surveyor probes of the early 1960s were originally designed to collect data to support numerous research goals within astronomy and space science. Once the Apollo program began, these probes were retooled to gather data important to the manned program. This was called human factors research and was a small part of the discipline called space biology. NASA's focus on human factors at the expense of broader research in space biology, space science, and astronomy brought harsh criticism from scientists.

In 1967 a committee appointed by President Lyndon Johnson recommended that the nation establish a well-rounded space program following Apollo with more emphasis on science and less emphasis on human exploration. NASA did conduct unmanned spaceflights geared toward general space biology. In 1962 the Bio-satellite program began with a series of three flights designed to test the rigors of space travel on subhuman beings. In 1969 *Biosatellite III* flew with a male pig-tailed monkey named Bonnie aboard. The mission had to be ended early when Bonnie got sick. He died soon after returning to Earth.

During the early 1970s NASA wanted to build on its Apollo success with another ambitious manned space program. The agency lobbied Congress to allow it to transfer funds designated for the Biosatellite program to the manned program. This angered many scientists. As one historian has noted NASA pursued a plan of action that terminated "a relatively inexpensive science-oriented project in favor of a relatively expensive, exploration-oriented manned program." This type of criticism was to plague NASA for decades to come.

NASA's FIRST SPACE STATION

As early as the 1960s NASA made plans to put a manned space station in orbit around the Earth. These plans took center stage at the agency when the Apollo program ended. For its next great project NASA envisioned an orbiting space station devoted to scientific research and a fleet of reusable space planes to carry humans to and from the station. In 1969 neither President Richard Nixon nor the U.S. Congress were interested in extending the Apollo program, let alone pursuing a new and costly endeavor. NASA found its budget cut year after year. Nevertheless, the agency devoted many of its resources to developing a new manned space program.

The first step was the temporary *Skylab* space station. This was a small scientific laboratory and solar observatory that could hold three crewmembers at a time. Three separate crews visited and lived on *Skylab* between May 1973 and February 1974. The first mission lasted twenty-eight days, the second fifty-nine days, and the last mission was eighty-four days in length. The *Skylab* program saved money by using rockets and spacecraft components left over from the Apollo program. *Skylab* was considered key to gathering data on the effects on humans of prolonged weightlessness and spaceflight. As it had with its earlier space efforts, the United States lagged behind the Soviet Union in this area. The first Soviet space station (*Salyut 1*) was put into orbit two years before *Skylab*.

THE SHUTTLE PROGRAM

In 1972 development got under way at NASA on a reusable space plane called a shuttle. This program was supposed to produce a finished product within five years, but it eventually took twice that long. The first shuttle did not launch until 1981. By this time the Soviet space station *Mir* had been in orbit for several years. The Soviet space program had pursued—but failed to develop—a reusable space plane. Transportation to and from *Mir* was accomplished using expendable rocket boosters.

NASA relied on a series of space shuttles to conduct most Earth orbit operations. Throughout the early 1980s shuttles carried satellites for government, military, and commercial clients.

On January 28, 1986, the space shuttle *Challenger* exploded seventy-three seconds after liftoff. Seven astronauts were killed. The shuttle fleet was grounded for more than two years as a result. An investigation revealed that a faulty joint in a rocket booster had allowed hot gases to escape that ignited and destroyed the vehicle.

Government investigators also found fault with the entire shuttle program. They complained that NASA managers emphasized schedule over safety. Before the *Challenger* disaster, shuttles carried commercial and military satellites (called payloads) into space. To reduce the scheduling pressure NASA decided to immediately cease carrying commercial payloads. Military payloads were quickly phased out as well, leaving only scientific payloads. The shuttle's commercial and military clients were forced to turn back to expendable rockets to launch their satellites into orbit.

The loss of the *Challenger* shuttle forced NASA to scale back shuttle operations. The tragedy brought harsh criticism of NASA from scientists and politicians alike. NASA's can-do culture, left over from the Apollo years, was blamed for making the agency over-confident and overly optimistic about its abilities to safely operate a major space program on a limited budget.

Between 1994 and 1998 NASA shuttles played a major role in a cooperative space venture between the United States and Russia. (When the Soviet Union dissolved into numerous individual republics in 1991, the largest and most powerful, Russia, carried on the old Soviet space program under the new name of Rosaviakosmos.) U.S. shuttles docked with the Russian space station *Mir* for joint scientific missions of astronauts and cosmonauts. A decade before, NASA had worked with its Soviet counterpart to develop mutual docking mechanisms on U.S. and Soviet spacecraft.

In October 1998 NASA achieved a public relations boost when Senator John Glenn flew into space aboard the shuttle *Discovery*. The seventy-seven-year-old Glenn was already a hero for his participation in the Mercury program of the early 1960s. In 1962 Glenn had been the third American in space and the first to complete an Earth orbit during a five-hour trip aboard *Friendship 7*. In 1998 his spaceflight lasted nine days. He became the oldest person ever to travel into space. NASA scientists conducted extensive medical tests before, during, and after his flight to monitor his well-being. They were particularly eager to learn about the effects of weightlessness on an older person. Prolonged weightlessness in space is known to weaken human bones, a condition also seen on Earth in older people suffering from osteoporosis.

NASA's shuttle program continued into the twenty-first century, and was again touched by tragedy when the space shuttle *Columbia* broke apart on February 1, 2003, during reentry over the western United States. Seven crewmembers were killed. Investigators found that a piece of foam had fallen off the shuttle's external fuel tank during lift-off and smashed into *Columbia*'s wing. The resulting damage allowed super-hot gases to enter the shuttle during reentry and tore it apart.

This second shuttle tragedy shook NASA to the core. The agency had made many promises to Congress and the American public about better shuttle safety and reliability following the 1986 disaster. NASA's capability to operate a manned space program again came under attack. A government investigation blamed NASA for continuing to perpetuate the can-do culture in the face of serious operational and budget problems within the shuttle program. The shuttle fleet was grounded for more than two years.

In July 2005 a shuttle successfully carried out the first of two planned "return-to-flight" missions. Although the shuttle returned safely to Earth, video images revealed that foam had again shed from the external tank during lift-off. Luckily, it fell harmlessly to the ground and did not damage shuttle components. NASA and the public faced the sobering realization that fixing the shuttle's safety problems had proved to be an elusive goal. The second return-to-flight was postponed indefinitely.

THE INTERNATIONAL SPACE STATION

In 1988 the United States and fifteen other nations embarked on a new space venture called the *International Space Station* (*ISS*). NASA and Rosaviakosmos collaborated throughout the 1990s to lead construction of an orbiting space station designed for prolonged inhabitation by scientists engaged in space research. They invited other countries to participate by contributing parts, components, and scientific facilities or sending researchers to the station.

Rosaviakosmos was anxious to play a major role in the *ISS* but had even less funding than NASA. Both agencies struggled to put U.S. and Russian modules into place and keep them operational. The station was scaled back in size and capability numerous times due to budget restrictions in both countries. The bulk of the heavy lifting required to put *ISS* modules into space was performed by space shuttles. The 2003 grounding of the fleet halted *ISS* construction. NASA continued to send astronauts to the *ISS* but they had to travel aboard Russian spacecraft to get there.

NASA's ROBOTIC SPACE PROGRAMS

Although NASA's crewed missions have historically received the most public attention, the agency has sent a number of unmanned (robotic) spacecraft into outer space. These machines have taken a number of forms and achieved some incredible milestones in space exploration. Satellites have been put into Earth orbit since the earliest days of NASA's space program to collect weather data or serve military purposes. During the 1960s and 1970s lunar probes were sent to the Moon to support the Apollo program. At the same time, NASA began launching robotic explorers that traveled to other planets. These were followed by sophisticated observatories and other robotic spacecraft placed in orbit around the Earth or Sun or sent to intercept asteroids. These projects are considered crucial to enhancing human understanding of the workings of the Earth, the surrounding solar system, and the universe at large.

Interplanetary Explorers

NASA has conducted a number of interplanetary robotic missions to further space science. The Mariner program began in 1962 and included ten spacecraft sent to gather data as they flew by Mercury, Venus, or Mars. In December 1962 *Mariner 2* became the first spacecraft to fly by another planet when it flew within 22,000 miles of Venus. In 1971 *Mariner 9* became Mars's first artificial satellite. It sent back more than 7,000 photos of the Red Planet.

The Mariner program provided valuable information that was used during the 1970s to conduct the Viking program. In 1976 twin Viking spacecraft were placed in Martian orbit and sent landers down to the planet's surface. On July 20, 1976, the *Viking 1* lander set down on Mars. It was the first safe landing of a spacecraft on another planet. The next year *Voyager 1* and *Voyager 2* were launched toward the far planets in the solar system.

They journeyed past Jupiter and Saturn and continued outward. *Voyager 2* flew by Uranus and Neptune. By 1998 *Voyager 1* had traveled farther than any human-made object in history. As of May 24, 2005, *Voyager 1* was reported to be more than 8.7 billion miles from the Sun and had crossed into "the solar system's final frontier" described in a NASA press release as "a vast, turbulent expanse, where the sun's influence ends and the solar wind crashes into the thin gas between stars." *Voyager 2* was similarly moving toward the outer boundary of the solar system as of 2006. NASA hopes it can maintain contact with them as they enter the new frontier of interstellar space.

In 1989 NASA launched two interplanetary missions: *Magellan* to Venus and *Galileo* to Jupiter. In August 1990 the *Magellan* lander set down on the Venutian surface. Five years later *Galileo* began orbiting Jupiter. It continued to send NASA data until September 2003, when it was destroyed on entry into Jupiter's atmosphere. Both missions provided new maps of planetary surfaces and important atmospheric data.

The 1990s were a tough decade for NASA's interplanetary craft. Three spacecraft sent to Mars were lost: the *Mars Observer* in 1992, the *Mars Climate Orbiter* in 1998, and the *Mars Polar Lander/Deep Space 2* of 1999. However, in 1997 the *Mars Global Surveyor* was put into Martian orbit. The same year *Mars Pathfinder* successfully put a lander, including the *Sojourner* rover, on the Martian surface. It explored an area called Ares Vallis in the planet's northern hemisphere.

During the early 2000s NASA had great success with interplanetary missions. In late 2001 *Mars Odyssey* arrived in Martian orbit and began mapping the planet. Two years later the *Mars Exploration* mission was launched including twin landing craft with rovers named *Spirit* and *Opportunity*. In January 2004 the rovers landed safely on Mars and began exploring its terrain.

In June 2004 the *Cassini* spacecraft reached Saturn. It was launched in 1997 as a joint mission between NASA, the European Space Agency (ESA), and the Italian space agency. This robotic spacecraft released a scientific probe called *Huygens* into the atmosphere of Saturn's largest moon, Titan. *Cassini* is scheduled to orbit Saturn through 2008.

During the 1990s NASA began the Discovery program. The goal of this program is to carry out numerous, relatively small and inexpensive space missions with specific objectives. Each mission must cost less than $299 million and proceed from initial development to launch within thirty-six months. NASA calls Discovery the "faster, better, cheaper" approach to space science. Discovery missions ongoing as of April 2006 have a number of goals, including investigation of comets, aster-

oids, and space weather. The *Messenger* spacecraft is projected to fly by Venus on October 24, 2006, at an altitude of 1,951 miles on its way to Mercury. After flying by Mercury in 2008 and 2009 to map the planet and gather data, the *Messenger* craft will enter Mercury's orbit in March 2011.

Earth and Sun Orbiters

NASA's space science program has been launching robotic spacecraft into orbit since the 1960s. These spacecraft have included numerous satellites put into Earth orbit to collect data about the planet's weather, oceans, and atmosphere. Other NASA satellites were designed to look outward into space.

During the 1980s the agency began a program called NASA's Great Observatories. There are four robotic spacecraft in this program: the *Hubble Space Telescope* (launched in April 1990), the *Compton Gamma Ray Observatory* (launched in April 1991), the *Chandra X-Ray Observatory* (launched in July 1999), and the *Spitzer Space Telescope* (launched in August 2003). These Earth-orbiting spacecraft contain highly advanced and sensitive instruments that allow scientists to study radiation emitted from nearby celestial bodies and distant galaxies. The space observatories provide a clear picture of the universe because they are located outside the interference of Earth's atmosphere.

Throughout its history NASA has operated a number of small observatories and satellites that incorporate sophisticated instruments such as scatterometers (a specialized form of radar for Earth study) and interferometers (for precise determinations of distance or wavelengths in space).

Numerous NASA science programs are geared toward studying the flow of energy from the Sun to the Earth. This energy includes a continuous flow of plasma called the solar wind. The solar wind and other emissions from the sun affect the magnetic properties of the space surrounding the Earth (or geospace). This magnetic phenomenon is known as "space weather."

The Explorers program is another NASA science program dedicated to operating small- to medium-sized space missions for a modest cost (less than $180 million per mission). This program includes satellites and observatories that gather data about the Earth, the Sun, and the surrounding universe.

NASA also works with international partners to perform space science missions. The International Solar-Terrestrial Physics (ISTP) Science Initiative is a collaborative effort between NASA, ESA, and Japan's space agency. Begun in the 1990s, the ISTP program uses satellites to gather information about space weather and its effects on geospace (the space around the Earth). *Ulysses* is a joint

NASA-ESA mission to study the Sun. Orbiting the Earth once every seventy-two hours, the *International Gamma-Ray Astrophysics Laboratory* (*INTEGRAL*) is a space observatory operated by the ESA in conjunction with NASA and the Russian space agency.

NASA's ORGANIZATION AND FACILITIES

Agency-level management takes place at NASA headquarters in Washington, D.C. People at this level interact with national leaders and NASA customers regarding overall agency concerns, such as budget, strategy, policies, and long-term investments. Headquarters is considered the centralized point of accountability and communication between NASA and people outside the agency. As of 2006 more than 1,200 personnel were employed at the Washington site.

During 2004 and 2005 NASA made major changes to its organizational structure to streamline the agency. The reorganization was designed to eliminate the so-called "stove pipe effect" in which individual facilities and enterprises within the agency operated too independently and did not communicate well with one another or with NASA headquarters. Many critics had blamed NASA's management structure for contributing to the *Challenger* and *Columbia* disasters. According to NASA's fiscal year 2005 performance and accountability report, the new system will ensure that "all parts of the Agency act as One NASA team to make decisions for the common good, collaborate across traditional boundaries, and leverage the Agency's many unique capabilities in support of a single focus: exploration."

NASA's organizational structure is shown in Figure 2.3. It includes four major divisions called Mission Directorates:

- Aeronautics Research—Devoted to research and development of new aeronautical technologies and aviation systems

- Exploration Systems—Responsible for biological research and technological development to support human and robotic exploration

- Science—Charged with ensuring that missions are planned to reap scientific benefits, analyzing scientific data, and facilitating cross-transfer between earth and space science findings

- Space Operations—Dedicated to directing launches and flight operations and related communications systems

NASA Facilities

Figure 2.4 shows the locations of NASA headquarters and various field facilities, including the ten major facilities called centers.

Each center supports multiple projects. Each center is also assigned a particular area of expertise for which it is supposed to build and maintain human resources, facilities, and other capabilities. NASA calls these "centers of excellence."

AMES RESEARCH CENTER. The Ames Research Center (ARC) is located in Mountain View, California. It was founded as an aeronautics research laboratory in 1939 adjacent to a military base later named Moffett Field. The base was closed in 1994 and its facilities and runways turned over to ARC. The center conducts research in astrobiology (the origin, evolution, distribution, and destiny of life in the universe), air traffic management, supercomputing, artificial intelligence, nanotechnology, and other areas of importance to space exploration. It also conducts wind tunnel testing and flight simulations. ARC is a center of excellence for information technology. As of 2006 it employed more than 2,500 people.

DRYDEN FLIGHT RESEARCH CENTER. The Dryden Flight Research Center (DFRC) is located at Edwards Air Force Base in Edwards, California. The base was the site of joint NACA military testing of high-speed experimental aircraft during the late 1940s. In 1959 the high-speed flight station at the base was designated a NASA flight research center. DFRC is NASA's primary installation for flight research. It also serves as a back-up landing site for the space shuttle. DFRC is a center of excellence for atmospheric flight operations. As of 2006 approximately 1,000 people were employed there.

GLENN RESEARCH CENTER. The Glenn Research Center (GRC) is located in Cleveland, Ohio, at Lewis Field adjacent to Cleveland Hopkins International Airport. It began in 1941 as NACA's Aircraft Engine Research Laboratory. GRC researches and develops technologies in aeropropulsion, aerospace power, microgravity science, electric propulsion, and communications technologies for aeronautics and space applications. Its facilities include the nearby Plum Brook field station at which large-scale testing is conducted. GRC is a center of excellence for turbomachinery (turbine-based machines). As of 2006 it employed more than 1,000 people.

GODDARD SPACE FLIGHT CENTER. The Goddard Space Flight Center (GSFC) is located in Greenbelt, Maryland, a suburb of Washington, D.C. It was founded in 1959 as NASA's first spaceflight center. GSFC is a major laboratory for developing robotic (unmanned) scientific spacecraft. The center also operates the Wallops Flight Facility near Chincoteague, Virginia, and the Independent Verification and Validation (IV&V) Facility in Fairmont, West Virginia. Wallops is NASA's principal installation for managing and implementing suborbital research programs. The IV&V facility was formed following the space shuttle *Challenger* accident to ensure that mission-critical software is safe and cost-effective.

FIGURE 2.3

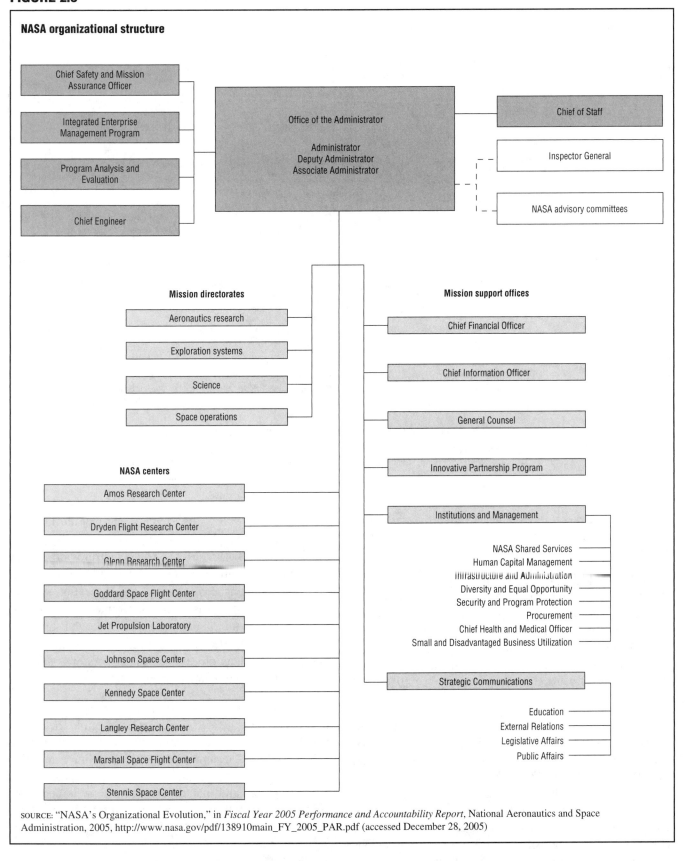

NASA organizational structure

Chief Safety and Mission Assurance Officer

Integrated Enterprise Management Program

Program Analysis and Evaluation

Chief Engineer

Office of the Administrator

Administrator
Deputy Administrator
Associate Administrator

Chief of Staff

Inspector General

NASA advisory committees

Mission directorates

Aeronautics research

Exploration systems

Science

Space operations

Mission support offices

Chief Financial Officer

Chief Information Officer

General Counsel

Innovative Partnership Program

Institutions and Management

NASA Shared Services
Human Capital Management
Infrastructure and Administration
Diversity and Equal Opportunity
Security and Program Protection
Procurement
Chief Health and Medical Officer
Small and Disadvantaged Business Utilization

Strategic Communications

Education
External Relations
Legislative Affairs
Public Affairs

NASA centers

Amos Research Center

Dryden Flight Research Center

Glenn Research Center

Goddard Space Flight Center

Jet Propulsion Laboratory

Johnson Space Center

Kennedy Space Center

Langley Research Center

Marshall Space Flight Center

Stennis Space Center

SOURCE: "NASA's Organizational Evolution," in *Fiscal Year 2005 Performance and Accountability Report*, National Aeronautics and Space Administration, 2005, http://www.nasa.gov/pdf/138910main_FY_2005_PAR.pdf (accessed December 28, 2005)

In 1966 NASA established the National Space Science Data Center (NSSDC) at GSFC. The NSSDC became the archive center for data from NASA's space science missions and continues to serve that purpose. Space science data from NASA missions are made available to researchers and, in some cases, to the general public.

FIGURE 2.4

NASA sites

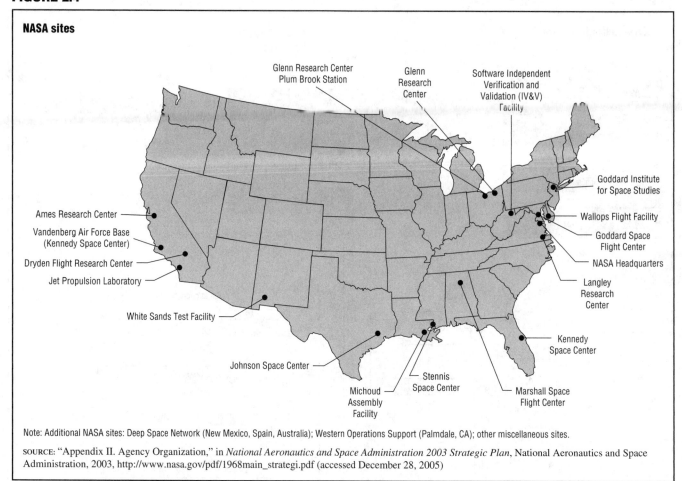

Note: Additional NASA sites: Deep Space Network (New Mexico, Spain, Australia); Western Operations Support (Palmdale, CA); other miscellaneous sites.

SOURCE: "Appendix II. Agency Organization," in *National Aeronautics and Space Administration 2003 Strategic Plan*, National Aeronautics and Space Administration, 2003, http://www.nasa.gov/pdf/1968main_strategi.pdf (accessed December 28, 2005)

GSFC is a center of excellence for earth science and physics and astronomy. As of 2006 more than 8,000 people were employed there.

JOHNSON SPACE CENTER. The Johnson Space Center (JSC) is located in Houston, Texas. It was established in 1961 to be the focus of the manned spaceflight program. At that time, it was known simply as the Manned Space-craft Center. In 1973 the Center was renamed the Lyndon B. Johnson Space Center in honor of the late president's support of NASA space programs during the 1950s and 1960s.

The JSC houses the program offices and mission control centers for the space shuttle and the International Space Station. JSC facilities are used for astronaut training and spaceflight simulations for both these programs. Aircraft used to train astronauts and to support the space shuttle program are stationed at nearby Ellington Field, a joint civilian and military airport operated by the City of Houston. JSC is a center of excellence for human operations in space. As of 2006 it employed more than 15,000 people.

KENNEDY SPACE CENTER. The Kennedy Space Center (KSC) is located on Merritt Island, Florida, adjacent to the Cape Canaveral Air Force Station. The air force station was the site of the Mercury and Gemini launches of the early 1960s. KSC was created specifically for the Apollo missions to the Moon. The center provides launch and landing facilities for the space shuttle program and performs maintenance, assembly, and inspection services on the spacecraft. It is also responsible for packaging components of the laboratory experiments that are used on the space shuttle. KSC is a center of excellence for launch and payload processing systems. As of 2006 more than 11,000 people worked there.

LANGLEY RESEARCH CENTER. The Langley Research Center (LRC) is located in Hampton, Virginia. In 1917 it was established as the country's first civilian aeronautics laboratory. LRC designs and develops military and civilian aircraft, conducts atmospheric flight research, and tests structures and materials in wind tunnels and other testing facilities. It is a center of excellence for structures and materials. As of 2006 LRC employed more than 3,500 people.

MARSHALL SPACE FLIGHT CENTER. The Marshall Space Flight Center (MSFC) is located near Huntsville, Alabama, on the Redstone Arsenal Site. During the 1950s a team of rocketry specialists led by Wernher von Braun worked at the arsenal site developing rockets for the U.S. military. In 1960 the Redstone Arsenal Site's space-related projects and personnel were transferred to the newly formed MSFC. The center developed the Saturn rockets used throughout the Apollo program. MSFC manages the manufacturing contracts for the space shuttle main engine, external tank, and reusable solid rocket motor. The center also conducts research in micro-gravity and space optics and develops programs for space shuttle payloads. It is a center of excellence for space propulsion. As of 2006 more than 6,000 people were employed there.

STENNIS SPACE CENTER. The Stennis Space Center (SSC) is located in Bay St. Louis, Mississippi. It was founded in 1961 as the static test facility for launch vehicles to be used in the Apollo program. SSC is home to the largest rocket propulsion test complex in the United States. It is NASA's primary installation for testing and flight-certifying rocket propulsion systems for the space shuttle and other space vehicles. The center also works with government and commercial partners to develop remote sensing technology. SSC is a center of excellence for rocket propulsion testing systems. As of 2006 it employed more than 1,500 people.

Other NASA Facilities

There are numerous facilities and installations that provide support to the field Centers and are either operated by NASA or under contract to NASA. Some of the major ones are described below.

JET PROPULSION LABORATORY. The Jet Propulsion Laboratory (JPL) is located in Pasadena, California. This facility is owned by NASA but operated under a contractual agreement by the California Institute of Technology. JPL began informally during the 1930s as a group of student rocket enthusiasts under the direction of Professor Theodore von Kármán, head of the university's Guggenheim Aeronautical Laboratory. These rocket scientists achieved funding for their projects from the U.S. Army, and by the 1940s they were investigating new technologies in aerodynamics and propellant chemistry under the name of the Jet Propulsion Laboratory. In 1958 JPL was transferred from Army jurisdiction to NASA.

Jet propulsion is no longer the primary focus at JPL. The facility now serves as NASA's primary operator of robotic exploration missions. It also manages and operates NASA's Deep Space Network. As of 2006 JPL employed approximately 5,000 people.

DEEP SPACE NETWORK. The Deep Space Network (DSN) is an international network of antennas that enables NASA mission teams to communicate with distant spacecraft. As shown in Figure 2.5 DSN communications complexes are situated at three locations around the world (roughly 120 degrees apart) in Goldstone, California, Robledo, Spain, and Tidbindilla, Australia.

This placement allows the JPL operations control center to maintain constant contact with spacecraft as the earth rotates.

WHITE SANDS TEST FACILITY. The White Sands Test Facility (WSTF) is located at Las Cruces, New Mexico, a remote desert location. WSTF provides services to military and government clients. It is NASA's primary facility for testing and evaluating rocket propulsion systems, spacecraft components, and hazardous materials used in space travel. WSTF supports the space shuttle and ISS programs.

NASA's WORKFORCE

People employed by federal agencies (excluding the military) are called civil servants. As of 2006 NASA employed approximately 16,650 full-time civil servants. Another 40,000 people supported NASA projects by working under contracts or grants handed out by the agency. The vast majority of these people work at or near NASA facilities.

At the height of Apollo development NASA employed nearly 36,000 civil servants. By the early 1990s this number had dropped to 24,000 and continued to decrease over the next several years. NASA reduced its workforce by offering employees cash bonuses to retire early and through normal attrition (not replacing workers that leave). During most of the 1990s the agency operated under a hiring freeze. One consequence of this was that very few young people entered the NASA workforce during that period.

NASA divides its civil service workforce into four main categories:

- Scientists and Engineers—Highly educated professionals that conduct aerospace research and development or perform biological, life science, or medical research or services. This category includes space scientists, biologists, aerospace engineers, physicians, nurses, and psychologists.

- Technicians—Technicians fall into two categories. Some are specialists that provide services such as drafting or photographic development. Others are skilled at particular trades (such as mechanics or electrical work).

- Professional Administrators—These employees operate non-technical functions such as management, legal affairs, public relations, and human resources.

- Clerical Workers—This includes secretarial, administrative, and clerical positions.

People engaged in technical work comprise nearly 60% of the agency's workforce.

FIGURE 2.5

NASA Deep Space Network communications complexes

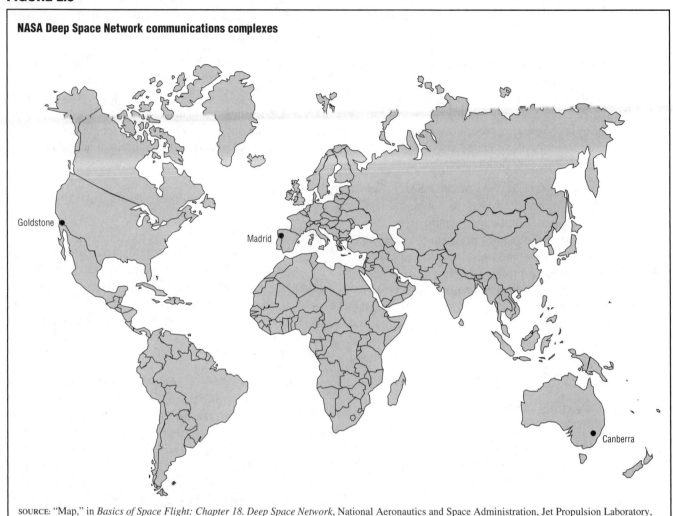

SOURCE: "Map," in *Basics of Space Flight: Chapter 18. Deep Space Network*, National Aeronautics and Space Administration, Jet Propulsion Laboratory, California Institute of Technology, February 2001, http://www2.jpl.nasa.gov/basics/basics.pdf (accessed December 28, 2005)

According to NASA's Human Resources department the vast majority of the agency's workforce is at least forty years old, and the average NASA employee is forty-seven years old. Approximately two-thirds of the NASA employees are male. Most male employees work in science and engineering professions. The majority of the female employees work in professional administrative positions. More than 75% of NASA employees are white.

Most NASA workers have college degrees. More than a third of the workforce have advanced degrees beyond the bachelor level. The average salary for a NASA employee is about $95,000 per year. Scientists and engineers are the highest paid, averaging $105,000 per year, while clerical employees are the lowest paid, earning around $45,000 per year.

Contractors and Grantees

More than 40,000 people support NASA services under contracts and grant arrangements. NASA's major contractors are manufacturing companies in the aerospace industry. United Space Alliance is a joint venture between the Boeing and Lockheed Martin corporations. As of 2005 it performed day-to-day operations of the space shuttle program and employed more than 10,000 people. Most worked at JSC and KSC. The California Institute of Technology is another major contractor. It operates JPL and employs approximately 5,000 people.

NASA also funds research projects at private institutions, such as universities, and encourages commercial investment in space research. In 1985 the U.S. Congress amended the National Aeronautics and Space Administration Act to direct NASA to "seek and encourage to the maximum extent possible the fullest commercial use of space." In response NASA developed a Space Partnership Development (SPD) Office, an industry-university-government collaboration. The SPD manages twelve Research Partnership Centers (RPCs) at universities and nonprofit institutions engaged in space research and product development. (See Figure 2.6.) Each dollar of NASA funding provided to an RPC must be matched by at least two dollars of non-NASA funding.

FIGURE 2.6

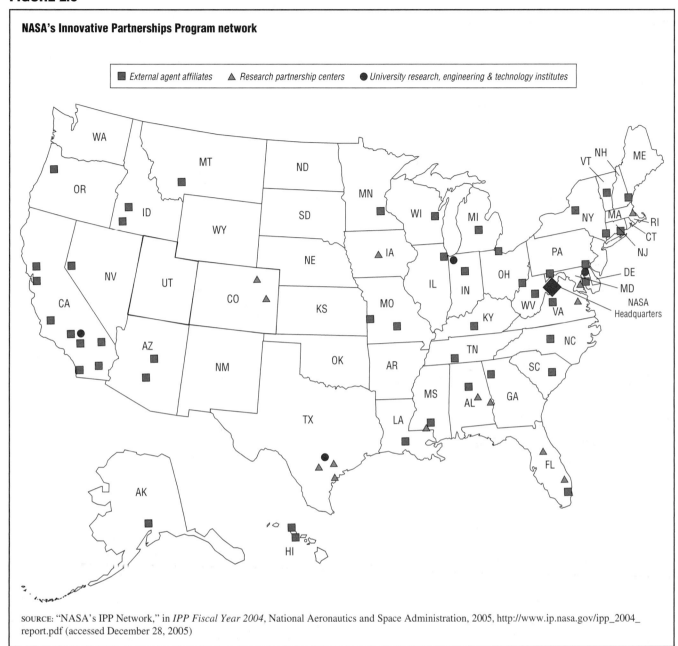

NASA's Innovative Partnerships Program network

■ *External agent affiliates*   ▲ *Research partnership centers*   ● *University research, engineering & technology institutes*

SOURCE: "NASA's IPP Network," in *IPP Fiscal Year 2004*, National Aeronautics and Space Administration, 2005, http://www.ip.nasa.gov/ipp_2004_report.pdf (accessed December 28, 2005)

In 2004 President George W. Bush urged NASA to expand its commercial partnerships to develop technology to support his new vision for the space agency for crewed missions to the Moon and Mars. NASA incorporated existing partnership arrangements into a new program called the Innovative Partnerships Program (IPP) under the direction of the Space Operations Mission Directorate. According to NASA the IPP worked with more than 150 partners (called external agents) during 2004 to develop technologies and products that will be useful to NASA and have commercial applications. Other programs operated under the IPP include the University Research, Engineering & Technology Institutes program, which funds cutting-edge research at selected universities and the Small Business Innovative Research and Small Business Technology

Transfer programs. The locations of major facilities involved in the IPP network are shown in Figure 2.6.

Astronauts

Astronauts are the most famous NASA workers. In 1959 the first group of seven astronauts was chosen from 500 candidates. All were military men with experience flying jets. At the time, spacecraft restrictions required that astronauts be less than 5 feet 11 inches tall. In the early days of the Apollo program all astronauts were chosen from the military services. This soon changed, and NASA began including civilian pilots with extensive flight experience. During the mid-1960s NASA expanded the astronaut corps to include non-pilots with academic qualifications in science, engineering, or medicine.

TABLE 2.4

Astronaut candidate selection process timeline, 2007–08

July 1st 2007	July–Aug. 2007	Aug.–Dec. 2007	Sept.–Dec. 2007	Spring 2008	Summer 2008
Cutoff date for receipt of new applications	Applications reviewed by Astronaut Candidate (ASCAN) Selection Rating Panel to determine highly qualified (HQ) applicants	Evaluation forms sent to supervisors and references of HQ applicants	Week-long interviews and medical examinations conducted	Applicants interviewed will be contacted by phone	New Astronaut Candidates report for duty to the Johnson Space Center in Houston, Texas
		Request for prescreening medical exam sent to civilian HQ applicants		New Astronaut Candidates publicly announced	
		HQ applications reviewed by ASCAN Selection Board to select the applicants to be interviewed		Applicants not selected notified by letter	

Note: This schedule is subject to change.

SOURCE: "Astronaut Candidate Selection Process Timeline," in *Astronaut Selection*, National Aeronautics and Space Administration, October 21, 2005, http://www.nasajobs.nasa.gov/astronauts/content/timeline.htm (accessed January 31, 2006)

In 1978 the first group of space shuttle astronauts was selected. For the first time the trainees included women and minorities. The unique environment aboard the space shuttle permitted even more opportunities for non-pilots to fly into space.

A typical shuttle crew includes a commander and a pilot. Both of these crewmembers are considered pilot astronauts. In addition there can be three to five other crewmembers called mission specialists or payload specialists. Mission specialists are NASA astronauts (typically scientists) with specific onboard responsibilities during a mission. Payload specialists can be scientists, engineers, and ordinary citizens from the private or commercial sector or foreign astronauts invited by NASA to participate in a shuttle mission.

Space shuttle commanders, pilots, and mission specialists are career NASA astronauts, as are commanders and flight engineers that serve aboard the *ISS*.

During the early 1980s NASA was enthusiastic about including private citizens on space shuttle flights. This was viewed as a way to better interest the public, and particularly children, in space travel. One of the most famous participants was Christa McAuliffe (1948–86), the first schoolteacher selected to go into space. On January 28, 1986, she died along with her crewmates when the shuttle *Challenger* exploded shortly after launch. This disaster ended NASA's policy of inviting private citizens on shuttle flights.

In 2002 NASA organized a new program to put a teacher in space called the Educator Astronaut program. Under this program qualified teachers were invited to apply to be full-time career astronauts. At the time NASA planned to select three to six Educator Astronauts for future space shuttle flights. Although the program contin-

ued following the 2003 *Columbia* disaster, its future is uncertain given the plans to eliminate the shuttle program.

The 1980s witnessed several firsts in NASA's astronaut corps. In June 1983 Sally Ride (1951–) became the first U.S. woman in space when she served as a mission specialist aboard the space shuttle *Challenger*. It was the shuttle's seventh mission. Two months later mission specialist Guion Bluford (1942–) became the first African-American in space as part of the shuttle's next mission.

Astronaut Selection

NASA accepts applications from astronaut candidates on an ongoing basis. Civilian candidates submit their applications directly to NASA. Candidates in the armed forces are pre-screened by the military. Every two years NASA conducts a review process to select a new group of astronauts. This process begins in odd-numbered years and follows a very specific format.

The next selection process begins on July 1, 2007, as shown in Table 2.4. The first day of July is the cutoff date for receipt of new applications. Throughout July and August the applications are reviewed by the Astronaut Candidate Selection Rating Panel. The panel narrows the field to those applicants considered highly qualified (HQ) and collects information about them through the remainder of the year. This information is used to select HQ applicants for extensive interviews and medical examinations. By spring of the following even-numbered year the selection process is complete. The names of the successful candidates are released to the media. Those selected begin training soon afterwards at Johnson Space Center in Texas. The training period lasts one to two years.

Pilot astronaut applicants must have at least 1,000 hours of command flying experience in jet aircraft.

Preference is given to pilots with flight test experience. Pilot candidates must also pass a stringent medical examination and be between sixty-four and seventy-six inches tall. Mission specialists are required to pass a less stringent medical examination and must be between fifty-eight-and-a-half and seventy-six inches tall.

Astronaut Pay Rates

Civilian astronauts employed by NASA are civil servants. They are paid salaries based on the federal government's pay scale called the General Schedule or GS. There are fifteen GS pay levels ranging from the lowest (GS-1) to the highest (GS-15). NASA's mission specialists fall within grades GS-11 through GS-13 depending on their education, experience, and qualifications. These grade scales cover a salary range between approximately $56,000 and $104,000 per year.

Active-duty military personnel selected to be NASA astronauts remain on the military payroll during their assignment to Johnson Space Center.

NASA's BUDGET

NASA is a federal government agency. For accounting purposes the federal government operates on a fiscal year (FY) that begins in October and runs through the end of September. Thus, fiscal year 2007 covers the time period of October 1, 2006, through September 30, 2007. Each year by the first Monday in February the President of the United States must present a proposed budget to the U.S. House of Representatives. This is the amount of money that the president estimates will be required to operate the federal government during the next fiscal year.

It can take many months for the House to debate, negotiate, and approve a final budget. Then, the U.S. Senate also must approve the budget. This entire process can take longer than a year, which means that NASA can be well into a fiscal year (or even beyond it) before knowing the exact amount of money appropriated for that year.

On February 6, 2006, NASA Administrator Mike Griffin outlined NASA's FY 2007 budget estimate at a news conference. The agency requested $16.8 billion, a 3.2% increase above the FY 2006 appropriated budget.

Table 2.5 shows the FY 2007 budget request broken down by mission directorate and theme. The space shuttle program is the single most expensive undertaking. It will cost more than $4 billion to operate for the year. Together, the space shuttle and ISS programs account for $6.2 billion (or 37% of the entire budget) for fiscal year 2007. They comprise the Space Operations Mission Directorate.

The Science Mission Directorate is the next most expensive category in the budget. NASA is requesting $5.3 billion to support robotic investigations of our solar system and beyond. Nearly $4 billion is devoted to the Exploration

TABLE 2.5

NASA budget request, fiscal year 2007

$ in millions	FY07 request
Exploration, science, & aeronautics	**10,524.4**
Science	5,330.0
Solar system exploration	1,602.0
Universe	1,517.1
Earth-Sun system	2,210.9
Exploration systems	3,978.3
Constellation systems	3,057.6
Exploration systems research & tech	646.1
Human systems research & tech	274.6
Aeronautics research	724.4
Cross-agency support programs	491.7
Education	153.3
IEMP*	108.2
Innovative partnerships	197.9
Shared capability assets	32.2
Exploration capabilities	**6,234.4**
Space operations	6,234.4
Space station	1,811.3
Space shuttle	4,056.7
Space & flight support	366.5
Inspector General	**33.5**
Total NASA	**16,792.3**

*IEMP=Integrated Enterprise Management Program

SOURCE: Adapted from "Budget Adjustments," in *Fiscal Year Budget Estimates 2007: Agency Summary*, National Aeronautics and Space Administration, February 6, 2006, http://www.nasa.gov/pdf/142543main_FY07%20Budget%20Agency%20Summary%20--%20FINAL%202-4-06.pdf (accessed February 6, 2006)

TABLE 2.6

NASA budget estimates, fiscal years 2006–11

$ in millions	FY2006	FY 2007	FY 2008	FY 2009	FY 2010	FY 2011
Total NASA	16,623	16,792	17,309	17,614	18,026	18,460
Percent change year to year		3.2%	3.1%	1.8%	2.3%	2.4%

SOURCE: "FY2007 Budget Request Summary," in *Fiscal Year Budget Estimates 2007: Agency Summary*, National Aeronautics and Space Administration, February 6, 2006, http://www.nasa.gov/pdf/142543main_FY07%20Budget%20Agency%20Summary%20--%20FINAL%202-4-06.pdf (accessed February 6, 2006)

Systems Mission Directorate. This money would fund research and development of new spacecraft and technologies to send human explorers to Earth's moon and to Mars.

Table 2.6 shows NASA's expectations for future budgets through FY 2011. The long-term plan assumes that shuttle program costs will decrease slowly until 2010 and disappear by 2012. The money saved by eliminating the shuttle program will give a funding boost to the new lunar and Martian exploration programs.

NASA's GOALS FOR THE FUTURE

NASA's stated overall goal for the future is to improve life on Earth, while extending human life to outer space and searching for other life in the universe. NASA believes that this goal will be achieved through three broad missions:

- Understanding and protecting Earth

- Exploring the universe and searching for life

- Inspiring young people to appreciate the importance of space exploration

In February 2004 NASA's goals for the twenty-first century were redefined in *A Renewed Spirit of Discovery: The President's Vision for U.S. Space Exploration*. In this document President George W. Bush articulated his goals for the nation's space program over the next few decades:

- Implementing an affordable space exploration program that includes robotic spacecraft and human explorers

- Putting astronauts on the Moon by the year 2020

- Developing new technologies and equipment needed to acquire data about potential destinations for human astronauts

- Promoting international and commercial participation in the exploration program

The president called for the space shuttle fleet to be retired by 2010. NASA's participation in the *ISS* would end in 2016 with the completion of specific research objectives at the station.

NASA's plan for achieving the president's mandate includes such ongoing missions as *Mars Rover*, which will be used as stepping-stones to future exploration missions. NASA plans to use other robotic spacecraft to test new technologies and gather data about the Moon and Mars before sending humans to explore them.

Human travel to the Moon and Mars will require development of new launch and crew vehicles. NASA no longer has any of the Saturn V rockets that lifted Apollo spacecraft into space. A new heavy-lift vehicle must be developed. A new crew exploration vehicle (CEV) is also needed. A space shuttle cannot serve this purpose, because it was designed only for low Earth orbit.

Under the new Exploration Systems Mission Directorate are major subdivisions called Exploration Systems Research and Technology, Human Systems Research and Technology, and Constellation Systems. The role of each is described below:

- Exploration Systems Research and Technology— Develop innovative technologies and capabilities with the help of contracts with NASA Centers, industry, and academic institutions. This includes development of nuclear electric power and propulsion technologies in partnership with the Department of Energy's Office of Naval Reactors. Oversee new technology transfer partnerships and the Centennial Challenges prize program in which cash prizes are awarded to non-NASA partners that develop innovative technologies.

- Human Systems Research and Technology—Perform research and development related to the health, living and working environments, safety, and activities of crewmembers.

- Constellation Systems—Develop the Crew Exploration Vehicle (CEV), Crew Launch Vehicle (CLV) and related infrastructure, transportation and communications systems, instrumentation, and robotic systems.

In September 2005 NASA presented its conceptual design for the spacecraft being developed under the Constellation Systems program. (See Figure 2.7.) The new CEV will be shaped like an Apollo capsule, but much larger. It will have room for up to six astronauts. The CEV will be blasted into space atop a long cylindrical rocket using a shuttle solid rocket booster as the first stage and a shuttle main engine as the second stage. The CEV will be capable of docking with the ISS. The CEV will parachute to dry land when it returns to Earth and will be reusable, except for the heat shield, which will be replaced for each trip. In addition, a heavy-lift spacecraft will be built to be powered by two longer solid rocket boosters and five shuttle main engines. This craft will carry heavy cargo to the ISS and later put components of the Moon and Mars missions into space.

Figure 2.8 shows NASA's long-range plans for human space exploration missions. Space shuttle operations are expected to cease by FY 2011. By that time development of the CEV and CLV should be nearly completed. NASA expects the first human CEV flight to the Moon to occur between 2015 and 2020. This will be followed by establishment of a lunar outpost and preparation for a crewed mission to Mars. NASA's ability to follow through with this long-range plan is dependent on Congressional approval of projected budgets and successful implementation of the new technologies that will be required for success.

FIGURE 2.7

Planned lunar launch vehicles for heavy cargo and crew transport

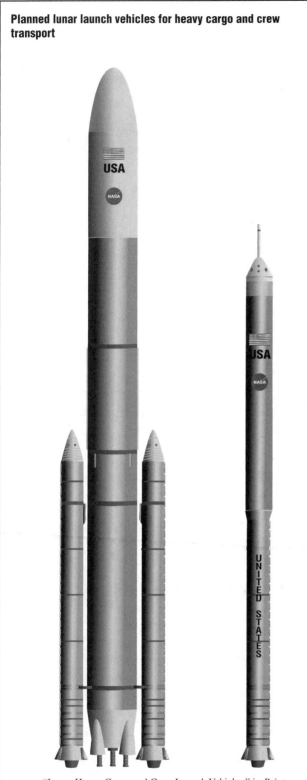

SOURCE: "Lunar Heavy Cargo and Crew Launch Vehicles," in *Print Materials: Launch Vehicles*, in *How We'll Get Back to the Moon*, National Aeronautics and Space Administration, September 22, 2005, http://www.nasa.gov/pdf/136545main_CEV_litho2_hi.pdf (accessed December 28, 2005)

FIGURE 2.8

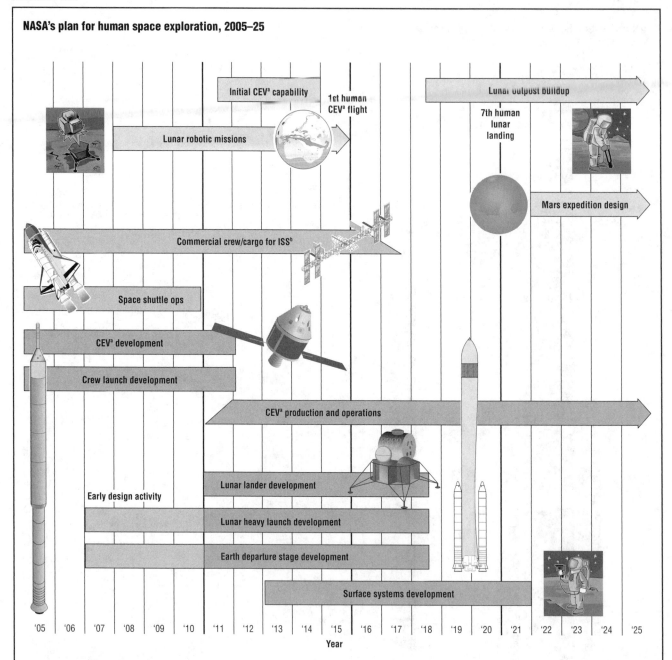

NASA's plan for human space exploration, 2005–25

Initial CEV[a] capability

1st human CEV[a] flight

Lunar outpost buildup

7th human lunar landing

Lunar robotic missions

Mars expedition design

Commercial crew/cargo for ISS[b]

Space shuttle ops

CEV[a] development

Crew launch development

CEV[a] production and operations

Early design activity

Lunar lander development

Lunar heavy launch development

Earth departure stage development

Surface systems development

'05 '06 '07 '08 '09 '10 '11 '12 '13 '14 '15 '16 '17 '18 '19 '20 '21 '22 '23 '24 '25

Year

Note: Specific dates and milestones not yet established. CEV/CLV[c] availablilty planned for no later than 2014, and potentially much sooner. Return to the Moon as early as 2018, but not later than 2020.
[a]CEV=Crew Exploration Vehicle
[b]ISS=International Space Station
[c]CLV=Crew Launch Vehicle

SOURCE: "NASA's Exploration Roadmap," in *Fiscal Year Budget Estimates 2007: Agency Summary*, National Aeronautics and Space Administration, February 6, 2006, http://www.nasa.gov/pdf/142543main_FY07%20Budget%20Agency%20Summary%20--%20FINAL%202-4-06.pdf (accessed February 6, 2006)

CHAPTER 3

SPACE ORGANIZATIONS, PART 2: U.S. MILITARY, FOREIGN, AND PRIVATE

Outer space, including the Moon and other celestial bodies, shall be free for exploration and use by all Nations.

—United Nations Treaty of 1967

Although the National Aeronautics and Space Agency (NASA) is the best-known space organization in the world, it is not the only one. The U.S. military and many foreign governments also have active space programs. In fact, the U.S. military program existed even before NASA was formed. Most modern military space ventures center around ballistic missiles and data-gathering satellites. These are unmanned projects. The United States officially holds the policy that it will not develop space weapons, only defensive systems. Some critics complain that the line between the two is growing vague. Increasingly the U.S. military is being criticized for developing spacecraft that could be considered weapons. For example, the Near Field Infrared Experiment is a satellite originally designed with a system capable of destroying other satellites in space. This capability was eventually dropped from the design.

Chief among the foreign governments with space programs is the Russian space program operated by an agency called Rosaviakosmos. The Russian agency continues the program begun by the Union of Soviet Socialist Republics decades ago. For about half of the twentieth century the Soviet Union engaged in a bitter Cold War rivalry for space supremacy with the United States. The Soviets achieved many milestones in space ahead of the United States, including the first manned spaceflight in 1961.

In 1991 the Soviet Union splintered into individual nations (including Russia) that were friendlier with the United States. Civilian space agencies in the United States and Russia struggled to carry on ambitious space programs as their funding was cut. They began working together on many space ventures. Eventually space programs sprang up in Europe, China, Japan, and other countries. This presented opportunities for new alliances in space.

In the past private organizations contributed to space exploration indirectly by promoting space programs and gathering together individuals interested in rocket science, physics, astronomy, space travel, or space commerce. In 2004 the private sector opened a new era in space exploration when the first privately funded manned vehicle traveled into space and back. Private space ventures are expected to grow quickly during the twenty-first century.

U.S. MILITARY SPACE PROGRAMS

The United States must win and maintain the capability to control space in order to assure the progress and preeminence of the free nations.

—Air Force Chief of Staff General Thomas White, 1959

The U.S. military had space aspirations long before spaceflight was possible. The three main branches of the military, the Army, Air Force, and Navy, began space programs following World War II. They sometimes collaborated, but more often they competed against each other to develop rockets, satellites, and manned space programs.

In 1958 President Eisenhower limited the military's role in space when he created NASA as a civilian agency. NASA was given responsibility for the nation's manned space programs. The military was allowed to pursue space projects that benefited national defense. Despite the separation, the two programs overlapped quite a bit. Even in the twenty-first century NASA is dependent on military resources to carry out human space exploration projects.

The Department of Defense (DOD) operates a comprehensive space program including a missile defense

FIGURE 3.1

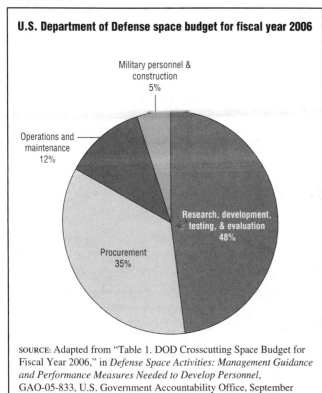

U.S. Department of Defense space budget for fiscal year 2006

Military personnel &
construction
5%

Operations and
maintenance
12%

Research, development,
testing, & evaluation
48%

Procurement
35%

SOURCE: Adapted from "Table 1. DOD Crosscutting Space Budget for Fiscal Year 2006," in *Defense Space Activities: Management Guidance and Performance Measures Needed to Develop Personnel*, GAO-05-833, U.S. Government Accountability Office, September 2005, http://www.gao.gov/new.items/d05833.pdf (accessed December 28, 2005)

system and communication, navigation, and spy satellites. For fiscal year 2006 DOD's space budget was $22.6 billion. By comparison, NASA's budget for fiscal year 2006 was $16.5 billion. As shown in Figure 3.1 nearly half of DOD's space-related budget was devoted to research, development, testing, and evaluation of space components.

World War II to 1955

The military space program began in earnest as World War II ended. In May 1945 a group of German rocket scientists led by Wernher von Braun (1912–77) surrendered to American forces. Under Operation Paperclip the U.S. Army signed a contract with von Braun's team and moved them to Fort Bliss, Texas, to work on America's rocket program. The Army also captured many German V-2 rocket parts. The von Braun team assembled the parts and launched rockets at the White Sands Proving Ground in New Mexico. On February 24, 1949, the team launched the first rocket from U.S. soil to travel beyond Earth's atmosphere and penetrate outer space. It was called *BUMPER Round 5*.

Meanwhile the U.S. Air Force had its own space program that included development of guided missiles and robotic aircraft at the Holloman Air Force Base near Alamogordo, New Mexico. As early as 1946 the Air Force was launching rockets into the upper atmosphere

that carried fruit flies, fungus spores, and small mammals. An Aeromedical Field Laboratory was established at the base as part of the Air Force's "Man in Space" program. The laboratory researched the new field of space biology and conducted high-altitude balloon flights with animals and humans.

By the early 1950s the Air Force was launching rockets to test the effects of weightlessness and radiation on mice and monkeys. Some of the animals survived the flights, while others perished. According to NASA historians in *The Human Factor: Biomedicine in the Manned Space Program to 1980* (http://history.nasa. gov/SP-4213/ch1.htm), at least four rhesus monkeys died when parachutes failed to open during descent of their spacecraft. In 1952 the Air Force ended its space biology program and turned toward ballistic missiles. However, by that time the Air Force had accumulated a wealth of knowledge and resources in the field of bioastronautics.

In 1950 the Army moved von Braun's rocket team from New Mexico to the Redstone Arsenal in Huntsville, Alabama. Four years later von Braun proposed that the Army launch an unmanned satellite into orbit using a Redstone missile as the main booster. The plan was eventually called Project Orbiter.

The Navy also pursued rocket research following World War II using captured German rockets. The Naval Research Laboratory (NRL) in Washington, D.C., equipped V-2 rockets with atmospheric probes and other scientific instruments. The NRL had a long and distinguished history in scientific research. It had been established in the 1920s at the urging of famous inventor Thomas Edison (1847–1931). The NRL invented the modern U.S. radar system and used V-2 rockets to obtain a far-ultraviolet spectrum of the sun and to discover solar x-rays. As the supply of V-2 rockets began to run out, the NRL developed its own rockets called Vikings and Aerobees.

1955 to 1958

In 1955 the United States decided to launch an unmanned satellite as part of the International Geophysical Year (IGY) project. The IGY was to run from July 1957 to the end of 1959. Various government agencies submitted proposals to develop the satellite. These included proposals from all three service branches: the Army's Project Orbiter, based on a Redstone rocket; an Air Force proposal, based on an Atlas rocket; and the Navy's Project Vanguard, based on a Viking missile. Project Vanguard was selected. The Naval Research Laboratory was delegated responsibility for developing the satellite and including a scientific experiment upon it.

The first test flights of Project Vanguard were conducted in December 1956 and May 1957. Although both tests were successful, the project proceeded slowly. In

October 1957 the Soviet Union successfully launched *Sputnik 1* into Earth orbit. It was the world's first artificial satellite. The United States was stunned that the Soviets had achieved this great milestone. The Department of Defense pressured the Navy to accelerate the Vanguard schedule. In early November 1957 the Soviets launched *Sputnik 2* with a dog named Laika aboard.

Meanwhile von Braun's team at the Redstone Arsenal had developed the Jupiter ballistic missile. Throughout the mid-1950s the Army had tried to convince the DOD that a Redstone or Jupiter rocket should be used to put a satellite into orbit. After *Sputnik 1* the DOD was ready to listen. In November 1957 the Army was authorized to pursue Project Explorer as a backup to Project Vanguard. A month later the first full-scale Vanguard launch attempt failed when the rocket exploded two seconds after lift-off.

On January 31, 1958, the Army successfully launched into space *Explorer 1*, the first U.S. satellite, using a Jupiter-C rocket. The satellite was nearly seven feet long and about six inches in diameter. It weighed thirty-one pounds. The scientific payload included temperature gauges and instruments to detect cosmic rays and the impacts of micrometeorites. The payload was developed under the direction of James Van Allen, a physics professor at the University of Iowa. Data from *Explorer 1* and the later *Explorer 3* satellite led to Van Allen's discovery of radiation belts around the Earth. In 1958 the existence of the belts was confirmed by the Soviet satellite *Sputnik 3*. (See Figure 3.2.)

On March 17, 1958, the Navy finally got its *Vanguard* satellite into orbit. Vanguard Test Vehicle 4 was launched at Cape Canaveral, Florida, and put the three-pound satellite into Earth orbit. The satellite was about the size of a grapefruit. It was the first orbiting satellite to be powered by solar energy. Solar cells also powered its radio until the radio failed in 1964. As of 2006 the silent *Vanguard* satellite continues to circle the Earth. It has remained in orbit longer than any human-made object in space.

The satellite successes of the 1950s encouraged the Air Force's space ambitions. The service began planning a manned spaceflight program called Dyna-Soar. This was to be an aircraft based on the X-15 experimental plane that could be launched into orbit by a missile, but glide back to Earth and land on an airstrip. Another project was called Man in Space Soonest (MISS). MISS called for a manned satellite to be launched by 1960, a manned laboratory to be in earth orbit by 1963, and a manned lunar landing to take place by 1965.

In June 1958 the Air Force announced a list of test pilots chosen to participate in the MISS project. (The list included only one pilot who eventually became an astronaut: Neil Armstrong.) These would have been the very

FIGURE 3.2

Sputnik 3

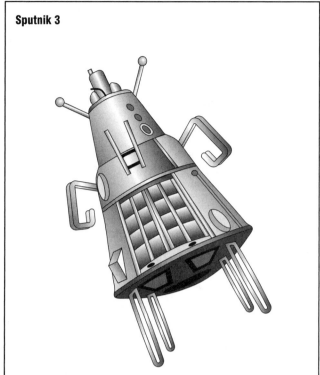

SOURCE: David P. Stern and Mauricio Peredo, "Sputnik 3," in *The Exploration of the Earth's Magnetosphere: Chapter 11. Explorers 1 and 3*, National Aeronautics and Space Administration, Goddard Space Flight Center, November 25, 2001, http://www.istp.gsfc.nasa.gov/Education/wexp13.html (accessed January 31, 2006)

first American astronauts. Four months later NASA was formed and took responsibility for manned spaceflights. Dyna-Soar and MISS were cancelled. Most of the would-be astronauts were given NASA assignments.

NASA Takes Over

Throughout the 1950s the Air Force had lobbied Congressional leaders to be given control of the nation's space program. The Air Force had excellent launch capabilities and extensive research and development capabilities in space science and bioastronautics.

According to NASA historians in *Beyond the Atmosphere: Early Years of Space Science* (http://history.nasa.gov/SP-4211/ch7-1.htm), President Eisenhower feared that militarizing the nation's space program would accelerate the nuclear arms race with the Soviet Union and locate too much political power within the military-industrial complex in the United States. Many scientists were also opposed to military control of the space program. They feared that weapon development and manned spaceflights would receive priority over scientific objectives. As a compromise, several prominent U.S. scientists urged Congress to divide the space program into two parts, with manned programs operated by the military and science programs operated by NASA. In 1958 when the agency was put into operation President Eisenhower decided to allow NASA to run the nation's space program.

Over the next few years most of the military's space programs, assets, and resources were turned over to NASA. The new agency was very dependent on military scientists with expertise in space science, particularly those of the Air Force. Even after NASA was created, Air Force officials continued to lobby political leaders for control over space programs. In January 1961 President-elect Kennedy received a report from his science adviser, Jerome Wiesner, that was very critical of NASA and its plans to develop manned space projects. Some observers interpreted the report as promoting military control of the nation's space program.

At the time, NASA was engaged in Project Mercury and planning the Apollo trips to the moon. The Air Force's Space Systems Division (SSD) proposed its own post-Mercury project called Lunex that promised to put three men on the moon by 1967. The SSD estimated the cost of the project at $7.5 billion (about $50.8 billion in 2006 dollars).

The Army's plan for a manned spaceflight was called Project Adam. It called for one astronaut to be sealed inside a capsule atop a ballistic missile for his ride in orbit. Although the Army had excellent launch capabilities and rocket technology, it lacked expertise in bioastronautics. Project Adam did not include any monitoring of the human during his spaceflight to gain medical knowledge. The Army also advocated a military outpost on the moon as part of Project Horizon. This ambitious plan included a dock and fueling station in orbit around Earth.

The U.S. Navy had its own plan for a manned spaceflight project called the Manned Earth Reconnaissance Project or Project MER. However, the Navy's space reputation was hurt by the poor performance of the Vanguard program. Also, the Navy was dependent on the Air Force for launch facilities and bioastronautics capabilities.

NASA had its share of influential supporters, including Overton Brooks (the chairman of the House Committee on Science and Astronautics) and Vice President Lyndon Johnson, both of Texas. Neither wanted the military to control the nation's space ventures. In March 1961 Brooks wrote the President a letter in which he pushed Kennedy to make clear his intentions on the matter. The President responded that he did not intend to "subordinate" NASA under military control. Kennedy increased NASA's budget and gave the agency responsibility for a manned lunar spaceflight.

NASA received Air Force help with many aspects of the early space programs. During the 1950s the Air Force

obtained infant chimpanzees and monkeys that were trained at the Holloman base for spaceflights. Many of the animals were not named. A rhesus monkey named Sam (after the Air Force School of Aviation Medicine) flew aboard a Mercury test flight in 1959.

Another chimp was named Ham (an acronym for Holloman Aero Medical). During 1961 NASA launched Ham and another "chimponaut" named Enos into outer space to orbit the Earth. The Air Force continued to run a space chimp colony until 1997. At that time twenty-one chimps were turned over to a chimpanzee rescue group in Florida.

Military and Intelligence Satellites

Following the formation of NASA the U.S. military focused most of its space resources on development of ballistic missiles and satellites. Satellites were designed for a variety of purposes, including communications, navigation, weather surveillance, and reconnaissance (spying).

During the late 1950s the Air Force worked with the Central Intelligence Agency (CIA) to develop a reconnaissance satellite capable of photographing installations on the ground in the Soviet Union from space. The project was code-named Corona. Publicly the U.S. called the satellite *Discoverer* and claimed that it conducted scientific research. More than 100 Corona missions were flown during the 1960s and early 1970s. The Soviet Union orbited its own spy satellites and also claimed that they were for scientific purposes.

Prior to the 1980s all satellites were launched aboard rockets called Expendable Launch Vehicles (ELVs). Once above Earth's atmosphere a satellite separated from its ELV, and the ELV burned up during reentry. During the 1970s the Air Force used a number of ELVs including the Scout, Thor, Delta, Atlas, and Titan rockets.

Development of the space shuttle introduced a new era in satellite deployment. The shuttle was reusable and included a crew of astronauts that could release, retrieve, and repair satellites as needed. The military was very excited about this prospect. During space shuttle development the DOD insisted that the vehicles be designed to carry heavy military satellites and be able to orbit the Earth along a polar path. Both requirements added substantially to the cost of the shuttle program and slowed its development.

The Air Force was given responsibility for developing a shuttle launch site at Vandenberg Air Force Base on the California coast. This would allow the shuttle to take off in a southerly direction toward the South Pole. The Air Force also developed a rocket for the shuttle program called the Interim Upper Stage (IUS). IUS boosters were designed to thrust satellites from the shuttle's typical orbit into higher orbits.

The first shuttle flight did not take place until April 1981. In June 1982 a shuttle carried a military satellite into orbit for the first time. Shuttles carried six subsequent DOD satellites into space during 1984 and 1985. Four of these satellites were SYNCOM communication satellites. The other two missions were classified.

When the shuttle was first proposed, NASA promised that it would fly frequently and routinely into Earth orbit and would meet the military's scheduling demands for satellite launches. It soon became apparent that this was not the case. The shuttle program was plagued by problems and flew only a few times each year. The DOD decided it could not rely completely on shuttles for the nation's military missions. In 1984 Air Force officials convinced Congress to fund development of a fleet of new ELVs for military missions. NASA protested strongly against this action, but was overruled.

The initiative turned out to be a good one. The explosion of the space shuttle *Challenger* shortly after liftoff in 1986 forced NASA to make drastic changes in the shuttle program. This had profound effects on the military's space ambitions. The *Challenger* explosion happened only months before the first planned launch of an Air Force shuttle from Vandenberg Air Force Base. The base's shuttle launch facilities were dismantled. Most of the related equipment was turned over to NASA. The DOD focused more resources on developing ELVs.

In September 1988 the space shuttle resumed flying. Between 1988 and 1992 shuttles carried less than ten military payloads into space. These were satellites that could not be launched aboard ELVs for some reason.

Star Wars

On March 23, 1983, President Ronald Reagan announced a new military space venture for the United States. He called it the Space Defense Initiative or SDI. Basically the plan called for the placement of a satellite shield in space that would protect the United States from incoming Soviet nuclear missiles. Reagan said that SDI would make nuclear weapons "impotent and obsolete."

Earlier that month Reagan had denounced the Soviet Union as the "focus of evil in the modern world." The Soviet news agency TASS responded that Reagan was full of "bellicose lunatic anti-communism." Reagan's SDI proposal heightened tensions between the two countries. The Soviets warned that it would set off a new and more dangerous arms race. Later that year the Soviet Union broke off nuclear arms negotiations in Geneva, Switzerland.

The media nicknamed the SDI proposal the "Star Wars" program. (*Star Wars* had been a hit 1977 movie featuring elaborate space weapons.) Many scientists publicly questioned whether SDI was technically feasible given the technologies of the times. Major newspapers openly ridiculed the idea. Politicians complained about the potential costs. Discovering whether SDI was even possible was expected to be immensely expensive. Some high-ranking government officials feared that SDI would start an arms race in space.

In March 1984 the DOD established a Strategic Defense Initiative Organization (SDIO). Later that year the Army successfully tested an interceptor missile as part of SDIO operations. The missile was launched from the Kwajalein Missile Range in the Marshall Islands. It flew above the atmosphere and then located and tracked a reentry missile that had been launched from Vandenberg Air Force Base in California. The interceptor missile homed in on the target using onboard sensors and computer targeting. It crashed into the target and destroyed it.

Reagan met with Soviet premier Mikhail Gorbachev for private talks during 1985 and 1986. Both times they argued about SDI. In a 1986 meeting in Reykjavik, Iceland, Premier Gorbachev offered to cut Soviet missile stocks if the United States would cease development of the SDI project. Reagan refused. By this time the military had developed a working concept for the space shield that included numerous small, computerized satellites. The concept was called "Brilliant Pebbles."

In 1989 President George H. W. Bush assumed office. He supported the SDI project, and research and development on it continued. Two years later, the United States entered the Gulf War against Iraq. By this time the Soviet Union had dissolved into a number of independent republics. The Cold War was over. During the administration of President Bill Clinton in 1993 the SDIO was redesignated the Ballistic Defense Missile Organization (BDMO). The new threat was considered to be limited-range missiles in the hands of unfriendly dictators and terrorists.

In 2002 the United States withdrew from the Anti-Ballistic Missile Treaty of 1972. This treaty with the Soviet Union (and later Russia) had strictly limited each nation's deployment of anti-ballistic missiles. Soon afterward, President George W. Bush converted the BDMO into the Missile Defense Agency (MDA).

The goal of the MDA is to intercept and destroy ballistic missiles along their flight path. There are three flight phases for an intercontinental ballistic missile (ICBM): boost phase, midcourse, and terminal phase. The boost phase occurs during the first three to five minutes after an ICBM is launched, when it is being powered by its engines. During the boost phase an ICBM can reach an altitude of up to 300 miles. The midcourse stage takes the ICBM on a trajectory above the

FIGURE 3.3

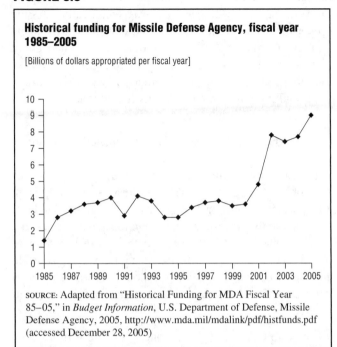

Historical funding for Missile Defense Agency, fiscal year 1985–2005

[Billions of dollars appropriated per fiscal year]

SOURCE: Adapted from "Historical Funding for MDA Fiscal Year 85–05," in *Budget Information*, U.S. Department of Defense, Missile Defense Agency, 2005, http://www.mda.mil/mdalink/pdf/histfunds.pdf (accessed December 28, 2005)

TABLE 3.1

Components of the U.S. missile defense system

Component	Description
Airborne laser	Plane-based laser destroys the missile during boost phase by heating its metal skin until it cracks, which causes the boosting missile to fail.
Aegis ballistic missile defense	Sea-based system intended to intercept short to medium range hostile missiles in the ascent and descent phase of midcourse flight.
Ground-based midcourse defense	Ground-based booster missile flies toward a target's predicted location and releases a "kill vehicle" on a path with the incoming target. The kill vehicle uses data from ground-based radars and its own on-board sensors to collide with the target, thus destroying both the target and the kill vehicle using only the force of the impact.
Kinetic energy interceptors	Ground-based interceptor capable of destroying incoming missiles while their booster rockets are still burning.
Sensors	Ground-, sea-, and space-based sensors detect and track threat missiles through all phases of their trajectory.
Terminal phase defense	Ground-based elements with the capability to shoot down short or medium range ballistic missiles in their final stages of flight, both inside and just outside of the atmosphere.

SOURCE: Adapted from *Fact Sheets*, U.S. Department of Defense, Missile Defense Agency, 2005, http://www.mda.mil/mdalink/html/factsheet.html, and "Appendix" in *A Historic Beginning: BMDS System Second Edition*, U.S. Department of Defense, Missile Defense Agency, 2005, http://www.mda.mil/mdalink/pdf/bmdsbook.pdf (accessed December 28, 2005)

atmosphere through space and can last up to twenty minutes. During this phase the missile can release countermeasures and decoys. Once the missile reenters Earth's atmosphere it is in the terminal phase of its flight. This can last from thirty seconds to a minute. Preferably interception and destruction would be done outside of Earth's atmosphere so that nuclear or biological warheads would be destroyed during reentry.

Missile Defense System

As of February 2006 the MDA continues development and testing of interceptor missiles and tracking systems. Things that were once considered science fiction fantasy are slowly becoming viable components in the DOD arsenal. This is due to technological advances and a large influx of money to the program. As shown in Figure 3.3 MDA funding has increased from just over $1 billion in 1985 to $9 billion in 2005.

Table 3.1 lists the components of the nation's ballistic missile defense systems that are in testing or operational stages. The only space-based system is called the Space Tracking and Surveillance System (STSS). This is a series of satellites in low Earth orbit that would detect and track ballistic missiles launched anywhere on the planet. Eventually the system will include space-based infrared sensors that will be able to distinguish missile warheads from other nearby objects (such as decoys, rocket casings, or space debris). The first launch of an STSS satellite is planned for 2007.

The nation's ballistic missile defense systems have evoked severe criticism from some in the scientific community. One of the most vocal critics is the Union of Concerned Scientists, an organization of independent scientists who research and analyze policy issues, such as the environment and missile development. In 2004 the USC issued a report titled *Technical Realities: An Analysis of the 2004 Deployment of a U.S. National Missile Defense System* (http://www.ucsusa.org/global_security/missile_defense/technical-realities-national-missile-defense-deployment-in-2004.html), which stated:

> The ballistic missile defense system that the United States will deploy later this year will have no demonstrated defensive capability and will be ineffective against a real attack by long-range ballistic missiles. The administration's claims that the system will be reliable and highly effective are irresponsible exaggerations. There is no technical justification for deployment of the system, nor are there sound reasons to procure and deploy additional interceptors.

The UCS advocates new nonproliferation treaties with Russia and China to prevent an arms race in space between the three countries.

In December 2005 the MDA reported that a test of the ground-based midcourse defense system against a "non-live" target had been successful. An interceptor launched and encountered a simulated target. The MDA noted that the system has been successful in five of ten tests conducted to date.

Space Weapons?

Historically the United States has focused on developing defensive, rather than offensive, space-based assets. It is a party to the Outer Space Treaty of 1967, which says that nations "may not place in orbit around the Earth any objects carrying nuclear weapons or any other kinds of weapons of mass destruction, install such weapons on celestial bodies, or station such weapons in outer space in any other manner." In addition, presidential space policy since the 1950s has focused on unarmed satellites to prevent a new arms race in space.

In May 2005 Tim Weiner reported in the *New York Times* that the Bush administration was considering a change in space policy to allow the development of space weapons ("Air Force Seeks Bush's Approval for Space Arms," May 18, 2005). The article claims that the Air Force wants the capability to develop offensive and defensive space assets to protect the country. An Air Force spokesperson denied that the new policy would militarize space, but would ensure the United States has "free access in space." According to Weiner, the potential policy change had already drawn objections from leaders in Canada, China, Russia, and the European Union. Critics fear that such a move would encourage countries like China and Russia to build their own space weapons. Proponents argue that the United States must develop new space capabilities to protect vulnerable satellites upon which the nation depends for communications, global positioning data, and military reconnaissance.

In the article, Weiner noted that a change in space policy would face numerous technical, financial, and diplomatic challenges. Critics complain that billions of dollars have been spent on missile defense systems that have not proven to be reliable. They fear that development of space weapons would be a wasteful expense and do little to combat the spread and very real threat of terrorist strikes.

Some critics complain that assets already being developed by the DOD are preludes to space weapons. For example, the Air Force is testing maneuverable microsatellites called the Experimental Satellite Series (XSS). These are small, mobile satellites that can be maneuvered up to other orbiting objects and take pictures of them.

The first microsatellite, XSS-10, was successfully tested in space in 2003 during a one-day trial. Although the Pentagon denies that the XSS satellites are space weapons, critics claim it would be relatively easy to convert their photographic capabilities to firepower. Then they could seek out and destroy targets in orbit, such as the satellites of unfriendly countries. Development of anti-satellite weapons or ASATs is highly controversial. In April 2005 the Air Force launched the much more sophisticated XSS-11 into space. This satellite is designed for a one-year lifetime during which it will approach up to eight objects in space. The XSS-11 is about the size of a washing machine and weighs around 300 pounds.

Another project under fire by critics is the MDA's Near Field Infrared Experiment. This satellite would carry infrared sensors capable of detecting and tracking ballistic missiles during their boost phase. The MDA's original proposal for NFIRE included a "kill vehicle" aboard the satellite that could be released to smash into a missile and destroy it in space. In 2004 the kill vehicle was cancelled. Officially the DOD cited technical reasons for the cancellation. However, critics believe that the agency backed down in the face of intense criticism for proposing an obvious space weapon. The test launch of NFIRE, which was supposed to take place in 2004, has been postponed until 2006 or 2007.

Other space weapons reportedly being considered by the DOD include the Common Aero Vehicle (CAV) and Hypervelocity Rods. The CAV would deliver high explosives from space with the capability to bomb targets thousands of miles from the United States. Hypervelocity rods are long metallic rods that would be delivered from space and—having built up tremendous speed during the long fall to Earth—smash into and destroy deep underground bunkers. They have been nicknamed "Rods from God" by the media.

U.S. Strategic Command

In 1985 the Reagan Administration established the U.S. Space Command to oversee military space operations. Its commander was also in charge of the North American Aerospace Defense command (NORAD). NORAD protects the air space of the United States and Canada. In 1992 President George H. W. Bush established the U.S. Strategic Command (StratCom) to oversee the nation's nuclear arsenal.

Following the terrorist attacks of September 11, 2001, President George W. Bush abolished U.S. Space Command and assigned its responsibilities to StratCom. StratCom is headquartered at Offutt Air Force Base in Nebraska. It is the command and control center for U.S. strategic forces, controls military space operations, and is responsible for early warning and defense against missile attacks.

StratCom has four space-related missions:

- Satellite launches and operations including telemetry, tracking, and command. Satellite launches take place at Cape Canaveral, Florida, and Vandenberg Air Force Base, California

- Armed forces support via use of communication, navigation, weather, missile warning, and intelligence satellites

- Protecting U.S. access to space and denying access to enemies
- Researching and developing space assets that can engage enemies from space. Such projects cannot presently be implemented due to long-standing U.S. policy against deploying orbiting weapons.

AIR FORCE SPACE COMMAND. Much of StratCom's space operations are carried out by the Air Force Space Command (AFSPC) headquartered at Peterson Air Force Base (AFB) in Colorado. The AFSPC has facilities at three other Colorado locations (Cheyenne Mountain Air Station, Schriever AFB, and Buckley AFB) and in Alaska, California, Florida, North Dakota, Wyoming, Montana, New Hampshire, and Greenland.

The AFSPC operates the Global Positioning System and launches and operates satellites that provide weather, communications, intelligence, navigation, and missile warning capabilities. The Command also provides services, facilities, and aerospace control for NASA operations. In 2004 AFSPC established the National Security Space Institute to provide education and training in space-based topics.

The AFSPC's Space Control Center maintains a database of more than 9,000 objects known to be in Earth orbit. These include operating and inoperative satellites, pieces of rockets, and other objects. When a space shuttle mission is taking place, the Center tracks the shuttle's path and establishes a safety zone twenty-five miles long around the vehicle, as shown in Figure 3.4. If the Center determines that an object is on a collision path with the shuttle, the Center notifies NASA so that evasive maneuvers can be performed.

DOD Manned Space Flight Support Office

In 1958 the U.S. government established the DOD Manned Space Flight Support Office (DDMS) to support NASA's manned spaceflight programs. The DDMS provided medical support and communications, tracking, and data capabilities, and recovered astronauts and space capsules after splashdown for all manned programs from Mercury (1959–63) through Skylab (1973–74).

When the space shuttle program began in the 1980s the DDMS assumed responsibility for astronaut rescue and recovery, payload security, and a variety of contingency services in the event of an emergency. Near the Kennedy Space Center in Florida the DDMS has at its disposal a number of Air Force and Navy resources including helicopters, tanker aircraft, ships, air traffic control facilities, and medical and search-and-rescue personnel. The DDMS also supports potential emergency landing sites in Spain, Morocco, and Gambia.

SPACE AGENCIES AROUND THE WORLD

NASA and Rosaviakosmos (the Russian Space Agency) operate the two most active space programs in

FIGURE 3.4

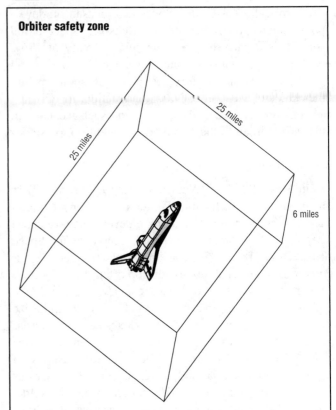

Orbiter safety zone

SOURCE: "Figure A. Orbiter Safety Zone," in *Columbia Accident Investigation Board: Report Volume I*, National Aeronautics and Space Administration, August 2003, http://www.nasa.gov/columbia/caib/PDFS/VOL1/PART01.PDF (accessed January 31, 2006)

the world. Rosaviakosmos evolved from the great space agency of the former Soviet Union. The Soviet space program achieved many important milestones in robotic and human spaceflight. Other nations with the resources to do so have ventured into space. Some have sent their astronauts aboard U.S., Soviet, or Russian spacecraft. Others have developed their own space vehicles and programs. This has created new opportunities for cooperation and competition among space-faring nations.

Table 3.2 lists the space agencies of various countries around the world. Major programs are described below.

RUSSIA

The Russian Space Agency is called Rosaviakosmos (RKA). It was officially created on February 25, 1992, by decree of the President of the Russian Federation. The RKA inherited the technologies, programs, and facilities of the Soviet Union space program.

Sergei Korolev

Sergei Korolev (1906–66) is considered the founder of the Soviet Union space program. Korolev was born in Zhitomir, a town in what is now the Ukraine. An engineer and aviator who began building rockets in the 1930s, he

TABLE 3.2

International space agencies

Country	Name	Acronym
Argentina	Comisión Nacional de Investigaciones Espaciales	CONAE
Australia	Australian Space Office	ASO
Austria	Österreichische Gesellschaft für Weltraumfragen Ges.m.b.H (Austrian Space Agency)	ASA
Belgium	Belgium Federal Science Policy Office	SPO
Brazil	Agência Espacial Brasileira	AEB
Bulgaria	Bulgarian Aerospace Agency	BASA
Canada	Canadian Space Agency	CSA
China	China National Space Administration	CNSA
Denmark	Dansk Rumforskningsinstitut (Danish Space Research Institute)	DSRI
Finland	National Technology Agency of Finland	Tekes
France	Centre National d'Etudes Spatiales	CNES
Germany	Deutschen Zentrum für Luft- und Raumfahrt	DLR
Hungary	Magyar Ürkutatási Iroda (Hungarian Space Office)	HSO
India	Indian Space Research Organisation	ISRO
Indonesia	National Institute of Aeronautics & Space	LAPAN
Israel	Israel Space Agency	ISA
Italy	Agenzia Spaziale Italiana	ASI
Japan	Japan Aerospace Exploration Agency	JAXA
Korea	Korea Aerospace Research Institute	KARI
Netherlands	Nationaal Lucht-en Ruimtevaartlaboratorium (National Aerospace Laboratory)	NAL
Norway	Norsk Romsenter (Norwegian Space Centre)	NSC
Poland	Space Research Centre	SRC
Portugal	Instituto Nacional de Engenharia e Tecnologia Industrial	INETI
Romania	Romanian Space Agency	ROSA
Russia	Rosaviakosmos	RKA
Spain	Instituto Nacional de Técnica Aeroespacial	INTA
Sweden	Swedish National Space Board	SNSB
United Kingdom	British National Space Centre	BNSC

SOURCE: Created by Kim Weldon for Thomson Gale, 2004

founded a rocket organization called *Gruppa Isutcheniya Reaktivnovo Dvisheniya* (Group for Investigation of Reactive Motion). Following World War II the government appointed Korolev to develop Soviet missile systems.

In August 1957 his team successfully tested the R-7, the world's first intercontinental ballistic missile (ICBM). The R-7 was powerful enough to carry a nuclear warhead to the United States or a satellite into outer space. In October 1957 an R-7 rocket carried *Sputnik 1* into orbit. It was the world's first artificial satellite. The Soviet Union had beaten the United States into space.

VOSTOK. Korolev's next challenge was to beat the United States to the Moon. In January 1959 the Soviet probe *Luna 1* flew past the moon. In September 1959 *Luna 2* was deliberately crashed into the lunar surface, making it the first manmade object to reach the Moon. A month later *Luna 3* took the first photographs of the far side of the moon. Korolev was already working on a spacecraft for manned missions. It was a modified R-7 called Vostok ("East" in English). Vostok included a sphere-shaped cosmonaut module that held one person. The module was too heavy for a parachute. Instead it included an ejection seat so that the cosmonaut could eject from the module following reentry and parachute to Earth by himself.

Throughout 1960 and early 1961 the Vostok was tested unmanned, with dogs, small mammals, and a mannequin aboard. Vostok flying dogs included Strelka, Belka, Pchelka, Mushka, Chernushka, and Zvezdochka. Many of the dogs died during these tests. The mannequin was nicknamed Ivan Ivanovich, which is the Russian equivalent of "John Doe."

On April 12, 1961, the Soviets launched the first man into space aboard *Vostok 1*. His name was Yuri Gagarin (1934–68). Gagarin was one of the twenty original cosmonauts selected by the Soviet Union in 1959 for manned spaceflights. Korolev played a major role in selecting and training the cosmonauts. In 1960 they began training at a sprawling new complex called Zvyozdny Gorodok (Star City) in the Russian countryside. Gagarin's flight into orbit lasted 108 minutes and reached an altitude of 203 miles. During descent, Gagarin parachuted safely from the module and landed in a rural field. The following month, on May 6, 1961, U.S. astronaut Alan Shepard became the first American in space.

There were five more Vostok flights from 1961 through 1963. *Vostok 2* carried Gherman Titov (1935–2000) to 17.5 orbits around Earth on August 6–7, 1961. *Vostok 3* and *Vostok 4* were launched only one day apart on August 11 and August 12 of 1962. *Vostok 3* carried Andriyan Nikolayev (1929–2004), and *Vostok 4* carried Pavel Popovich (1930–). The two cosmonauts landed within minutes of each other on August 15, 1962. In June 1963 *Vostok 5* and *Vostok 6* also conducted a joint operation. *Vostok 5* launched on June 14th with Valeri Bykovsky (1934–) aboard. It was followed two days later by *Vostok 6* with Valentina Tereshkova (1937–) aboard. Tereshkova was the first woman in space and had been personally selected for the task by Korolev. The two cosmonauts returned to Earth on June 19, 1963.

The Vostok program and the U.S. Mercury project both took place between 1961 and 1963. The Soviet cosmonauts beat the U.S. astronauts into space and spent much more time there. The longest Mercury flight lasted only one day and ten hours. The longest Vostok flight lasted nearly five days.

VOSKHOD. In 1964 the Soviets began testing a multipassenger spacecraft called Voskhod, which means "sunrise" in English. The Voskhod module had a parachute descent system that eliminated the need for ejection seats. On October 12, 1964, *Voskhod 1* carried three men into space: the pilot, Vladimir Komarov (1927–67), Boris Yegorov, a physician (1937–94), and Konstantin Feoktistov, a scientist (1926–). Their flight lasted just over twenty-four hours and circled the Earth sixteen times. A few months later *Voskhod 2* was put into orbit with two cosmonauts aboard: Aleksey Leonov (1934–) and Pavel Belyayev (1925–70). On March 18, 1965, Leonov conducted the first extravehicular activity (space walk) in history. It lasted about ten minutes.

Despite these successes, the *Voskhod 2* mission was plagued by life-threatening problems. Leonov's spacesuit and the vehicle's airlock and reentry rockets malfunctioned. The crew module spun out of control during reentry and landed in heavy woods far from its intended landing point. At the time, a number of crewed Voskhod missions were planned for the 1960s, including one with an all-female crew. However, the problems of *Voskhod 2* and the death of Korolev in January 1966 shook the Soviet space agency. All of these planned missions were cancelled.

N-1 ROCKET. During his lifetime Sergei Korolev was relatively unknown outside the Soviet Union. The Soviets were very secretive about national affairs and provided scant information to the foreign media. This was particularly true for the inner workings of the Soviet space program. It was only following Korolev's death that the Western world learned about his many contributions to space travel. These included numerous rockets and launch vehicles, satellites and probes of different types, and manned spacecraft. His most famous spacecraft was the Soyuz ("Union" in English). Modified versions of Soyuz rockets are still used by Rosaviakosmos in the twenty-first century.

Korolev is also remembered for his one great failure, the N-1 rocket. This was supposed to be the superbooster that would launch a Soviet spacecraft called the L1 (or Zond) to the moon. Korolev's design team created the L1 from a modified Soyuz spacecraft. The N-1 superbooster was similar in scope to von Braun's Saturn V rocket. Korolev worked on the N-1 project from 1962 until his death in 1966, but never achieved an operational rocket. His successors continued the work after his death, but were not successful.

On July 3, 1969, an unmanned N-1 rocket exploded only seconds before lift-off. The resulting fireball was so huge it destroyed the launch facilities. Thirteen days later a U.S. Saturn V rocket launched *Apollo 11* on its way to the moon.

The Soviet space program was shrouded in secrecy. Successes were publicized, while failures and plans were not. Although the Soviets had ambitions to land a man on the moon, this goal was never announced publicly. It was only years later that the West learned about the failed Soviet moon program in Sergei Leskov's article "How We Didn't Get to the Moon" (*Izvestiya*, August 18, 1989). Most observers in the United States assumed that the Soviet Union was aiming for the moon, but this was not certain. In fact, as early as 1963 NASA critics in the United States asserted that the "moon race" was a hoax advanced by the U.S. government to further its own aims.

After the United States reached the moon first the Soviets insisted that they had never intended to go there. According to James E. Oberg in "Yes, There Was a Moon Race" (*Air Force Magazine*, April 1990), "Examination of newly disclosed evidence about one of the most intense phases of the superpower rivalry makes plain that U.S. actions came in response to an authentic Soviet challenge."

Firsts in Space

Despite losing the moon race, the Soviet space program achieved many firsts in space during the 1950s and 1960s:

- *Sputnik 1*—The first artificial satellite in orbit (October 4, 1957)

- *Sputnik 2*—The first space passenger, Laika the dog, spent seven days in orbit (November 3, 1957)

- *Luna 2*—The first artificial object to reach a celestial body (September 14, 1959)

- *Vostok 1*—Yuri Gagarin is the first person to orbit the Earth (April 12, 1961)

- *Vostok 2*—Gherman Titov is the first person to spend a full day in orbit (August 6–7, 1961)

- *Vostok 3* and *Vostok 4*—First spaceflight including two spacecraft in orbit at once (August 11–15, 1962)

- *Vostok 6*—Valentina Tereshkova is the first woman in space (June 16, 1963)

- *Voskhod 1*—First spaceflight including three people (October 12–13, 1964)

- *Voskhod 2*—Aleksei Leonov is the first person to take a space walk (March 18, 1965)

The Soviet space program also experienced a tragic first in space. On April 23, 1967, the Soviet space agency launched the first manned Soyuz rocket with Vladimir Komarov aboard. A day later the flight ended in tragedy when the module's parachute failed during descent. Komarov was killed. He was the first human to die during a spaceflight. On June 29, 1971, three more cosmonauts died when their *Soyuz 11* spacecraft depressurized during descent after visiting the *Salyut 1* station. Their names were Georgiy Dobrovolskiy (1928–71), Vladislav Volkov (1935–71), and Viktor Paysayev (1933–71). At that time cosmonauts did not wear spacesuits during launch and reentry. This was later changed to provide them greater safety.

A New Focus

The Soviet's Moon program continued well into the 1970s. However, neither the N-1 nor a competing rocket called the Proton ever became dependable enough for manned launches. During the early 1970s the Soviets

FIGURE 3.5

Apollo-Soyuz rendezvous and docking test project

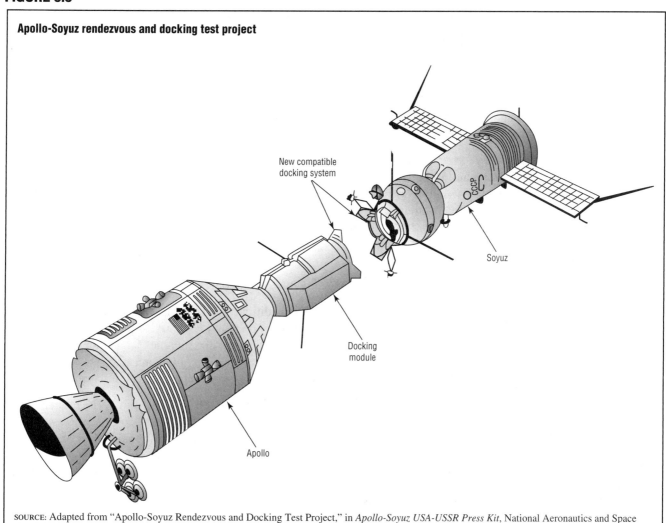

New compatible
docking system

Soyuz

Docking
module

Apollo

SOURCE: Adapted from "Apollo-Soyuz Rendezvous and Docking Test Project," in *Apollo-Soyuz USA-USSR Press Kit*, National Aeronautics and Space Administration, June 10, 1975, http://www-lib.ksc.nasa.gov/lib/archives/apollo/pk/soyuz.PDF (accessed January 31, 2006)

concentrated on perfecting their Soyuz rockets and building a space station. Like NASA, the Soviet space agency had always envisioned an orbiting space station as the next step after a lunar visit.

On April 19, 1971, space station *Salyut 1* was launched into orbit. It was another first for the Soviet Union. The United States space station *Skylab* would not launch for another two years. Between 1971 and 1982 the Soviets put seven Salyut stations into orbit, one after another. These were designed to be temporary stations. Some of them fell out of orbit only months after being launched.

The last station, *Salyut 7*, stayed in orbit for nearly nine years, from April 1982 to February 1991. It hosted ten crews of cosmonauts that spent a total of 861 days in space. The Soviet space program gained invaluable experience in long-duration exposure to weightlessness. The Salyut program was also notable in that cosmonauts and scientists from Cuba, India, and France were invited to visit the stations.

In 1972 the Soviet Union and the United States agreed to work together to achieve a common docking system for their respective spacecraft. This would permit docking in space of U.S. and Soviet spacecraft during future missions. On July 17, 1975, a Soviet Soyuz spacecraft carrying two cosmonauts docked with an Apollo spacecraft carrying three astronauts. (See Figure 3.5.)

The crewmembers conducted a variety of scientific experiments during the two-day docking period. Both spacecraft returned to Earth safely. The Apollo-Soyuz Rendezvous and Docking Test Project was the first union of spacecraft from two different countries.

By 1976 the Soviet space program was engrossed in another new project called Buran, a reusable space plane modeled after the U.S. shuttle. Buran means "snowstorm" in English. The Buran program (like the U.S. shuttle program) was plagued by development, cost, and scheduling problems. Although an unmanned Buran was successfully orbited in November 1988, the program was halted soon afterwards due to funding cuts.

The 1980s were a tense time in U.S.-Soviet relations. The Soviets were at war with Afghanistan and cracking down on dissidents in Poland. In 1983 the military shot down a Korean jet liner that allegedly veered into Soviet air space. More than sixty Americans were among the 269 passengers killed. The Soviet Union felt threatened by President Reagan's so-called "Star Wars" proposal to build a satellite shield. Throughout the decade, arms talks with the United States failed repeatedly.

On February 8, 1987, the Soviet space agency launched the space station *Mir* into orbit. Unlike the temporary Salyut stations, *Mir* was designed to last for years and to be continuously inhabited. Dozens of cosmonauts, astronauts, and space tourists visited the station during its fifteen-year lifetime in space. On March 22, 1995, cosmonaut Valeri Polyakov (1942–) returned to Earth after spending 437 days and eighteen hours aboard *Mir*. It was a record that remained unbroken in early 2006.

Rosaviakosmos Takes Over

During the early 1990s the Soviet Union splintered into a number of individual republics. The largest of these is Russia. In 1992 the new Russian government established a space agency called Rosaviakosmos to take over the space programs of the old Soviet Union. Russia and the United States began a new era of cooperation in space. In 1993 the two countries agreed to work together to build an *International Space Station* (*ISS*). Between 1994 and 1998 U.S. shuttles carried astronauts and cosmonauts into orbit together and to missions on the *Mir* station. In 1998 *ISS* construction began when the Russians placed the first module (Zarya) into orbit. Construction is expected to take place at least through 2010.

Rosaviakosmos controls all of the country's nonmilitary spaceflights. Military space ventures are controlled by Russia's Military Space Forces (VKS). The two agencies share control of the Baikonur Cosmodrome in Kazakhstan and the Gagarin Cosmonaut Training Center at Star City, Russia. The Plesetsk Cosmodrome launch facility in northern Russia is under the control of the VKS.

EUROPE

The European Space Agency (ESA) was formed in 1973 from two existing organizations, the European Space Research Organisation and the European Launcher Development Organisation. The ESA includes seventeen member states:

- Austria
- Belgium
- Denmark
- Finland
- France
- Germany
- Greece
- Ireland
- Italy
- Luxembourg
- The Netherlands
- Norway
- Portugal
- Spain
- Sweden
- Switzerland
- United Kingdom

In addition, the ESA has agreements with Canada, Czech Republic, and Hungary to participate as members in some projects. Although the ESA is independent of the European Union (EU), it maintains close ties with the EU and the two organizations share a joint space strategy.

ESA headquarters are located in Paris, France. Other ESA facilities include the European Space Research and Technology Centre (ESTEC) in Noordwijk, Netherlands; the European Space Operations Centre (ESOC) in Darmstadt, Germany; the European Astronauts Centre (EAC) in Cologne, Germany; the European Space Research Institute (ESRIN) in Frascati, Italy; and liaison offices in Belgium, Russia, and the United States. The ESA operates a launch base in French Guiana, near the equator in South America. As of February 2005 (the latest date available) the ESA employed just over 1,900 people.

Each member state funds mandatory ESA activities based on that country's gross national product. Mandatory activities include space science programs and the agency's general budget. In addition, the ESA operates optional projects in which countries may choose to participate and fund.

During the 1970s the ESA developed the Ariane rocket to launch satellites into orbit. Ariane is the French name for the Greek goddess Ariadne. According to Greek mythology Ariadne gave a thread to Theseus that helped him find his way out of the labyrinth of the Minotaur. One of the remarkable features of the rocket is that it can carry two satellites at once. The Ariane 4 series of rockets was used between 1988 and 2003. According to the ESA, the agency controlled 50% of the commercial satellite launch business during this period. By the end of 2005 Ariane rockets had achieved 169 successful launches.

As of February 2006 the ESA was participating in the *ISS* project and three interplanetary missions: Venus Express, Mars Express, and the Huygens Probe.

The Venus Express was launched on a Soyuz/Fregat rocket from Baikonur, Kazakhstan, in November 2005. It was scheduled to arrive at Venus in April 2006 and assume an orbit around the planet. It is designed to perform sophisticated atmospheric studies and measure surface temperatures on Venus.

The Mars Express was launched on a Soyuz/Fregat rocket from Baikonur, Kazakhstan, in June 2003 and went into orbit around Mars in December 2003. A landing vehicle named *Beagle 2* left the orbiter to head for the planet's surface. ESA lost contact with the lander, and it was presumed lost. However, the orbiter continued to circle the Red Planet and collect data with its scientific instruments.

The ESA was also responsible for the Huygens Probe mission to Saturn. The probe was launched aboard the NASA spacecraft *Cassini* in October 1997. On December 25, 2004, *Cassini* released the probe for a three-week journey to the surface of Titan, Saturn's moon. It penetrated the thick cloud cover that hides the moon and touched down on January 14, 2005. The probe sampled Titan's atmosphere and provided the first photographs ever of its surface. The probe is named after Christiaan Huygens (1629–95), the Dutch astronomer who discovered Saturn's rings and Titan. Huygens was the first probe to ever land on a celestial body in the outer solar system.

In December 2005 the ESA announced a new space endeavor called the Aurora Exploration Programme. It will include robotic missions to Mars that will return samples to Earth. The first launch is expected in 2011.

CHINA

China's space program is overseen by the China National Space Administration and operated by the China Aerospace Science and Technology Corporation (CASC). The CASC is a state-run enterprise that develops and produces rockets, spacecraft, and related products. It has conducted satellite launches since 1970. CASC launch sites include Jiuquan in the Gobi desert, Taiyuan in northern China, and Xichang in southeastern China.

Tsien Hsue-shen

The Chinese space program began in the late 1950s under the direction of rocket engineer Tsien Hsue-shen (1911–). Tsien was born in China, but immigrated to the United States during the 1930s, where he attended the Massachusetts Institute of Technology and California Institute of Technology (Cal Tech). He was a key member of the rocketry club at Cal Tech that evolved into NASA's Jet Propulsion Laboratory. He was also instrumental in the U.S. program to acquire and apply German rocket technology at the end of World War II. In 1950 Tsien was accused of being a communist spy and had his security clearance revoked. At the time he was pursuing U.S. citizenship.

In 1955, after five years under virtual house arrest, Tsien was deported to China. There he was put in charge of the nation's budding space program. He joined the Communist Party in 1958. Under his leadership China developed very successful satellite and missile systems. These included the anti-ship missile called Haiying (Sea Eagle) by the Chinese and dubbed Silkworm by the Western media. During the Cold War, China sold Silkworms to a number of third-world countries considered unfriendly to the United States. Tsien also led development of the Changzheng (Long March) rockets that became the primary launch vehicle of the Chinese space program.

During the late 1960s Tsien fell out of favor with the Chinese leadership and was removed from his post. This disgrace resulted in Tsien receiving little credit within China for his accomplishments. However, the Western world considers him the father of the Chinese space program. In her 1995 biography of Tsien, *Thread of the Silkworm*, Iris Chang asserts that deporting the brilliant rocket scientist was "one of the most monumental blunders committed by the United States."

The Long March into Space

On April 24, 1970, the first Chinese satellite was launched into Earth orbit. It was propelled into space by a Long March rocket. Oddly enough the powerful rocket was named in memory of a 1930s retreat across China by thousands of communist soldiers. They were fleeing from the national forces they had hoped to overthrow. The communist soldiers were led by Mao Tse-tung. During a civil war in the 1940s Mao's troops were victorious, and Mao assumed power over the People's Republic of China. He held this position until his death in 1976. China's first satellite was called *Mao-1* in his honor.

Since the 1970s China has conducted numerous satellite launches using Long March rockets. During the 1990s development began on capsules capable of carrying animals, and later humans, into space. In 1999 the first such spacecraft, *Shenzhou 1*, successfully completed fourteen orbits around Earth. (Shenzhou means "Divine Vessel" in English.) The Shenzhou series was updated with newer, more powerful Long March rockets throughout the early 2000s.

On October 15, 2003, China conducted its first human spaceflight. A Chinese "taikonaut" named Yang Liwei (1965–) was launched aboard *Shenzhou 5*. Liwei spent twenty-one hours and twenty-three minutes in space and completed fourteen orbits. He became a national hero. On October 12, 2005, *Shenzhou 6* carried two taikonauts into space: Fei Junlong (1965–) and Nie Haisheng (1964–). They spent just over four days orbiting the Earth before touching down safely in Inner Mongolia.

In 2003 China announced its intention to send a robotic orbiter to the moon before the end of the decade. The mission will be called Chang'e-1 in honor of the Chinese moon goddess. As of February 2006 the launch was scheduled for 2007. Future Chinese plans call for development of a space station in Earth orbit, interplanetary robotic probes, and a crewed lunar landing. The country has shown keen interest in participating in international space ventures and has such agreements with Russia, Brazil, and the European Space Agency. In 2003 the U.S. Congressional Research Service estimated that China spends approximately $2 billion a year on its space program.

JAPAN

The Japan Aerospace Exploration Agency (JAXA) was created on October 1, 2003, by merging the Institute of Space and Astronautical Science (ISAS), the National Space Development Agency of Japan (NASDA), and the National Aerospace Laboratory of Japan (NAL). JAXA is headquartered in Tokyo and has more than a dozen field facilities across Japan.

The first Japanese satellite, *Osumi*, was launched into space on February 12, 1970, by a Lambda-4 rocket. *Osumi* remained in space for more than three decades and was destroyed in 2003, as it reentered Earth's atmosphere. It was the first of many satellites launched by the nation. Japanese launch vehicles for lightweight satellites are named after letters in the Greek alphabet. In 2001 a new heavy-lift rocket called the H-II became Japan's primary launch vehicle for heavier spacecraft. Two years later an H-II malfunctioned soon after lift-off and had to be destroyed, along with the two satellites it was carrying. A lengthy safety review followed the incident. The H-II was not used again until February 2005, when it successfully launched a weather satellite into space.

JAXA's major ongoing projects as of 2006 include the solar orbiter *Nozomi*, the asteroid sampler *Hayabusa*, and development of the *Kibo* laboratory module for the ISS. *Nozomi* (which means "hope" in English) was launched in 1998 by a Mu rocket and was to go into orbit around Mars in December 2003. An equipment failure prevented this from happening. Instead, JAXA was forced to put the spacecraft into a solar orbit.

Hayabusa (which means "falcon" in English) was launched in May 2003 by a Mu rocket to intercept the asteroid Itokawa. The asteroid orbits the sun between Earth and Mars and is about 2,300 feet by 1,000 feet in size. It is named after Dr. Hideo Itokawa, who is considered the founder of Japan's space program. The robotic explorer was designed to land on the asteroid, take a surface sample, and return to Earth by 2007. In November 2005 JAXA lost contact with the spaceship during the touchdown procedure. Although contact was regained

after a few days, it is not known for sure if *Hayabusa* was able to collect dust particles. In December 2005 JAXA announced that thruster problems were going to delay *Hayabusa*'s return to Earth until 2010. It is supposed to land in a desolate region of the Australian Outback.

The *Kibo* laboratory facility was originally supposed to be flown to the *ISS* in 2004 or 2005. It is a heavy component and must be transported by the space shuttle. Continuing problems with the space shuttle fleet have delayed the *Kibo* transport mission until 2008 or 2009.

JAXA also participates in a number of scientific satellite projects with international partners. Future Japanese space projects include missions to the moon, Venus, and Mercury. The *SELenological and ENgineering Explorer* (*SELENE*), a lunar orbiter, is scheduled to launch in 2007. The PLANET-C mission is planned for launch in 2008 and will put an orbiter around Venus a year later. Mercury orbiters are under development for a mission in the early 2010s.

PRIVATE SPACE ORGANIZATIONS

Private space organizations have played a major role in advancing space exploration. As far back as the 1920s groups of scientists, hobbyists, and other enthusiasts were gathering together to share their passion for rocket science and space travel. Many of the early groups were absorbed by government and military space organizations or evolved into aerospace manufacturing businesses. Private groups continue to advance spaceflight by researching and developing new technologies, operating commercial space enterprises, promoting public interest in space, and influencing government decisions on the future of spaceflight.

During the 1990s new avenues arose for private parties to participate in space endeavors. The Russian government allowed high-paying "space tourists" to travel to the *International Space Station* for brief stays. The first non-governmental launch facilities were developed for commercial satellites.

In 2004 a major milestone was achieved in space exploration—the first manned spacecraft developed and launched by a commercial enterprise traveled into space. This opens a whole new realm of space travel opportunities to private citizens.

Early European Organizations

One of the first private space organizations was a German group called *Verein für Raumschiffahrt* (VfR), or Society for Spaceship Travel. The VfR was formed in July 1927 in Berlin by a group of scientists and authors interested in rocket research. In particular, they wanted to raise money to finance rocket experiments being conducted by Professor Hermann Oberth (1894–1989) at the University of Munich. During the early 1930s the group sponsored rocket research projects around Germany. The VfR included many famous members, including Wernher von Braun. The group disbanded in 1933 as the Nazi Party gained power in Germany.

The 1930s witnessed the formation of private space organizations throughout western and eastern Europe. In the Soviet Union there was *Gruppa Isutcheniya Reaktivnovo Dvisheniya* (Group for Investigation of Reactive Motion). Sergei Korolev was one of its founding members. He went on to become the chief designer of the Soviet space program. The British Interplanetary Society (BIS) was founded in October 1933. This group of scientists and intellectuals is credited with advancing many important theories used in spaceflight, including a design for a lunar landing vehicle that was incorporated into the Apollo Program. As of 2006 the BIS was very active and published several influential journals.

American Institute of Aeronautics and Astronautics

In April 1930 a group of American scientists, engineers, and writers interested in space exploration formed the American Interplanetary Society. The founders included G. Edward Pendray (1901–87; inventor of the time capsule), David Lasser (1902–96; an engineer and technical writer who advocated space travel) and Laurence Manning (1899–1972; a science fiction writer). In 1934 the name of the group was changed to the American Rocket Society (ARS). By this time the members were predominantly rocket scientists who specialized in the research, design, and testing of liquid-fuelled rockets. The American Rocket Society featured many prominent members including Robert Goddard (1892–1945), whose theories and experiments were instrumental in the development of rocket science during the early twentieth century.

During World War II several ARS members started a company called Reaction Motors, Inc., to support the war effort. The company later developed rocket engines used in the famous X-series planes. Over the decades, the company evolved into ATK Thiokol Propulsion, the manufacturer of the space shuttle's rockets.

In 1932 a group of American aeronautical engineers and scientists formed the Institute of Aeronautical Sciences. The name was later changed to the Institute of Aerospace Sciences (IAS). Although originally focused on Earth-bound aviation, the IAS grew increasingly interested in spaceflight. In 1963 the IAS merged with the ARS to become the American Institute of Aeronautics and Astronautics (AIAA).

As of 2006 the AIAA had more than 31,000 members and was the largest professional society in the world devoted to aviation and spaceflight. Its stated purpose is "to advance the arts, sciences, and technology of aeronautics and astronautics and to promote the professionalism of those engaged in these pursuits." The AIAA has published hundreds of books and hundreds of thousands of technical papers throughout its history.

The Planetary Society

The Planetary Society is a nonprofit space advocacy group based in Pasadena, California, that is funded by donations from its members. It was founded in 1980 by scientists Carl Sagan (1934–96), Bruce Murray (1931–), and Louis Friedman (1940–).

Its stated purpose is as follows: "The Planetary Society creates ways for the public to have active roles in space exploration. We develop innovative technologies, like the first solar sail spacecraft, we fund astronomers hunting for hazardous asteroids and planets orbiting other stars, we support radio and optical searches for extraterrestrial life, and we influence decision makers." The Society claims to be the largest space interest group on Earth. It operates an educational Web site: http://www.planetary.org/home/.

The Planetary Society funds projects that support its goals and educate the public about space travel. It also encourages its members and the public to contact government leaders regarding space exploration projects. During the 1980s the Society waged a campaign to encourage the U.S. Congress to restore funding for NASA's Search for Extra-Terrestrial Intelligence (SETI) project. In the early 1990s the battle was over NASA's planned postponement of the Mars Observer mission. In late 2003 and early 2004 Planetary Society members sent more than 10,000 postcards to Congressional leaders to protest funding cuts for NASA's planned mission to Pluto. According to the Planetary Society, all three of these campaigns were successful in that government funding was restored to the projects.

In 1999 the Society started the SETI@home project in which private citizens could allow their home computers to be used to analyze data recorded by a giant radio telescope as part of SETI. NASA operated a SETI program for a short time in the early 1990s, but it was cancelled due to lack of funding. By the time the SETI@home project ended in December 2005, more than five million people had participated. The project was turned over to the University of California at Berkeley, which plans to continue it under the Berkeley Online Infrastructure for Network Computing (BOINC).

In the early 2000s the organization launched an extensive project called Red Rover Goes to Mars, to coincide with NASA's Mars Exploration missions. The project included an essay contest for students that resulted in the names used for the Mars Rovers: *Spirit* and *Opportunity*. The contest was sponsored by the Planetary Society and the Lego toy company.

The two also funded creation of DVDs that were mounted to the Rovers for the missions. The DVDs were specially crafted out of silica glass (instead of plastic) and contain the names of nearly four million people who asked NASA to be listed. Each DVD surface features a drawing of an "astrobot" saying "Hello" to Mars. The spacecraft safely landed on Mars in January 2004. Photos transmitted to NASA by the rovers after landing showed that the DVDs survived the journey. The Rovers are designed to remain on Mars and not return to Earth.

Other components of the Red Rover Goes to Mars project included a contest in which the winning students visited mission control during the Mars missions and a classroom project in which students built models of the Mars Rover and the Martian landscape.

On June 21, 2005, the Planetary Society launched its first spacecraft, *Cosmos 1*—on a mission to test a solar sail in orbit around Earth. A solar sail is a novel technology that could power spaceflight in the future. It is composed of giant ultra-thin silvery blades that unfurl after launch to reflect sunlight. The electromagnetic radiation of sunlight exerts force on the objects upon which it shines. This force is fairly strong in outer space due to the absence of atmospheric friction, and it could potentially push a solar sail in much the same way that the wind pushes sailing ships on the Earth's oceans.

Cosmos-1 was built in Russia with funding and technical support from the Planetary Society. Each blade of the solar sail was forty-seven feet long. The sail was to be launched by the Russian Navy from a submarine. The mission was cosponsored by the media company Cosmos Studios through a contract with the Russian Space Agency. It was the first space mission ever funded by a private space interest organization.

Unfortunately *Cosmos-1* was lost soon after launch when its Russian-supplied Volna rocket failed to fire properly. The Planetary Society hopes to raise the money needed to fund a second solar sail.

Other programs being funded by the Planetary Society as of 2006 include an exosolar planet search and the Gene Shoemaker Near-Earth Object Grant Program. Exosolar planets are planets lying outside of our solar system. Searches are conducted by a robotically controlled telescope at the Kitt Peak Observatory in Arizona. The Planetary Society also provides grants to private observers around the world who help track small asteroids orbiting near the Earth.

Space Entrepreneurs

Commercial enterprises have played an important role in space exploration through the decades. Government and military space programs would not have been possible without the contributions of labor and technology from companies in the aerospace business and related fields. Communication corporations were among the first to see the potential of satellites to grow and revolutionize their businesses. Demand for satellite launches from the commercial sector helped fund and drive many advances in rocket science and launch technology.

For decades satellites could only be launched at state-operated facilities. The 1990s witnessed the birth of commercial satellite launching organizations in several countries. One of the most unusual is called Sea Launch Company, LLC. The company formed in 1995 and included U.S., Russian, Ukrainian, and Norwegian companies engaged in the aerospace business. The consortium modified an ocean oil-drilling platform into a rocket launch platform and placed it in the middle of the Pacific Ocean along the equator. Since the first successful launch in 1999, more than a dozen commercial satellites have been put into orbit from the sea-based facility.

The 1990s also witnessed the first space tourists. The Soviet (and later Russian) space agency allowed private citizens to visit the *Mir* space station and the *ISS* for fees ranging from $15 to $30 million per tourist. Some of the trips were arranged through a private U.S. company called Space Adventures, Ltd. Formed in 1998 by aerospace engineer Peter Diamandis (1961–), the company offers customers opportunities in space tourism and related entertainment areas, such as "zero gravity" experiences. The demand by private citizens for space travel is expected to grow substantially during the twenty-first century.

A New Way to Explore Space—Commercial Suborbital Flights

In 2004 a major milestone in space exploration was achieved when the first nongovernmental manned spacecraft traveled to space and back. The spacecraft was called *SpaceShipOne*, and it was funded by private investor Paul G. Allen (1953–), cofounder of the Microsoft Corporation. In 2001 Allen contracted a California design firm called Scaled Composites to develop a reusable space vehicle capable of carrying at least one passenger to sub-orbital space. Aside from re-engaging the public's interest and passion in space exploration, Allen and *SpaceShipOne* set out to win the Ansari X Prize. This prize was offered by a group of private investors called the X Prize Foundation, created by Peter Diamandis. The Ansari family was the prime funder of the $10 million prize, which was available to any nongovernmental group that could achieve the following:

- Build a spaceship and fly three people (or at least one person plus the equivalent weight of two people) into space (defined as an altitude of 100 kilometers or 62.14 miles)

- Return safely to Earth
- Repeat the feat with the same spaceship within two weeks

On June 21, 2004, test pilot Mike Melvill (1941–) of Scaled Composites became the first person to pilot a privately built plane into space when he took *SpaceShipOne* to an altitude of 62.2 miles during a test flight. On September 29, 2004, he achieved an altitude of 63.9 miles. Only five days later pilot Brian Binnie (1953–) took the same plane to an altitude of 69.6 miles to win the Ansari X Prize.

The flights were conducted from an airstrip in Mojave, California. A carrier plane called *White Knight* carried *SpaceShipOne* to an altitude of approximately 47,000 feet and released it. A rocket motor aboard the spaceship was fired to propel it vertically into space. The pilots experienced about three minutes of weightlessness at the height of their journeys. During reentry the wings of *SpaceShipOne* were maneuvered to provide maximum drag and slow its descent. The spacecraft was glided back to the airstrip and landed like a plane. The flight sequence is shown in Figure 3.6.

The *White Knight* was a manned twin-turbojet carrier aircraft designed to fly to high altitudes carrying a payload of up to 8,000 pounds. It was named after two U.S. Air Force pilots (Robert White and William "Pete" Knight) who earned their astronaut wings flying the experimental X-15 aircraft during the early 1960s.

SpaceShipOne used a unique hybrid rocket motor fueled by liquid nitrous oxide (laughing gas) and solid hydroxy-terminated polybutadiene (HTPB, a major con-stituent of the rubber used in tires). The individual fuel components are nontoxic and are not hazardous to transport or store. They do not react when mixed together unless a flame is supplied. In *SpaceShipOne* the nitrous oxide was gasified prior to combustion.

In July 2005 Burt Rutan (1943–; president of Scaled Composites) and Sir Richard Branson (1950–; founder of the Virgin Group) announced the formation of a new aerospace production company called The Spaceship Company. Using original technology licensed from Allen, the company plans to build a small fleet of spacecraft based on the designs of the *White Knight* and *SpaceShipOne*. In 2004 Branson created Virgin Galactic, which bills itself as the world's first commercial spaceline. Virgin Galactic signed an agreement to become the first "launch customer" for The Spaceship Company aircraft.

Virgin Galactic plans to sell sub-orbital spaceflights to tourists for approximately $200,000 per flight. In December 2005 the company announced its agreement with the State of New Mexico to build a spaceport near the White Sands Missile Range. The company intends to begin commercial spaceflights from an airstrip in Mojave, California, in 2008 and switch its operations to the New Mexico site in 2009 or 2010, when the spaceport is complete.

According to an MSNBC news report (Alan Boyle, "Virgin Unveils Space Plans for New Mexico," December 13, 2005), Virgin Galactic has already registered more than 38,000 people interested in taking a spaceflight, and nearly 100 people have pledged to pay the full fare upfront to be among the first to fly. The article notes that several other companies have plans to develop and launch commercial spacecraft.

FIGURE 3.6

Flight sequence for SpaceShipOne

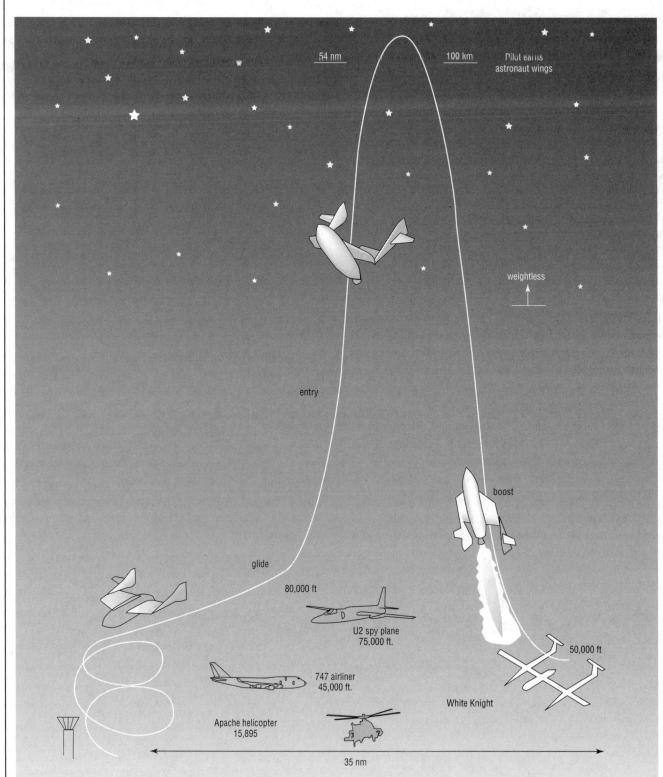

54 nm   100 km   Pilot earns
astronaut wings

weightless

entry

boost

glide

80,000 ft

U2 spy plane
75,000 ft.

50,000 ft

747 airliner
45,000 ft.

White Knight

Apache helicopter
15,895

35 nm

Note: nm=nautical mile

SOURCE: "SpaceShipOne Flight," in *PDF Graphics: Space Flight Profile* in *June 21 Space Flight Electronic Press Kit*, Scaled Composites, LLC, June 21, 2004, http://scaled.com/projects/tierone/data_sheets/PDF/SS1_flight_profile.pdf (accessed December 28, 2005) © 2004 Mojave AerospaceVentures LLC; SpaceShipOne is a Paul G. Allen Project.

CHAPTER 4
THE SPACE SHUTTLE PROGRAM

It will revolutionize transportation into near space, by routinizing it.

—President Richard Nixon, January 5, 1972

The space shuttle was supposed to make space travel a routine and frequent occurrence. Its conceivers envisioned shuttles regularly transporting humans and cargo back and forth between the Earth and a fleet of orbiting space stations. The shuttle was expected to be much cheaper than previous spacecraft, because it would be reusable. This would mean low operational and maintenance costs and a quick turnaround time between flights. It was predicted to bring in lots of money by hauling satellites into space for paying customers. The shuttle was going to be part of a massive transportation system and open up space the way the railroad opened up the frontier of the American West during the nineteenth century.

This vision never became a reality. Shuttle flights did not become routine, common, or frequent. Over the twenty-five-year period from 1981 through 2005, space shuttles flew only 114 times, averaging less than five flights per year. Two shuttles exploded, killing fourteen crewmembers. In addition to the human cost, the program experienced high operational and maintenance costs. Long turnaround times prevented the shuttle from flying frequently. However, the flights that took place did achieve many accomplishments. They put probes and observatories into space and were essential for building the *International Space Station (ISS).*

Nevertheless, many people believe that the United States has wasted too much time and money on a shuttle program that does not deliver what it promised. In January 2004 President George W. Bush announced his own vision for the nation's space program. It focuses on trips to the moon and Mars and calls for ending the shuttle program by 2010. The dream of routine access to space remains an elusive one.

THE POST-APOLLO VISION

In the early 1960s NASA planners envisioned a space station program as the next step after Apollo. It was assumed that the United States would establish large space stations in orbit around Earth and possibly outposts on the moon. In fact, NASA hoped to put at least one twelve-person space station in Earth orbit by 1975. This would require a new type of reusable space plane to carry cargo and personnel to and from the station.

These grand plans did not mesh with the political, cultural, and technological realities of the times. By the late 1960s the nation was heavily engaged in the Vietnam War. Domestic unrest and social issues dominated the political agenda into the early 1970s. Richard Nixon was president of the United States from 1969 to 1974. According to historians, Nixon was not interested in pursuing any grand and expensive vision for space exploration. In a March 1970 statement on space policy Nixon noted, "We must build on the successes of the past, always reaching out for new achievements. But we must also recognize that many critical problems here on this planet make high priority demands on our attention and our resources" (T. A. Heppenheimer, "The Space Shuttle Decision: NASA's Search for a Reusable Space Vehicle," 1999, http://history.nasa. gov/SP-4221/ch9.htm.) NASA's budget was severely cut, and plans for space stations were put on hold.

NASA did not give up on the shuttle program. They began promoting the project as a transport business, rather than an exploratory adventure. NASA officials argued that a shuttle could haul government and commercial satellites into space in a cost-effective manner because it would be reusable. The shuttle astronauts could service and repair these satellites as needed. The shuttle was touted as an investment. It would make money from commercial customers and save the government money on launching satellites for weather, science, and military purposes.

The argument was successful. In 1971 NASA was given a $5 billion budget over a five-year period for development of a shuttle program. This was later increased to $5.5 billion. NASA assured the White House that each shuttle would be good for 100 flights and each flight would have an average cost of $7.7 million. The planners agreed that the shuttle program would have to operate about fifty missions a year to satisfy demand for satellite launches. It was expected that the shuttle program would be operational by the end of the decade.

Historians such as T. A. Heppenheimer in *The Space Shuttle Decision* (1999) suggest that President Nixon had strong political motives to approve the space shuttle program. A presidential election was coming up in 1972, and he was anxious to gain favor in states like Florida and Texas that would benefit from new NASA projects. Also, the Soviet Union had already put a space station (*Salyut 1*) in orbit during 1971. The last Apollo mission was scheduled for 1972. On January 5, 1972, President Nixon announced to the nation that NASA would build a new Space Transportation System (STS) based around a new vehicle called a space shuttle.

SPACE SHUTTLE DESIGN AND DEVELOPMENT

Various space shuttle designs had been evolving since the 1950s. The U.S. Air Force had examined several options based on a reusable manned space plane that could be maneuvered in flight and glided to a landing. The best-known program was called DynaSoar (short for Dynamic Soaring). The DynaSoar concept included an expendable launch vehicle to carry a space plane out of Earth's atmosphere.

NASA engineers began designing a spacecraft much different from those used during the Apollo program. Apollo capsules and command modules were launched inside long cylindrical rockets. The thrust needed to get these vehicles off the ground was through the center of gravity of each rocket. The rockets were fueled by liquids: kerosene and liquid hydrogen-oxygen.

The shuttle design was completely different. At first engineers hoped to develop a fully reusable vehicle. Budget constraints soon made it obvious that this was not going to be possible. NASA designed a three-part vehicle for the shuttle:

- A reusable space plane called an orbiter

- An expendable external liquid fuel tank for the orbiter's three main engines

- A reusable pair of external rocket boosters containing a powdered fuel

Figure 4.1 shows the major components of the space shuttle design. The launch sequence called for the fuel tank and the rocket boosters to be jettisoned away from

FIGURE 4.1

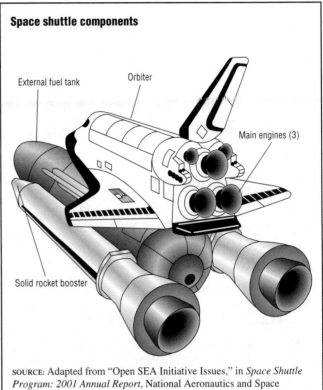

Space shuttle components

SOURCE: Adapted from "Open SEA Initiative Issues," in *Space Shuttle Program: 2001 Annual Report*, National Aeronautics and Space Administration, 2001, http://spaceflight.nasa.gov/shuttle/reference/2001_shuttle_ar.pdf (accessed January 31, 2006)

the orbiter during the ascent phase. The rocket boosters were designed to be recovered and refilled with fuel for the next launch. The external tank was to be jettisoned above Earth's atmosphere and burn up during reentry.

The orbiter holds the crew compartment and payload bay. (See Figure 4.2 for a typical orbiter layout.) The payload bay measures sixty feet by fifteen feet. The shuttle was designed to transport the orbiter into space 115 to 690 miles above the Earth's surface. This is considered low-Earth orbit or LEO.

The orbiter had to be capable of maneuvering while in space and during landing. The early designs were based on the Air Force's X-series of high-performance aircraft. Unlike the Apollo capsules, the orbiter was intended to be reusable. It had to land on the ground, rather than splash down into the ocean. At first engineers included jet engines on the orbiter for use within Earth's atmosphere. These proved to be too expensive and too heavy for the structure and were eliminated. Instead, the orbiter was designed to glide through the air to its landing site.

The orbiter was built to carry a crew of seven under normal circumstances, for a typical mission time of seven days in space. The maximum mission time for this crew number is thirty days (assuming that adequate supplies have been packed). The orbiter was designed to hold up to ten people in an emergency.

FIGURE 4.2

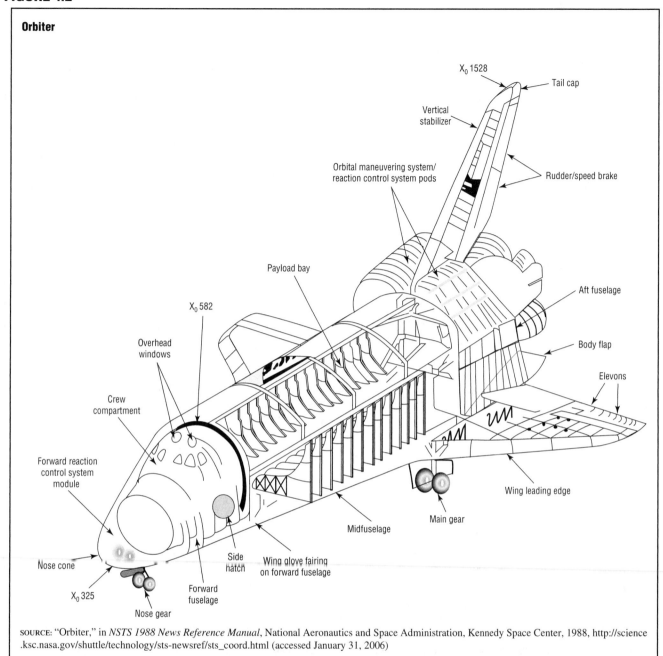

Orbiter

SOURCE: "Orbiter," in *NSTS 1988 News Reference Manual*, National Aeronautics and Space Administration, Kennedy Space Center, 1988, http://science .ksc.nasa.gov/shuttle/technology/sts-newsref/sts_coord.html (accessed January 31, 2006)

One of the most difficult design problems for the orbiter was a thermal protection system that could be reused. Previous spacecraft had been well protected from the intense heat of reentry, but their thermal protection materials were rendered unusable after one reentry. At first designers hoped to cloak the orbiter in metal plates that could withstand high temperatures. This proved to be too heavy. So the orbiter was built out of light-weight aluminum, and its underside was covered with high-tech thermal blankets and tiles. More than 24,000 individual tiles had to be applied by hand. These light-weight tiles are made of sand silicate fibers mixed with a ceramic material.

The new spacecraft had to be light enough to get off the ground, but large enough to carry military payloads that weighed substantially more than what shuttle engineers had expected. The Department of Defense also wanted the shuttle to be able to fly polar orbits (that is, orbits crossing over the North and South Poles). This meant that a launching facility on the West Coast was required, so that the shuttle could launch in a southerly direction toward the South Pole. In April 1972 it was decided that the Air Force would build this facility at Vandenberg Air Force Base in California.

The primary launch facility was authorized at Kennedy Space Center (KSC) in Florida. The KSC location

FIGURE 4.3

Space shuttle launch sites

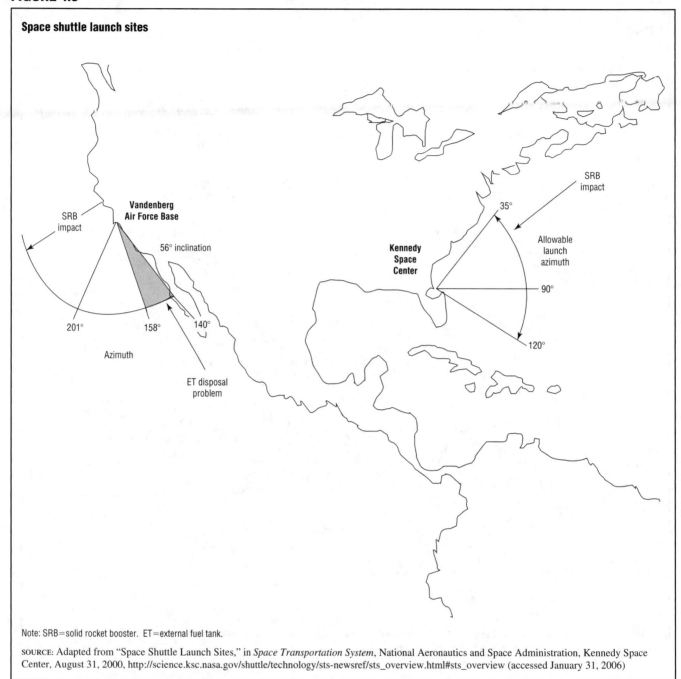

Note: SRB=solid rocket booster. ET=external fuel tank.

SOURCE: Adapted from "Space Shuttle Launch Sites," in *Space Transportation System*, National Aeronautics and Space Administration, Kennedy Space Center, August 31, 2000, http://science.ksc.nasa.gov/shuttle/technology/sts-newsref/sts_overview.html#sts_overview (accessed January 31, 2006)

was chosen to allow the shuttle to launch in an easterly direction over the Atlantic Ocean and assume an equatorial orbit (an orbit along lines of latitude near or at the equator.)

Figure 4.3 shows the original launch azimuths (angles) planned for Vandenberg and KSC. The azimuths were chosen so that launch trajectories did not cross over heavily populated areas or foreign soil. They also ensured that any parts jettisoned from the shuttle during ascent would fall harmlessly into the ocean. It was assumed that the shuttle would land at Vandenberg, KSC, or the Air Force's White Sands Testing Facility in New Mexico.

The Department of Defense insisted that the shuttle be designed to return to Vandenberg after only one polar orbit. This was a technological challenge because it meant that the shuttle had to fly more than 1,000 miles to the east during reentry. Engineers call this the "cross-range requirement." In order to meet this requirement the shuttle was given delta wings (symmetrical triangular wings designed for subsonic and supersonic flight) and an enhanced thermal protection system.

NASA had to meet the design demands of the military in order to keep the project alive. However, this added substantially to the development costs for the spacecraft.

TABLE 4.1

Space shuttle statistics

	Overall shuttle	Orbiter
Length	184.2 feet	122.17 feet
Height	76.6 feet	56.67 feet
Wingspan	—	78.06 feet
Approximate weight		
• Gross lift-off, which will vary depending on payload weight and onboard consumables	4.5 million pounds	—
• Nominal end of mission landing with payload, which will vary depending on payload return weight	—	230,000 pounds
Thrust (sea level)		—
• Solid rocket boosters	3,300,000 pounds of thrust each in vacuum	—
• Orbiter main engines	—	393,800 pounds of thrust each at sea level at 104 percent
Cargo bay		
• Length	—	60 feet
• Diameter	—	15 feet

SOURCE: "Table," in *Space Transportation System*, National Aeronautics and Space Administration, Kennedy Space Center, August 31, 2000, http://science .ksc.nasa.gov/shuttle/technology/sts-newsref/sts_overview.htm l#sts_overview (accessed January 31, 2006)

Most of the design work took place during the mid-1970s. This was a time of high inflation for the U.S. economy. High inflation means that the purchasing power of the dollar goes down. NASA would be designated funds in one year, but by the time those funds were received in the next year, their practical value had decreased.

The original date for the first space shuttle launch was to be March 1978. This date was postponed several times due to budget and equipment problems. The shuttle's main engines and thermal protection tiles proved to be particularly troublesome. In 1979 President Jimmy Carter reassessed the need for the space shuttle program and considered canceling it. Historians believe he decided to continue shuttle development because the United States wanted to launch intelligence satellites to monitor the Soviet Union's nuclear missile program (http://anon.nasa-global.speedera.net/ anon.nasa-global/CAIB/CAIB_lowres_chapter1.pdf.) The White House and the Congress put their support behind the space shuttle. In early 1981 NASA declared that development was complete. The shuttle was "finished" and was only 15% over its original budget.

Table 4.1 lists major design parameters of the shuttle and the orbiter. Figure 4.4 shows various views of the space shuttle orbiter.

SPACE SHUTTLE FLIGHT PROFILE

The ten panels of Figure 4.5 illustrate the major steps in a space shuttle flight from launch to landing.

Launch

The countdown to launch begins approximately four days before lift-off. During this time numerous systems checks are conducted on the spacecraft and its components. The flight crew is taken to the orbiter approximately 2.5 hours prior to lift-off and strapped into their seats.

The shuttle is launched in a vertical position, with its nose pointing up, as shown in "Shuttle launch" in Figure 4.5. At 6.6 seconds prior to launch, the three main engines at the rear of the orbiter are ignited. These engines burn fuel contained in the external fuel tank (ET). The ET includes two separate compartments. Liquid hydrogen is kept in one compartment, and liquid oxygen in the other.

When the countdown reaches zero, the solid rocket boosters (SRBs) are ignited. The SRBs are metal housings filled with solid fuel (aluminum powder and other dry chemicals). Ignition of the SRBs provides the powerful push needed to lift the spacecraft off the ground and overcome the effects of Earth's gravity during ascent. The ride for the crew is very rough and bumpy while the SRBs are firing. However, the acceleration load on the humans is designed to stay below three Gs. In other words, the force of gravity "pushing" against the crew members as the shuttle accelerates is only three times the force of gravity on Earth.

Approximately two minutes after lift-off, the shuttle reaches a vertical distance of twenty-eight miles and travels at about 3,000 miles per hour. At this point the SRBs are jettisoned away from the vehicle. Their fuel has been consumed. The SRBs are equipped with parachutes that open after the boosters have fallen a specified distance. The SRBs splash into the ocean and are retrieved for reuse.

The ride becomes much smoother for the shuttle crew after the SRBs are jettisoned. The shuttle's main engines continue to fire until they have used up all of the fuel in the external tank. This occurs at 8.5 minutes after lift-off. At this point the shuttle is above Earth's atmosphere and traveling at a speed of five miles per second. The main engines are shut down and, seconds later, the external fuel tank is jettisoned away from the vehicle. The tank burns up during atmospheric reentry. Only the orbiter is left to continue the journey.

Orbit

The shuttle assumes a low-Earth orbit, typically 150 to 250 miles above sea level, where it travels at about 17,600 miles per hour. It takes the craft approximately forty-five minutes after lift-off to reach its orbit.

The orbiter includes a series of small engines that allow the flight crew to maneuver while in space. These engines comprise the orbital maneuvering system (OMS) and the reaction control system (RCS).

FIGURE 4.4

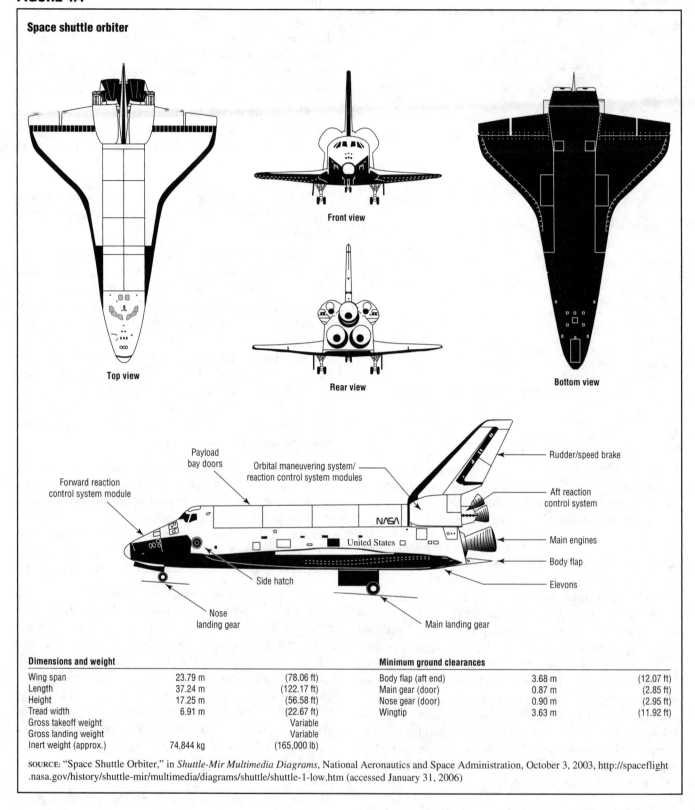

Space shuttle orbiter

Top view

Front view

Rear view

Bottom view

Payload bay doors

Orbital maneuvering system/ reaction control system modules

Forward reaction control system module

Rudder/speed brake

Aft reaction control system

Main engines

Body flap

Elevons

NASA

United States

Side hatch

Nose landing gear

Main landing gear

Dimensions and weight			Minimum ground clearances		
Wing span	23.79 m	(78.06 ft)	Body flap (aft end)	3.68 m	(12.07 ft)
Length	37.24 m	(122.17 ft)	Main gear (door)	0.87 m	(2.85 ft)
Height	17.25 m	(56.58 ft)	Nose gear (door)	0.90 m	(2.95 ft)
Tread width	6.91 m	(22.67 ft)	Wingtip	3.63 m	(11.92 ft)
Gross takeoff weight		Variable			
Gross landing weight		Variable			
Inert weight (approx.)	74,844 kg	(165,000 lb)			

SOURCE: "Space Shuttle Orbiter," in *Shuttle-Mir Multimedia Diagrams*, National Aeronautics and Space Administration, October 3, 2003, http://spaceflight .nasa.gov/history/shuttle-mir/multimedia/diagrams/shuttle/shuttle-1-low.htm (accessed January 31, 2006)

The OMS engines are mounted on both sides of the upper aft fuselage. They provide the thrust needed to make major orbital maneuvers, for example, to move the shuttle into orbit, change orbits, and rendezvous with other spacecraft in orbit. Such instances are called orbit maneuver burns, because the engines are temporarily ignited to achieve them. The last illustration in the first column of Figure 4.5 shows the shuttle undergoing an orbit maneuver burn. The RCS engines are located along either side of the orbiter's tail and on its nose. They provide small amounts of thrust for delicate and exacting maneuvers.

FIGURE 4.5

A space shuttle mission profile

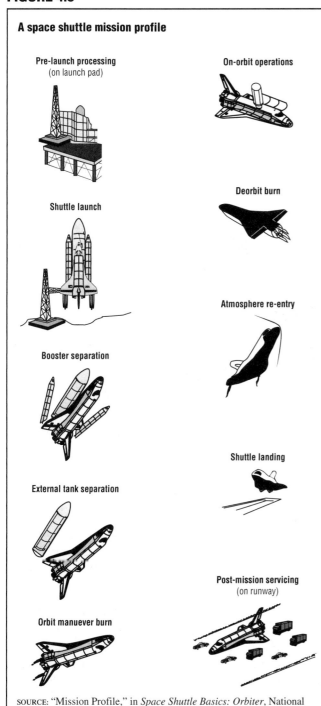

Pre-launch processing
(on launch pad)

Shuttle launch

Booster separation

External tank separation

Orbit manuever burn

On-orbit operations

Deorbit burn

Atmosphere re-entry

Shuttle landing

Post-mission servicing
(on runway)

SOURCE: "Mission Profile," in *Space Shuttle Basics: Orbiter*, National Aeronautics and Space Administration, June 25, 2003, http://spaceflight .nasa.gov/shuttle/reference/basics/orbiter/index.html (accessed January 31, 2006)

During orbit the space shuttle crew performs a variety of tasks depending on the mission requirements. The shuttle was designed to carry payloads into space and to serve as a short-term laboratory for science experiments.

The shuttle can carry satellites or heavy equipment needed for space station construction in its large payload bay. Some satellites are intended for LEO, while others orbit at much higher distances. Shuttle crews can deploy, retrieve, and service LEO satellites from their spacecraft. Satellites that require higher orbits can also be deployed from the shuttle. These satellites have built-in propulsion systems that boost them into their orbits once they are a safe distance away from the shuttle.

The payload bay is equipped with a fifty-foot-long robotic arm called the remote manipulator system (RMS). The RMS is also called the Canadarm, because it was developed by Canadian companies. A crewmember operates the RMS from the orbiter flight deck. The RMS is used to move things in and out of the cargo bay and on and off the *International Space Station* and to grab and position satellites or even space-walking astronauts. In 2005 a fifty-foot-long extension to the RMS was added to the shuttle. The Orbiter Boom Sensor System (OBSS) is equipped with sensors and imaging systems and allows the crew to scan most of the outside of the orbiter for any damage.

The shuttle is equipped with specialized laboratories in which crewmembers can conduct experiments related to astronomy, earth sciences, medicine, and other fields. Most of these experiments take place in pressurized modules specifically designed for shuttle flights. Similar containers called multi-purpose logistics modules (MPLMs) are used as cargo vessels to transport equipment and supplies to and from the *ISS*. Figure 4.6 shows how the MPLM nicknamed Raffaello and other payloads were configured in the payload bay for a shuttle flight.

A space shuttle crew normally consists of five members: a commander, a pilot, and three mission specialists. These are all NASA personnel. The commander has onboard responsibility for the mission, the crew, and the vehicle. The pilot assists the commander in operating and controlling the shuttle and may help deploy and retrieve satellites using the RMS. Mission specialists work with the commander and pilot and have specific responsibilities relating to shuttle systems, crew activities, consumables, scientific experiments, and/or payloads. Mission specialists are trained to perform EVAs (space walks) and to operate the RMS.

In addition to the commander, pilot, and mission specialists, there may be one or two "guest" crewmembers called payload specialists. Payload specialists are not considered NASA astronauts. They perform specialized functions related to payloads and may be nominated by private companies, universities, foreign payload sponsors, or NASA. Payload specialists can also be foreign astronauts recommended by foreign space agencies.

The crew spends time in the crew module. This 2,325-cubic-foot module is pressurized and maintained at a comfortable temperature to provide what is called a "shirt-sleeve environment." The crew module includes the flight deck, the middeck/equipment bay, and an airlock. The airlock contains two spacesuits and space for

FIGURE 4.6

The configuration of the space shuttle Discovery's payload bay for STS-114

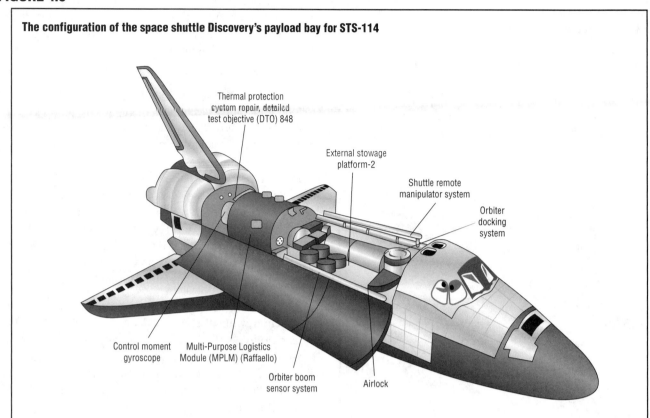

SOURCE: "STS-114 Payload Bay Configuration," in *The Space Shuttle's Return to Flight: Mission STS-114 Press Kit*, National Aeronautics and Space Administration, July 2005, http://www.nasa.gov/pdf/112301main_114_pk_july05.pdf (accessed December 28, 2005)

two crewmembers to put on and take off these suits. Spacesuits are required during EVA activities.

The flight deck is the top level of the crew module. (See Figure 4.7.) This is where the commander and pilot spend most of their time during a mission. During launch and reentry they sit in the two seats facing the front of the orbiter with the commander on the left and the pilot on the right. The orbiter can be piloted from either seat. Two other crewmembers sit behind these seats further back in the flight deck. Any other crew members sit in the mid-deck section during launch and reentry.

The mid-deck of the crew cabin includes stations for meals, personal hygiene, and sleeping. This area includes the waste management system, a table, and stowage space for gear. In an emergency three additional seats can be placed in the mid-deck crew cabin for reentry. This allows the shuttle to carry ten crewmembers back to Earth. Such a contingency might be needed to rescue astronauts from the *International Space Station*.

Reentry and Landing

In order to reenter Earth's atmosphere the shuttle has to decrease its speed by a substantial amount. This is performed via a deorbit burn in which the shuttle is turned upside down with its tail toward the direction it wants to go. Firing of the OMS engines slows the spacecraft down. It then flips over and reenters the atmosphere with the nose pointed up at an angle, shown as "Atmosphere re-entry" in Figure 4.5. This ensures that the well-protected underside of the orbiter takes the brunt of reentry heat.

Reentry is a very dangerous time for the shuttle. Any failure of the thermal protection system could allow superhot gases to enter the orbiter. Reentry begins about seventy-six miles above the Earth's surface. Following reentry, the shuttle glides through the air to its landing site.

Emergency Flight Options

The shuttle program includes a variety of flight options in the event of an emergency. If there is a problem with the main engines up to four minutes after lift-off, the shuttle can undergo a procedure called Return to Launch Site (RTLS) abort. The SRBs and external tank are jettisoned and the orbiter is maneuvered into position to glide back to the launch site. If an RTLS abort is not possible, there is also the option to land the orbiter at an overseas location. This is called a Transatlantic Abort Landing or TAL. There are three TAL landing locations along the western coast of Europe and Africa: near

FIGURE 4.7

Shuttle crew module layout

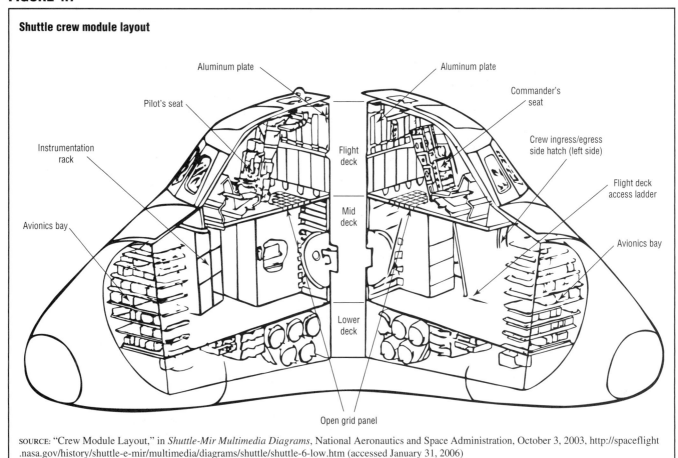

SOURCE: "Crew Module Layout," in *Shuttle-Mir Multimedia Diagrams*, National Aeronautics and Space Administration, October 3, 2003, http://spaceflight .nasa.gov/history/shuttle-e-mir/multimedia/diagrams/shuttle/shuttle-6-low.htm (accessed January 31, 2006)

Moron in Spain, at Ben Guerur in Morocco, and at Dakar in Senegal.

If the orbiter launches successfully but cannot reach its intended orbit, then an Abort-to-Orbit procedure is followed. This means that the spacecraft assumes a lower orbit than planned. In the event that the orbiter cannot maintain any orbit, it returns to Earth for reentry and landing. It may travel once around the Earth before it does so. This option is called the Abort Once Around.

The final emergency flight option is called the contingency abort. This procedure is undertaken if the orbiter cannot land on a landing strip for some reason. It calls for the orbiter to be put into a glide and the crew to use the in-flight escape system. This includes a pole that is extended out the side hatch door. The crewmembers can then slide along the pole to the end and parachute to the ground.

SPACE SHUTTLE PROGRAM ORGANIZATION

The Space Shuttle Program (SSP) is administered and operated by NASA, with the help of thousands of contract employees. Figure 4.8 shows the locations of key NASA and contractor facilities involved in the SSP. Strategic management of the program is handled at NASA's headquarters in Washington, D.C. This is where major decisions are made about future missions.

Johnson Space Center (JSC) in Houston, Texas, is home to the operational offices of the program. This office administers the Space Flight Operations Contract, a six-year, $7 billion contract originally signed in 1996 between NASA and United Space Alliance (a joint venture between the Boeing and Lockheed Martin corporations). United Space Alliance performs day-to-day operation of the space shuttle program. The original contract included two two-year extension options, both of which have been exercised by NASA. The estimated worth of the first extension was $2.8 billion, while the second extension was valued at approximately $3.6 billion. The contract expires at the end of September 2006; however, a new contract is expected to be developed to take its place. As of 2005 United Space Alliance employed more than 10,000 people. Most of these people work at JSC, the Kennedy Space Center (KSC) in Florida, and the Marshall Space Flight Center (MSFC) in Huntsville, Alabama.

JSC also hosts the mission control center, astronaut training, and shuttle simulation facilities. KSC supplies the shuttle launch and landing facilities; maintains and

FIGURE 4.8

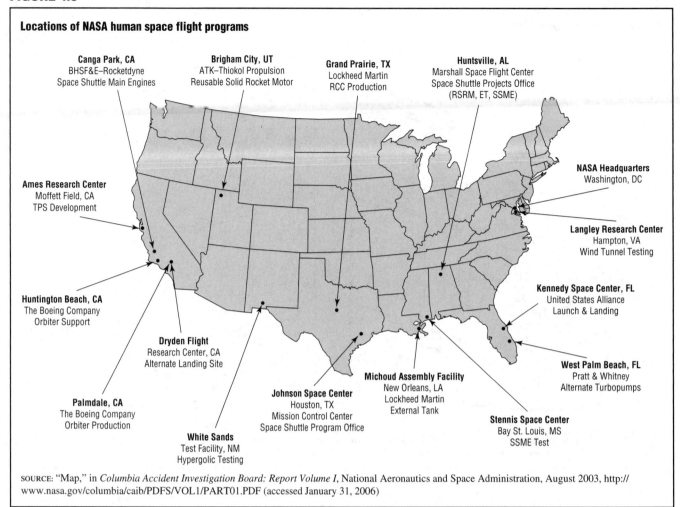

Locations of NASA human space flight programs

Canga Park, CA
BHSF&E–Rocketdyne
Space Shuttle Main Engines

Brigham City, UT
ATK–Thiokol Propulsion
Reusable Solid Rocket Motor

Grand Prairie, TX
Lockheed Martin
RCC Production

Huntsville, AL
Marshall Space Flight Center
Space Shuttle Projects Office
(RSRM, ET, SSME)

NASA Headquarters
Washington, DC

Ames Research Center
Moffett Field, CA
TPS Development

Langley Research Center
Hampton, VA
Wind Tunnel Testing

Huntington Beach, CA
The Boeing Company
Orbiter Support

Kennedy Space Center, FL
United States Alliance
Launch & Landing

Dryden Flight
Research Center, CA
Alternate Landing Site

West Palm Beach, FL
Pratt & Whitney
Alternate Turbopumps

Palmdale, CA
The Boeing Company
Orbiter Production

Johnson Space Center
Houston, TX
Mission Control Center
Space Shuttle Program Office

Michoud Assembly Facility
New Orleans, LA
Lockheed Martin
External Tank

White Sands
Test Facility, NM
Hypergolic Testing

Stennis Space Center
Bay St. Louis, MS
SSME Test

SOURCE: "Map," in *Columbia Accident Investigation Board: Report Volume I*, National Aeronautics and Space Administration, August 2003, http://www.nasa.gov/columbia/caib/PDFS/VOL1/PART01.PDF (accessed January 31, 2006)

overhauls the orbiters; packages components for the orbiter laboratories; and assembles, tests, and refurbishes motors for the solid rocket boosters (SRBs). Most of the contractor personnel working at KSC fall under the Space Flight Operations Contract administered at JSC.

Manufacturing contracts for the shuttle program are overseen by NASA at the MSFC. Major contractors include Boeing, United Technologies Corporation's Pratt & Whitney Rocketdyne, Lockheed Martin, and ATK Thiokol Propulsion. These companies manufacture the Space Shuttle Main Engines (SSME) and SSME turbopumps, the external tank (ET), solid rocket motors (RSRM), and reinforced carbon-carbon (RCC) panels for the thermal protection system (TPS). MSFC is also involved in the research and development of payloads that fly on the shuttles.

The shuttles' main engines and external tanks are tested at NASA's Stennis Space Center in Bay St. Louis, Mississippi. The Dryden Flight Research Center is located at Edwards Air Force Base in California. This is the back-up landing site for the shuttle.

Other NASA Centers assist the SSP by developing or testing shuttle components or fuels at their facilities. The shuttle thermal protection system is developed at Ames Research Center in Moffett Field, California. The highly toxic fuels called hypergols that are used to run the orbiter's maneuvering system and reaction control system are tested at the White Sands Test Facility in New Mexico. The orbiter structure is tested in wind tunnels at the Langley Research Center in Hampton, Virginia.

Since the 1990s the number of workers dedicated to the SSP has declined sharply from more than 30,000 in 1993 to fewer than 12,000 in 2005. The vast majority were contractor employees working for United Space Alliance under the Space Flight Operations Contract.

SPACE SHUTTLE MISSIONS

On April 12, 1981, *Columbia* became the first shuttle to fly into space. The flight's purpose was to test the shuttle's systems, and the mission lasted only two days. It was considered a huge success. Three more test flights were conducted during 1981 and 1982, all with the orbiter *Columbia*. On July 4, 1982, President Ronald Reagan

TABLE 4.2

Orbiter vehicles

Orbiter name	NASA code number	Date completed	Date of first launch	Named after	Note
Enterprise	OV-101	September 1976	Not applicable	The starship Enterprise in the television series "Star Trek"	Used for testing only during the 1970s, never launched into space
Columbia	OV-102	March 1979	April 12, 1981	A ship captained by American explorer Robert Gray during the 1790s	Destroyed during reentry, February 1, 2003
Challenger	OV-99	July 1982	April 4, 1983	A British Naval research vessel that sailed during the 1870s	Destroyed shortly after launch, January 28, 1986
Discovery	OV-103	November 1983	August 30, 1984	A ship captained by British Explorer James Cook during the 1770s	First shuttle to dock with the International Space Station (1999)
Atlantis	OV-104	April 1985	October 3, 1985	A research vessel used by the Woods Hole Oceanographic Institute in Massachusetts from 1930 to 1966	First shuttle to dock with the Russian spacecraft Mir (1995)
Endeavour	OV-105	May 1991	September 12, 1992	A ship captained by British Explorer James Cook during the 1760s	Built to replace Challenger. Endeavour was the first shuttle to fly to the International Space Station (1998)

SOURCE: Adapted from *Orbiter Vehicles*, National Aeronautics and Space Administration, February 1, 2003, http://science.ksc.nasa.gov/shuttle/resources/orbiters/orbiters.html (accessed January 31, 2006)

announced that shuttle testing was completed. The next flight of the shuttle was to begin its operational phase.

Space shuttle flights including the orbiters *Columbia, Challenger, Discovery*, and *Atlantis* carried out twenty-four missions before disaster struck.

On January 28, 1986, the space shuttle *Challenger* broke apart only seventy-three seconds after lift-off. All seven crewmembers were killed, including the commander, Francis "Dick" Scobee (1939–86), the pilot, Michael Smith (1945–86), Judith Resnik (1949–86), Ronald McNair (1950–86), Ellison Onizuka (1946–86), Gregory Jarvis (1944–86), and Christa McAuliffe (1948–86). McAuliffe was a teacher who had been selected for the mission by NASA to capture the imagination of America's schoolchildren.

An investigation of the accident revealed that a faulty joint and seal in a solid rocket booster allowed hot gases to escape from the booster and ignite the hydrogen fuel. The resulting explosion tore the shuttle apart. The tragedy brought intense scrutiny and criticism of the shuttle program from government investigators.

President Reagan appointed a panel called the Rogers Commission to investigate the accident. In June 1986 the Rogers Commission issued its findings in a report that was extremely critical of NASA. The report complained that the decision to launch the *Challenger* was flawed due to poor communication. The managers making the launch decision did not have access to all information. For example, they were not aware that some contractor engineers were very concerned about the cold weather forecast for the morning of the launch. They feared that cold

temperatures might compromise the integrity of the SRB seals. These fears were downplayed by NASA officials and not passed on to those making the launch decision.

In addition to problems specific to the *Challenger* accident, the Commission blamed NASA for fostering an overall culture that put schedule ahead of safety concerns. To reduce the scheduling pressure, it was decided that the shuttle would immediately cease carrying commercial satellites and phase out military missions as soon as possible. The Air Force had hoped to stage the first shuttle launch ever from Vandenberg Air Force Base in 1986. The *Challenger* accident and the resulting decision to cease carrying military payloads put an end to these plans. The launch facilities at Vandenberg were dismantled and abandoned. Most of the equipment was transferred to NASA facilities.

A number of organizational changes were made within NASA in response to the *Challenger* accident. Shuttle management was moved from Johnson Space Center to NASA headquarters in Washington, D.C. In addition, NASA created a new office in charge of safety, reliability, and quality assurance. The entire orbiter fleet was grounded and upgraded with new equipment and systems. A new orbiter named *Endeavour* was built to replace *Challenger*. A White House committee later estimated that the shuttle disaster cost the nation approximately $12 billion. This included the cost of building a new orbiter.

Table 4.2 provides general information about each orbiter in the shuttle fleet.

The shuttle flew again on September 29, 1988, with the successful launch of *Discovery* thirty-two months after the *Challenger* accident. Space shuttles flew eighty-seven successful missions between 1988 and 2002. Then, tragedy struck again. On February 1, 2003, the space shuttle *Columbia* broke apart during reentry over the western United States. Seven crewmembers were killed: the commander, Rick Husband (1957–2003), the pilot, William McCool (1961–2003), David Brown (1956–2003), Kalpana Chawla (1961–2003), Michael Anderson (1959–2003), Laurel Clark (1961–2003), and Ilan Ramon (1954–2003). Ramon was a colonel from the Israeli Air Force who traveled on the shuttle as a guest payload specialist. Following the accident, the shuttle fleet was grounded for more than two years.

THE *COLUMBIA* ACCIDENT

Immediately after the *Columbia* disaster, President George W. Bush appointed a panel to investigate what happened. The panel was called the *Columbia* Accident Investigation Board (CAIB) or the Gehman Board. In June 2003 the CAIB released its report, in which it concluded that the most likely cause of the accident was a damaged thermal protection tile on the orbiter's left wing. Video clips of the launch showed a large piece of foam falling off the external tank and striking the left wing eighty-two seconds after lift-off. This piece of foam fell a distance of only fifty-eight feet. However, the space shuttle was traveling very fast when this occurred, so the foam struck with extreme force.

NASA engineers knew about the foam strike, but were unsure whether it had caused any damage. While the *Columbia* was in orbit, some engineers suggested that high-resolution photographs be taken of the orbiter using Department of Defense satellites or NASA's ground-based telescopes. This suggestion was overruled by NASA officials who believed that the foam strike did not endanger mission safety.

During reentry to Earth's atmosphere, one or more damaged thermal tiles along the left wing likely allowed hot gases to breach the shuttle structure. Aerodynamic stresses then tore it apart. Debris from the shuttle was found spread along a corridor across southeastern Texas and into Louisiana.

The CAIB report was extremely critical of the entire shuttle program and complained that NASA shuttle managers had once again become preoccupied with schedule, rather than safety. Beginning in 1998 the shuttle program was under tremendous pressure to meet construction deadlines for the *ISS*. Nearly every shuttle flight undertaken between 1999 and 2003 was in support of the *ISS*.

The CAIB recommended a number of major changes within the shuttle program and within NASA management. One of the recommendations was that NASA develop a means for the shuttle crew to inspect the orbiter

while docked at the *ISS* and repair any damage discovered. Such a procedure might have saved the *Columbia* crew. Implementation of the so-called "Safe Haven" program was recommended prior to any future shuttle flight.

THE RETURN TO FLIGHT

Soon after publication of the CAIB report, the NASA administrator appointed a Return to Flight (RTF) Task Group to assess the progress of the agency at implementing CAIB recommendations before shuttle flights were resumed. The RTF Task Group was an independent advisory group comprising more than two dozen non-NASA employees with expertise in engineering, science, planning, budget, safety, and risk management. Its members were granted access to NASA facilities and meetings as the agency regrouped and developed new safety strategies.

The most critical technical issue was debris shedding from the ET during ascent and subsequent damage to the orbiter's thermal protection system. The primary focus was on eliminating ET debris and using devices to detect debris impacts. The procedures were changed for applying foam insulation to the ET, and quality control and inspection programs were expanded. Equipment changes were implemented to provide a smoother surface for foam application and to impede ice formation.

NASA decided that the first two shuttle flights after *Columbia* would be "test" flights to assess the effectiveness of new safety changes. The orbiter *Discovery* was selected for the first RTF mission. More than 100 cameras were installed on exterior spacecraft surfaces and at ground locations to provide an array of observation angles during ascent. The fifty-foot-long orbiter boom sensor system (OBSS) was installed on the end of the shuttle remote manipulator system (SRMS) to allow visual inspection of the wing tips and most of the orbiter underbelly while in flight. (See Figure 4.9.) A team of image analysts was assembled at JSC to inspect the images for any signs of damage. Dozens of sensors were installed on the wing edges of *Discovery* to take temperature readings and record the time and location of any debris impacts.

On July 26, 2005, *Discovery* launched from KSC for a fifteen-day mission. The orbiter, with a seven-member crew onboard, docked with the *ISS* and unloaded equipment there. Three spacewalks were conducted including one in which astronauts tested new repair techniques for the thermal protection system. The shuttle landed safely at Edwards Air Force Base on August 9, 2005. NASA proclaimed the first RTF a success. However, camera footage showed that foam debris had shed from the ET during shuttle ascent. Luckily, the debris did not hit the orbiter. NASA and the public realized that the hazard that

FIGURE 4.9

Orbiter boom sensor system

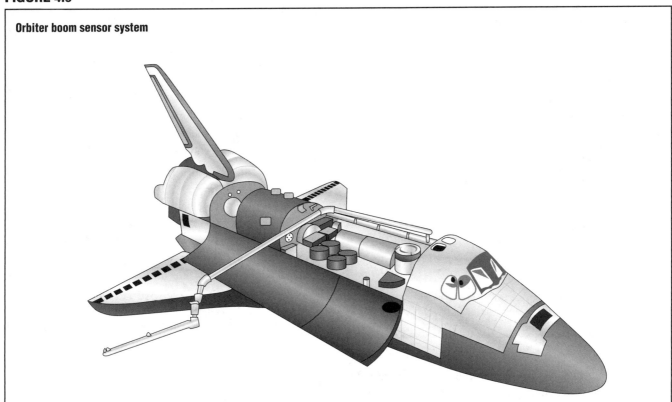

SOURCE: "A Computer-Generated Image of the OBSS in Operation while Attached to the Shuttle Remote Manipulator System," in *Final Report of the Return to Flight Task Group*, National Aeronautics and Space Administration, July 2005, http://www.nasa.gov/pdf/125343main_RTFTF_final_081705.pdf (accessed December 28, 2005)

had doomed *Columbia* had not been eliminated, but merely avoided by chance this time. The second RTF test flight was postponed indefinitely and is not expected to occur until late 2006.

The final report of the RTF Task Force was released to the public after the landing of *Discovery*. However, its findings were communicated to NASA directors prior to launch. The Task Force reported that NASA had "met the intent" of twelve of the fifteen most critical recommendations made by the CAIB. (See Table 4.3.) The other three recommendations were considered "so challenging" that NASA was unable to comply with them prior to RTF. The three problem areas were:

• External Tank Debris Shedding

• Orbiter Hardening

• TPS Inspection and Repair

The report noted, "It has proven impossible to completely eliminate debris shedding from the External Tank. The hard fact of the matter is that the External Tank will always shed debris, perhaps even pieces large enough to do critical damage to the Orbiter." Technical and time limitations also prevented NASA from successfully hardening orbiter surfaces to prevent damage from debris impacts and from proving that a damaged thermal protection system could be repaired while a shuttle was in orbit.

ACCOMPLISHMENTS OF THE SPACE SHUTTLE PROGRAM

A historical summary of all space shuttle missions conducted as of February 2006 is presented in Table 4.4.

NASA refers to each shuttle flight using an STS number. STS stands for Space Transportation System. Thus, STS-1 was the first shuttle flight into space. NASA assigns numbers to space shuttle flights in the order in which they are planned (or manifested). There is typically a period of several years between the time a mission is planned and the time of its scheduled launch. During this period priorities can change, and missions are often reshuffled or cancelled. This explains why the STS numbers in Table 4.4 do not always match the flight order number. For example, *Columbia*'s flight in 2003 was called STS-107, yet it was actually the 113th flight of a space shuttle. The missions numbered STS-108 through STS-113 wound up launching before STS-107, because they moved up in priority as launch time approached.

Shuttle flights deployed more than fifty satellites for military, governmental, and commercial clients. In addition,

TABLE 4.3

Status of CAIB recommendations, 2005

CAIB number	CAIB recommendation	Return to flight status
3.2-1	External tank debris shedding	Not met
3.3-1	Reinforced carbon-carbon non-destructive inspection	Met
3.3-2	Orbiter hardening	Not met
3.4-1	Ground-based imagery	Met
3.4-2	High-resolution images of external tank	Met
3.4-3	High-resolution images of Orbiter	Met
4.2-1	Solid rocket booster bolt catcher	Met
4.2-3	Two person close-out inspection	Met
4.2-5	Kennedy Space Center foreign object debris definition	Met
6.2-1	Consistency with resources (schedule pressures)	Met
6.3-1	Mission management team improvements	Met
6.3-2	National Imagery and Mapping Agency agreement	Met
6.4-1	Thermal protection system inspection and repair	Not met
9.1-1	Detailed plan for organizational change	Met
10.3-1	Digitize closeout photos	Met

SOURCE: Adapted from "The Following Table Summarizes the Task Group's Assessment of the CAIB Return-to-Flight Recomendations," in *Final Report of the Return to Flight Task Group*, National Aeronautics and Space Administration, Return to Flight Task Group, July 2005, http://www.nasa.gov/pdf/125343main_RTFTF_final_081705.pdf (accessed December 28, 2005)

three interplanetary craft were launched from shuttles: the *Magellan* spacecraft that traveled to Venus, the *Galileo* spacecraft that traveled to Jupiter, and the *Ulysses* spacecraft that traveled to the sun. Shuttles also deployed important observatories into space, including the *Hubble Space Telescope*, *Gamma Ray Observatory*, the *Diffuse X-Ray Spectrometer*, and the *Chandra X-Ray Observatory*.

The shuttle has carried more than three million pounds of cargo and more than 600 crewmembers into space. Hundreds of scientific experiments were conducted in orbit. Shuttle crews also serviced and repaired satellites as needed, particularly the *Hubble Space Telescope*. Between 1995 and 1998 shuttles docked nine times with the Russian space station *Mir*. Flights to construct the *International Space Station* (*ISS*) began in 1998. Shuttles carried major pieces of the *ISS* into space and traveled to the station seventeen times through the end of 2005.

Despite these accomplishments, the shuttle has not met many of the original goals that NASA set for the program. NASA planners had promised that the shuttle would fly dozens of times per year. As shown in Figure 4.10, the most shuttle flights ever accomplished in one year was nine flights in 1985. For the twenty-five-year period from 1981 to 2005 the shuttle averaged fewer than five flights per year.

NASA also promised that each shuttle orbiter would be good for 100 flights. Figure 4.11 shows the number of flights achieved by each orbiter in the shuttle fleet as of February 2006. *Discovery* has made thirty-one flights, the most of any orbiter. *Challenger* made only ten flights before it was lost. *Columbia* made twenty-eight flights during its lifetime.

There are only three orbiters left in the fleet: *Discovery, Atlantis*, and *Endeavour*. As of 2006 the *Discovery* shuttle is more than twenty years old, and the other two are not much younger. Most of the original facilities and infrastructure built on the ground for the space shuttle program are more than three decades old. To make matters worse Hurricane Katrina inflicted severe damage to two crucial SSP facilities during the summer of 2005—the Michoud Assembly Facility in New Orleans, Louisiana, and Stennis Space Center in Bay St. Louis, Mississippi.

CAIB's 2003 report was extremely critical of the space shuttle program overall. While the panel acknowledged the shuttle as an "engineering marvel" with a wide range of abilities in Earth orbit, it nevertheless concluded that "the shuttle has few of the mission capabilities that NASA originally promised. It cannot be launched on demand; does not recoup its costs; no longer carries national security payloads; and is not cost-effective enough, nor allowed by law, to carry commercial satellites. Despite efforts to improve its safety, the shuttle remains a complex and risky system."

THE FUTURE OF THE SPACE SHUTTLE PROGRAM

In January 2004 President Bush announced a new vision for the future of the nation's space program. It calls for NASA to send astronauts to the moon by 2020 and to Mars after that. This would require a completely new spacecraft. The space shuttle was not designed to fly farther than a few hundred miles from Earth. The space shuttle program would be ended by 2010, assuming that existing U.S. commitments to build the *ISS* are completed by then. The nearly $6 billion spent each year on the shuttle and *ISS* programs would be transferred to the new projects, which would also be allocated new funds.

In October 2005 NASA announced it planned to fly nineteen more space shuttle missions between 2006 and 2010. The first of these missions, the second return-to-flight, was scheduled for the second half of 2006 using the orbiter *Discovery*.

During grounding of the space shuttle fleet American *ISS* crewmembers were transported aboard Soyuz rockets by the Russian space agency, Rosaviakosmos. NASA was unable to pay Rosaviakosmos for this service due to the Iran Nonproliferation Act of 2000, which forbids payment of "extraordinary" amounts of money from the United States to Russia until it is proven that Russia is not sharing with Iran any technology related to missiles or weapons of mass destruction. In order to raise badly needed funds, Rosaviakosmos charged "space tourists" millions of dollars to fly to the *ISS*. In November 2005 the U.S. Senate approved amendments to the Iran Nonproliferation Act allowing NASA to pay the Russian space agency until 2012 for launches supporting the *ISS*.

TABLE 4.4

Space shuttle missions

Flight order	STS* number	Orbiter name	Primary payload	Launch date	Landing date
1	STS-1	Columbia	Shuttle systems test	4/12/1981	4/14/1981
2	STS-2	Columbia	OSTA-1	11/12/1981	11/14/1981
3	STS-3	Columbia	Office of Space Science-1 (OSS-1)	3/22/1982	3/30/1982
4	STS-4	Columbia	DOD and Continuous Flow Electrophoresis System (CFES)	6/27/1982	7/4/1982
5	STS-5	Columbia	Canadian satellite ANIK C-3; SBS-C	11/11/1982	11/16/1982
6	STS-6	Challenger	TDRS-1	4/4/1983	4/9/1983
7	STS-7	Challenger	Canadian satellite ANIK C-2; PALAPA B1	6/18/1983	6/24/1983
8	STS-8	Challenger	India satellite INSAT-1B	8/30/1983	9/5/1983
9	STS-9	Columbia	Spacelab-1	11/28/1983	12/8/1983
10	STS-41-B	Challenger	WESTAR-VI; PALAPA-B2	2/3/1984	2/11/1984
11	STS-41-C	Challenger	LDEF deploy	4/6/1984	4/13/1984
12	STS-41-D	Discovery	SBS-D; SYNCOM IV-2; TELSTAR	8/30/1984	9/5/1984
13	STS-41-G	Challenger	Earth Radiation Budget Satellite (ERBS); OSTA-3	10/5/1984	10/13/1984
14	STS-51-A	Discovery	Canadian communications satellite TELESAT-H; SYNCOM IV-1	11/8/1984	11/16/1984
15	STS-51-C	Discovery	DOD	1/24/1985	1/27/1985
16	STS-51-D	Discovery	Canadian satellite TELESAT-I; SYNCOM IV-3	4/12/1985	4/19/1985
17	STS-51-B	Challenger	Spacelab-3	4/29/1985	5/6/1985
18	STS-51-G	Discovery	MORELOS-A; Arab satellite ARABSAT-A; AT&T satellite TELSTAR-3D	6/17/1985	6/24/1985
19	STS-51-F	Challenger	Spacelab-2	7/29/1985	8/6/1985
20	STS-51-I	Discovery	American satellite ASC-1; AUSSAT-1; SYNCOM IV-4	8/27/1985	9/3/1985
21	STS-51-J	Atlantis	DOD	10/3/1985	10/7/1985
22	STS-61-A	Challenger	D-1 spacelab mission (first German-dedicated spacelab)	10/30/1985	11/6/1985
23	STS-61-B	Atlantis	MORELOS-B; AUSSAT-2; RCA americom satellite SATCOM KU-2	11/26/1985	12/3/1985
24	STS-61-C	Columbia	RCA americom satellite SATCOM KU-1	1/12/1986	1/18/1986
25	STS-51-L	Challenger	TDRS-B; SPARTAN-203	1/28/1986	Vehicle broke apart 73 seconds after liftoff
26	STS-26	Discovery	TDRS-C	9/29/1988	10/3/1988
27	STS-27	Atlantis	DOD	12/2/1988	12/6/1988
28	STS-29	Discovery	TDRS-D	3/13/1989	3/18/1989
29	STS-30	Atlantis	Magellan	5/4/1989	5/8/1989
30	STS-28	Columbia	DOD	8/8/1989	8/13/1989
31	STS-34	Atlantis	Galileo; SSBUV	10/18/1989	10/23/1989
32	STS-33	Discovery	DOD	11/22/1989	11/27/1989
33	STS-32	Columbia	SYNCOM IV-F5; LDEF retrieval	1/9/1990	1/20/1990
34	STS-36	Atlantis	DOD	2/28/1990	3/4/1990
35	STS-31	Discovery	HST deploy	4/24/1990	4/29/1990
36	STS-41	Discovery	Ulysses; SSBUV; INTELSAT Solar Array Coupon (ISAC)	10/6/1990	10/10/1990
37	STS-38	Atlantis	DOD	11/15/1990	11/20/1990
38	STS-35	Columbia	ASTRO-1	12/2/1990	12/10/1990
39	STS-37	Atlantis	Gamma Ray Observatory (GRO)	4/5/1991	4/11/1991
40	STS-39	Discovery	DOD; Air Force Program-675 (AFP675); Infrared Background Signature Survey (IBSS); Shuttle Pallet Satellite-II (SPAS-II)	4/28/1991	5/6/1991
41	STS-40	Columbia	Spacelab Life Sciences-1 (SLS-1)	6/5/1991	6/14/1991
42	STS-43	Atlantis	TDRS-E; SSBUV	8/2/1991	8/11/1991
43	STS-48	Discovery	Upper Atmosphere Research Satellite (UARS)	9/12/1991	9/18/1991
44	STS-44	Atlantis	DOD; Defense Support Program (DSP)	11/24/1991	12/1/1991
45	STS-42	Discovery	IML-1	1/22/1992	1/30/1992
46	STS-45	Atlantis	ATLAS-1	3/24/1992	4/2/1992
47	STS-49	Endeavour	Intelsat VI repair	5/7/1992	5/16/1992
48	STS-50	Columbia	USML-1	6/25/1992	7/9/1992
49	STS-46	Atlantis	TSS-1; EURECA deploy	7/31/1992	8/8/1992
50	STS-47	Endeavour	Space lab-J	9/12/1992	9/20/1992
51	STS-52	Columbia	USMP-1; Laser Geodynamic Satellite-II (LAGEOS-II)	10/22/1992	11/1/1992
52	STS-53	Discovery	DOD; Orbital Debris Radar Calibration Spheres (ODERACS)	12/2/1992	12/9/1992
53	STS-54	Endeavour	TDRS-F; Diffuse X-ray Spectrometer (DXS)	1/13/1993	1/19/1993
54	STS-56	Discovery	ATLAS-2; SPARTAN-201	4/8/1993	4/17/1993
55	STS-55	Columbia	D-2 spacelab mission (second German-dedicated spacelab)	4/26/1993	5/6/1993

TABLE 4.4

Space shuttle missions (CONTINUED)

Flight order	STS* number	Orbiter name	Primary payload	Launch date	Landing date
56	STS-57	Endeavour	SPACEHAB-1; EURECA retrieval	6/21/1993	7/1/1993
57	STS-51	Discovery	Advanced Communications Technology Satellite (AOTO)/Transfer orbit stage (TOS)	9/12/1993	9/22/1993
58	STS-58	Columbia	Spacelab SLS-2	10/18/1993	11/1/1993
59	STS-61	Endeavour	1st HST servicing	12/2/1993	12/13/1993
60	STS-60	Discovery	WSF; SpaceHab-2	2/3/1994	2/11/1994
61	STS-62	Columbia	USMP-2; Office of Aeronautics and Space Technology-2 (OAST-2)	3/4/1994	3/18/1994
62	STS-59	Endeavour	SRL-1	4/9/1994	4/20/1994
63	STS-65	Columbia	IML-2	7/8/1994	7/23/1994
64	STS-64	Discovery	LIDAR In-space Technology Experiment (LITE); Spartan-201	9/9/1994	9/20/1994
65	STS-68	Endeavour	SRL-2	9/30/1994	10/11/1994
66	STS-66	Atlantis	ATLAS-03	11/3/1994	11/14/1994
67	STS-63	Discovery	SpaceHab-3; Mir rendezvous	2/3/1995	2/11/1995
68	STS-67	Endeavour	ASTRO-2	3/2/1995	3/18/1995
69	STS-71	Atlantis	First shuttle-Mir docking	6/27/1995	7/7/1995
70	STS-70	Discovery	TDRS-G	7/13/1995	7/22/1995
71	STS-69	Endeavour	Spartan 201-03; WSF-2	9/7/1995	9/18/1995
72	STS-73	Columbia	USML-2	10/20/1995	11/5/1995
73	STS-74	Atlantis	Second shuttle-Mir docking	11/12/1995	11/20/1995
74	STS-72	Endeavour	Space Flyer Unit (SFU); Office of Aeronautics and Space Technology flyer (OAST-flyer)	1/11/1996	1/20/1996
75	STS-75	Columbia	TSS-1 reflight; USMP-3	2/22/1996	3/9/1996
76	STS-76	Atlantis	Third shuttle-Mir docking; SpaceHab	3/22/1996	3/31/1996
77	STS-77	Endeavour	SpaceHab; SPARTAN (inflatable antenna experiment)	5/19/1996	5/29/1996
78	STS-78	Columbia	Life and Microgravity Spacelab (LMS)	6/20/1996	7/7/1996
79	STS-79	Atlantis	Fourth shuttle-Mir docking	9/16/1996	9/26/1996
80	STS-80	Columbia	Orbiting and Retrievable Far and Extreme Ultraviolet Spectrograph-Shuttle Pallet Satellite II (ORFEUS-SPAS II)	11/19/1996	12/7/1996
81	STS-81	Atlantis	Fifth shuttle-Mir docking	1/12/1997	1/22/1997
82	STS-82	Discovery	Second HST servicing	2/11/1997	2/21/1997
83	STS-83	Columbia	MSL-1	4/4/1997	4/8/1997
84	STS-84	Atlantis	Sixth shuttle-Mir docking	5/15/1997	5/24/1997
85	STS-94	Columbia	MSL-1 reflight	7/1/1997	7/17/1997
86	STS-85	Discovery	Cryogenic Infrared Spectrometers and Telescopes for the Atmosphere-Shuttle Pallet Satellite-2 (CRISTA-SPAS-2)	8/7/1997	8/19/1997
87	STS-86	Atlantis	Seventh shuttle-Mir docking	9/25/1997	10/6/1997
88	STS-87	Columbia	USMP-4, Spartan-201 rescue	11/19/1997	12/5/1997
89	STS-89	Endeavour	Eighth shuttle-Mir docking	1/22/1998	1/31/1998
90	STS-90	Columbia	Final spacelab mission	4/17/1998	5/3/1998
91	STS-91	Discovery	Ninth and final shuttle-Mir docking	6/2/1998	6/12/1998
92	STS-95	Discovery	John Glenn's Flight; SPACEHAB	10/29/1998	11/7/1998
93	STS-88	Endeavour	First ISS flight	12/4/1998	12/15/1998
94	STS-96	Discovery	1st ISS docking	5/27/1999	6/6/1999
95	STS-93	Columbia	Chandra X-Ray Observatory	7/22/1999	7/27/1999
96	STS-103	Discovery	HST repair - 3A	12/19/1999	12/27/1999
97	STS-99	Endeavour	Shuttle Radar Topography Mission (SRTM)	2/11/2000	2/22/2000
98	STS-101	Atlantis	ISS assembly flight 2A.2a	5/19/2000	5/29/2000
99	STS-106	Atlantis	ISS assembly flight 2A.2b	9/8/2000	9/20/2000
100	STS-92	Discovery	ISS assembly flight 3A, Z1 truss and PMA 3	10/11/2000	10/24/2000
101	STS-97	Endeavour	ISS assembly flight 4A, P6 truss	11/30/2000	12/11/2000
102	STS-98	Atlantis	ISS assembly flight 5A, U.S. Destiny laboratory	2/7/2001	2/20/2001
103	STS-102	Discovery	ISS assembly flight 5A.1, crew exchange, Leonardo multi-purpose logistics module	3/8/2001	3/21/2001
104	STS-100	Endeavour	ISS assembly flight 6A, Canadarm2, Raffaello multi-purpose logistics module	4/19/2001	5/1/2001
105	STS-104	Atlantis	ISS assembly flight 7A, Quest airlock, high pressure gas assembly	7/12/2001	7/24/2001
106	STS-105	Discovery	ISS assembly flight 7A.1, crew exchange, Leonardo multi-purpose logistics module	8/10/2001	8/22/2001
107	STS-108	Endeavour	ISS flight UF-1, crew exchange, Raffaello Multi-Purpose Logistics Module, STARSHINE 2	12/5/2001	12/17/2001
108	STS-109	Columbia	HST servicing mission 3B	3/1/2002	3/12/2002
109	STS-110	Atlantis	ISS flight 8A, S0 (s-zero) truss, mobile transporter	4/8/2002	4/19/2002
110	STS-111	Endeavour	ISS flight UF-2, crew exchange, mobile base system	6/5/2002	6/19/2002

TABLE 4.4

Space shuttle missions (CONTINUED)

Flight order	STS* number	Orbiter name	Primary payload	Launch date	Landing date
111	STS-112	Atlantis	ISS flight 9a, S1 (s-one) truss	10/7/2002	10/16/2002
112	STS-113	Endeavour	ISS flight 11a, P1 (p-one) truss	11/23/2002	12/7/2002
113	STS-107	Columbia	SpaceHab-DM research mission, Freestar module,	1/16/2003	Vehicle broke up during reentry 2/1/03
114	STS-114	Discovery	ISS assembly flight LF1, external stowage platform-2, Raffaello Multi-Purpose Logistics Module	7/26/2005	8/9/2005

*STS=Space Transportation System
Notes:
Acronyms:

ATLAS	Atmospheric Laboratory for Applications and Science
AUSSAT	Australian Satellite
DOD	Department of Defense
EURECA	European Retrievable Carrier
HST	Hubble Space Telescope
IML	International Microgravity Laboratory
ISS	International Space Station
LDEF	Long Duration Exposure Facility
MORELOS	Mexican Satellite
MSL	Microgravity Science Laboratory
OSTA	Office of Space and Terrestrial Applications
PALAPA	Indonesian Satellite
SBS	Satellite Business Systems
SRL	Space Radar Laboratory
SSBUV	Shuttle Solar Backscatter Ultraviolet
SYNCOM	Synchronous Communication Satellite
TDRS	Tracking and Data Relay Satellite
TSS	Tethered Satellite System
USML	United States Microgravity Laboratory
USMP	U.S. Microgravity Payload
WSF	Wake Shield Facility

SOURCE: Adapted from *Past Missions*, National Aeronautics and Space Administration, 2004, http://spaceflight.NASA.gov/shuttle/archives (accessed December 28, 2005)

Space shuttle difficulties affect other ongoing missions. NASA originally planned to send shuttle astronauts in 2006 to service the *Hubble Space Telescope* (*HST*). The *HST* orbits in space about 350 miles above the Earth's surface. Scientists believe that it can last only until 2008 without servicing. Then it will lose its orbit and fall to Earth.

In January 2004 NASA administrator Sean O'Keefe announced that the planned *HST* servicing mission would be canceled due to safety concerns. The *HST* orbits far from the *ISS*. A shuttle sent to service the *HST* would not be able to make it to the *ISS* in an emergency.

NASA does not plan to launch *HST*'s replacement (the *James Webb Space Telescope*) until 2011. Cancellation of the shuttle servicing mission will likely mean that the *HST* will fall out of orbit before the new telescope is in place. This would mean a gap of several years in which scientists would not have access to a space telescope. This caused an uproar in the scientific community and resulted in intense lobbying for NASA to reinstate the mission. In August 2005 a new NASA administrator, Mike Griffin, announced that the agency was reconsidering the *HST* servicing mission assuming that the return-to-flight missions are successful.

FIGURE 4.10

Number of shuttle flights per year, 1981–2005

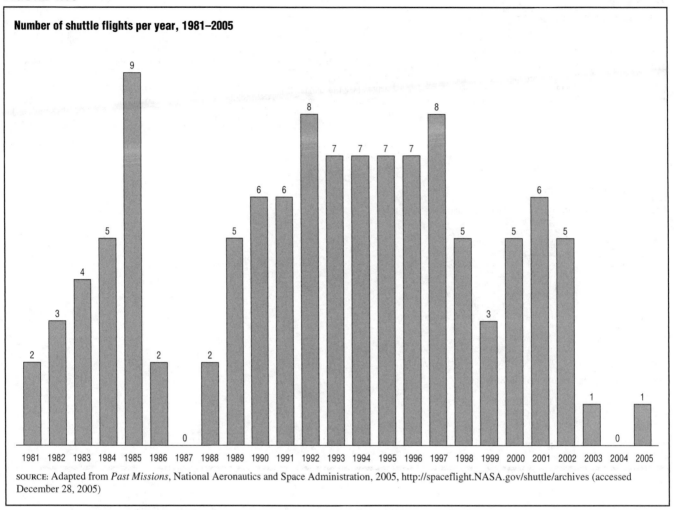

SOURCE: Adapted from *Past Missions*, National Aeronautics and Space Administration, 2005, http://spaceflight.NASA.gov/shuttle/archives (accessed December 28, 2005)

FIGURE 4.11

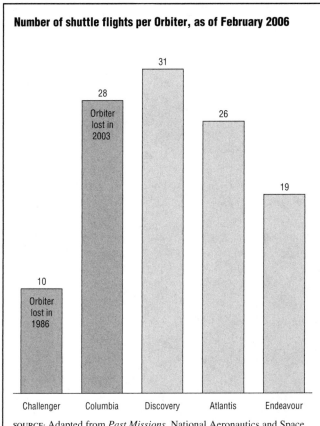

Number of shuttle flights per Orbiter, as of February 2006

SOURCE: Adapted from *Past Missions*, National Aeronautics and Space Administration, 2005, http://spaceflight.NASA.gov/shuttle/archives (accessed December 28, 2005)

CHAPTER 5
THE INTERNATIONAL SPACE STATION

I am directing NASA to develop a permanently manned space station and to do it within a decade.

—President Ronald Reagan, January 25, 1984

A space station is an orbiting structure designed to accommodate visiting crewmembers for an extended period of time. In 1984 the U.S. government envisioned building a continuously manned space station in which scientists would conduct long-term research in a microgravity environment. The station was to be large and spacious, with room for up to ten crewmembers at a time. The U.S. space shuttle was going to be the workhorse that carried cargo and astronauts to the station and back on a routine basis.

To save money on this expensive undertaking, the United States invited other countries to participate. Eventually fifteen countries did so, including Russia, which assumed a major role in the project. The space station became an international venture. It also became very expensive. The design was changed several times in order to bring costs down. The development phase alone dragged on for more than a decade. In 1998 construction finally began on the *International Space Station* (*ISS*). After five years a small portion of the *ISS* was in orbit around Earth. Then the space shuttle *Columbia* exploded. The entire shuttle fleet was grounded while NASA tried to figure out what went wrong. *ISS* construction was halted as well because only shuttles are powerful enough to haul heavy pieces of the *ISS* into space. This was a serious setback for the project.

In January 2004 U.S. President George W. Bush set forth a new space agenda for the United States—sending astronauts to the moon and Mars. The plan calls for aborting the shuttle program and ceasing *ISS* construction as soon as possible. That would mean a space station much smaller than expected, with limited research capabilities. Such a prospect is disappointing to many scientists around the world. However, others believe that even

a downsized *ISS* is a major step on humanity's journey into outer space.

EARLY VISIONS OF A SPACE STATION

The first serious proposal for a space station was made in 1923 by the scientist Hermann Oberth (1894–1989). Although born in Transylvania, Romania, he lived and worked in Germany. Oberth is considered one of the fathers of rocket science. His doctoral dissertation was titled *Die Rakete zu den Planetenräumen* (*The Rocket into Interplanetary Space*).

Oberth proposed building an orbiting structure called a *weltraumstation* (space station) that would serve as a launching and refueling station for spacecraft engaged in deep space travel. Six years later he expanded upon his ideas in *Wege zur Raumschiffart* (*Methods of Achieving Space Flight*). Oberth's writings had a profound effect on the young Wernher von Braun (1912–77), who later became a key developer of American rocket science.

In 1929 Austrian engineer Hermann Noordung (1892–1929) published his idea for an orbiting space station. Noordung's spacecraft was wheel-shaped and primarily designed to be an observatory and scientific laboratory.

In 1952 German rocket scientist Wernher von Braun published a drawing of his vision of a space station. It was a wheel-shaped structure that would orbit 1,075 miles above the Earth and serve a variety of purposes. Von Braun envisioned the station aiding navigation and weather forecasting on Earth and serving as a military outpost, spaceport, and a launching platform for ventures into deeper space.

According to NASA historians, the von Braun team encouraged NASA to build a space station prior to sending a man to the moon. President Kennedy decided that the Apollo program should receive priority. However, a space station was always considered the next step after Apollo.

The U.S. Air Force pursued its own version of a space station during the 1960s. The Manned Orbiting Laboratory (MOL) included a large laboratory module that was reached by a Gemini-type spacecraft launched aboard a Titan rocket. The military hoped to use the MOL for reconnaissance missions and for weather observation. The U.S. government spent more than one billion dollars researching and developing the MOL. The project suffered constant budget overruns and schedule delays and was finally cancelled in 1969. By that time, unmanned reconnaissance satellites were available that could do much of what the MOL was to accomplish. Military astronauts who had been training in the MOL program were transferred to NASA.

THE AMERICAN *SKYLAB*

Long before an Apollo spacecraft landed on the moon, NASA planners were looking ahead to their next great project. The Apollo Applications Program (AAP) began in 1963 with a plan to use leftover Apollo hardware in some kind of orbiting station including a laboratory, workshop, and space telescope. When the *Apollo 20* mission was cancelled in 1970, the AAP inherited a Saturn V rocket. They used the rocket as the launch vehicle for a newly developed station called *Skylab*.

The *Skylab* program had two primary goals:

- Prove that humans could live and work in space for extended periods of time

- Expand knowledge of solar astronomy using a space-based telescope

The program was composed of four flights, as shown in Table 5.1.

The station was designed with two solar panels that were folded flat against the rocket during launch. Once in orbit, they were to open up like wings and harness the sun's energy to provide electricity for the station. On May 14, 1973, the unmanned *Skylab* station was launched into orbit. It was damaged during liftoff when a protective shield came loose and smashed against the solar panels, ripping one of them off and damaging the other.

A team of three *Skylab* astronauts was to have launched the next day. However, their flight was delayed for ten days as engineers assessed the damage to the station. The astronauts, called the *Skylab 2* crew, finally launched on May 25, 1973. They successfully docked with the station and began repairing its damaged components. Crewmembers deployed a temporary sail-like shield to replace the torn-off solar panel. Their mission lasted just over twenty-eight days, a new record for Americans in space. This record was bested by the astronauts of *Skylab 3* and *Skylab 4*.

The *Skylab* station weighed nearly 100 tons and was about the size of a small three-bedroom house. It included a two-level workshop.

The *Skylab 3* mission included two spiders named Anita and Arabella. The spiders were part of an experiment suggested by a high school student named Judith Miles from Lexington, Massachusetts. She wondered if spiders would be able to spin their webs in microgravity. NASA scientists seized upon the idea and sent the spiders into space in cages equipped with still cameras and television cameras. The public became enthralled in hearing about the two spiders.

Neither spider adjusted well to their new environment. Arabella's initial webs were sloppy and lopsided. After a few days the spider seemed to get her space legs and began spinning webs of Earth-like patterns. Both spiders died during the mission, apparently of dehydration. Their bodies were turned over to the Smithsonian Institution and are still kept there in 2006.

TABLE 5.1

Skylab statistics

	Skylab 1	Skylab 2	Skylab 3	Skylab 4
Launch date	5/14/1973	5/25/1973	7/28/1973	11/16/1973
Launch vehicles	Saturn V	Saturn 1B	Saturn 1B	Saturn 1B
Orbital parameters	268.1 × 269.5 miles	268.1 × 269.5 miles	268.1 × 269.5 miles	268.1 × 69.5 miles
Orbital inclination	50 degrees	50 degrees	50 degrees	50 degrees
Orbital period (approximate)	93 minutes	93 minutes	93 minutes	93 minutes
Distance orbit	26,575 miles	26,575 miles	26,575 miles	26,575 miles
Crew's mission distance		11.5 million miles	24.5 million miles	34.5 million miles
Crew's number of revolutions		404	585	1,214
Crew's mission duration		28 days 49 min	59 days 11 hrs 9 min	84 days 1 hr 16 min
Crew's experiment time		392 hr	1,081 hr	1,563 hr
Crew's EVA time		6 hr 20 min	13 hr 43 min	22 hr 13 min

SOURCE: Adapted from "Skylab Statistics," in *Skylab Program Overview*, National Aeronautics and Space Administration, Kennedy Space Center, December 12, 2000, http://www-pao.ksc.nasa.gov/kscpao/history/skylab/skylab-stats.htm (accessed January 31, 2006)

Skylab was not designed for long-term use. It had no method of independent reboost to keep it from falling out of orbit. On July 11, 1979, the station reentered Earth's atmosphere and broke apart over the Pacific Ocean.

Despite its early mechanical problems, the *Skylab* program was considered a great success. The total number of hours spent in space by *Skylab* astronauts was greater than the combined totals of all space flights made up to that time. NASA gained valuable knowledge about human performance under microgravity conditions.

SOVIET AND RUSSIAN SPACE STATIONS

When the Soviets realized that they could not beat the Americans to the moon during the 1960s, they turned their attention to other space goals. In 1971 they put the first of many Soviet space stations into orbit around Earth. Soviet and Russian cosmonauts spent the next three decades gaining valuable experience in long-duration space flight.

The Salyut Series

On April 19, 1971, the Soviet space station called *Salyut 1* was launched from the Baikonur Cosmodrome in what is now Kazakhstan. Salyut means "salute" in English. The station was cylindrically shaped and approximately twelve meters long (about thirty-nine feet) and four meters (thirteen feet) wide at its widest point. It was placed into orbit approximately 200 kilometers (124.3 miles) above Earth.

The station was built so that Soviet scientists could study the long-term effects on humans of living in space. A crew of three cosmonauts flew aboard *Soyuz 10* to the station a few days after the station was placed in orbit. They were unable to dock with it and returned to Earth. In June 1971 the Soviet spacecraft *Soyuz 11* successfully docked with the station, and three cosmonauts inhabited it for twenty-four days. They were killed as they returned to Earth, when a valve opened on their spacecraft and allowed it to depressurize. At that time, cosmonauts did not wear pressurized space suits during launch or reentry.

The Soviet space agency cancelled future flights to the station and began an extensive redesign of the Soyuz spacecraft. In October 1971 *Salyut 1* fell into Earth's atmosphere and was destroyed. In total the Soviets put seven Salyuts into orbit as shown in Table 5.2. These stations were visited by cosmonauts and scientists from a number of countries, including France, India, and Cuba. In 1984 three Soviet cosmonauts spent 237 days aboard *Salyut 7*. This was a new record for human duration in space. *Salyut 7* was deorbited in February 1991.

Mir

In February 1986 the Soviet Union launched a new space station into orbit. It was called *Mir*. Although this word is often translated into English as "peace" or "world," it has a deeper meaning in Russian culture and history. It refers to a type of village established in the harsh and cold Russian countryside during the 1800s in which the villagers shared limited resources and worked together to survive.

Mir was to be Russia's first continuously occupied space station. Although originally planned to stay in orbit for five years, *Mir* survived for fifteen years. It finally tumbled to Earth in 2001.

Russian cosmonauts repeatedly set and broke space duration records aboard the *Mir* station. Vladimir Titov (1947–) and Musa Manarov (1951–) reached the one-year milestone when they completed 366 days in space in 1988. By 1995 the record was 438 days, set by Valeri Polyakov (1942–). This record still stands as of 2006.

The *Mir* is also famous for its non-governmental inhabitants. The Soviet space program suffered financial difficulties beginning in the 1980s. To raise funds, the space agency sold seats on *Mir* to a variety of foreign astronauts and adventurers. In 1990 a Japanese journalist named Tohiro Akiyama became the first citizen of Japan to fly in space and the first private citizen to pay for a space flight. Akiyama's television network paid $28 million to send the man on a seven-day mission to *Mir*. In 1991 a British chemist named Helen Sharman (1963–)

TABLE 5.2

The Salyut series of Soviet space stations

Name	Launch date	Deorbit date	Total crew occupancy time	Note
Salyut 1	April 1971	October 1971	24 days	Three cosmonauts died on their return to Earth.
Salyut 2	April 1973	April 1973	0 days	Unmanned. Station fell apart soon after reaching orbit.
Salyut 3	June 1974	January 1975	15 days	Hosted 1 crew. One unsuccessful docking.
Salyut 4	December 1974	February 1977	92 days	Hosted 2 crews and 1 unmanned craft. One abort.
Salyut 5	June 1976	August 1977	67 days	Hosted 2 crews. One unsuccessful docking.
Salyut 6	September 1977	July 1982	676 days	Hosted 16 crews and 1 unmanned craft.
Salyut 7	April 1982	February 1991	861 days	Hosted 10 crews.

SOURCE: Created by Kim Weldon for Thomson Gale, 2004

spent eight days in space after winning a contest sponsored by a London bank.

Shuttle-*Mir* Missions

As early as 1978 NASA proposed a joint U.S.-Soviet mission to a Salyut station. NASA engineers discussed possible ways to dock a U.S. space shuttle with the station and hoped to put a scientific payload on board the station. Scientific hopes were overshadowed by international politics. In 1979 the Soviet Union began a war in Afghanistan. Two years later the Soviet government imposed martial law in Poland to suppress dissenters. The U.S. response to both incidents was a sharp reduction in cooperative efforts between the two countries. The Soviet empire began to dissolve during the late 1980s and was officially ended in 1991, when it separated into numerous independent countries. The largest of these was Russia, which inherited most of the Soviet space program.

In June 1992 U.S. President George H. W. Bush and Russian President Boris Yeltsin signed a document called the "Agreement between the United States of America and the Russian Federation Concerning Cooperation in the Exploration and Use of Outer Space for Peaceful Purposes." NASA and the Russian space agency Rosaviakosmos (which had been recently created) worked out a plan for joint shuttle-*Mir* missions. Both agencies considered this a prelude to a joint U.S.-Russian space station. In fact, the shuttle-*Mir* program was officially called "Phase 1" at NASA. Phase 2 was to be assembly of a space station. Phase 3 was to be operation of a space station with gradual addition of scientific and operational capabilities.

The agency set four goals for the Phase 1 program:

- Learn to work with international partners
- Reduce the risks associated with developing and building a space station
- Gain American experience in long-duration missions
- Perform research in life sciences, microgravity, and environmental programs

In 1993 U.S. President Bill Clinton met with Yeltsin and agreed to continue cooperative efforts in space exploration. On February 3, 1994, the space shuttle *Discovery* (STS-60) launched for the first time ever with a cosmonaut aboard. Exactly a year later the shuttle was launched again. This time the shuttle flew near *Mir*. In June of 1995, STS-71, with the orbiter *Atlantis*, docked with *Mir*. The shuttle was carrying four cosmonauts in addition to its American crew.

Between 1994 and 1998 space shuttles docked ten times with *Mir*. Figure 5.1 depicts a shuttle docked to the *Mir* space station. American astronauts logged nearly 1,000 days of orbit time during the Phase 1 program. One of them, Shannon Lucid (1943–), set the women's record for space flight duration—188 days. A summary of all Phase 1 accomplishments is given in Table 5.3,

Mir Mishaps

When the first Americans arrived at *Mir* in 1995, the station had already been in orbit for nine years. They found a cramped and crowded spacecraft bulging with hoses, cables, and scientific equipment. Every closet and storage space was crammed full. Some gear and tools floated around, because there was no space left to stow or fasten them. Over the years, water droplets had escaped from environmental control systems and now clung to delicate electronics. *Mir*'s systems were plagued by computer crashes and battery problems. The cosmonauts spent the vast majority of their time doing repair and maintenance tasks.

On February 23, 1997, a fire broke out aboard *Mir* when a cosmonaut lit a lithium perchlorate candle. Flames one-foot long shot out of the unit and ignited the canister. At the time, there were six men aboard the station—four Russians, a German, and an American. The fire quickly filled the spacecraft with smoke. The Russians ordered everyone to evacuate the station. However, the fire blocked access to one of the two Soyuz capsules that served as their lifeboats. Only three men would be able to escape if the hull was breached.

The men fought the flames with towels and a few working fire extinguishers. Many of the ship's fire extinguishers malfunctioned or were bolted down and could not be released. After fifteen minutes the fire died, apparently snuffed out by lack of oxygen in its immediate area. The crew had donned respirators and floated quietly, barely moving for hours and waiting for the ship's ventilation system to remove the smoke.

Russian mission control downplayed the fire to the public and American officials, telling them it was a minor and isolated event. In truth, there had been a similar occurrence several years before in which a candle had burst into flames. Neither the most recent crew nor the public had been informed of that incident. Secrecy had always been a hallmark of the Soviet space program, and this culture persisted in the Russian space program of the 1990s.

The fire in February 1997 was followed by even more problems aboard *Mir*. Only a week later, a camera failed during a docking exercise and the station was nearly rammed by a supply ship. In late March the cooling system failed. The temperature rose to 95° Fahrenheit on the station, and it was permeated by an odor of antifreeze. High carbon dioxide levels forced the crew to limit their physical activity.

FIGURE 5.1

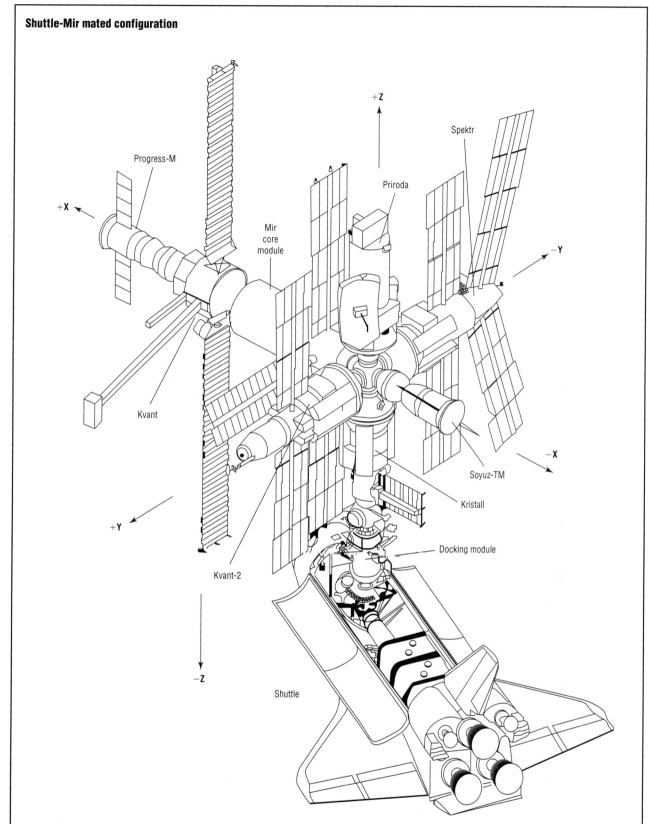

Shuttle-Mir mated configuration

SOURCE: "Shuttle-Mir Mated Configuration," in *Shuttle-Mir Multimedia Diagrams*, National Aeronautics and Space Administration, October 3, 2003, http://spaceflight.nasa.gov/history/shuttle-mir/multimedia/diagrams/shutmir-low.htm (accessed January 31, 2006)

TABLE 5.3

Shuttle-Mir timeline, 1994–98

Year	1994	1995	1996
Month	J F M A M J J A S O N D	J F M A M J J A S O N D	J F M A M J J A S O N D
Russian flights/ progress resupply		Mir 18/19 (Soyuz TM 21 70) 3/14–9/11; Spektr launched 5/20; Progress TM-27 4/9; Mir 20 (Soyuz TM 22 71) 9/3–2/29	Mir 21 (Soyuz TM 23 72) 2/21–9/2; Priroda launched 4/23; Progress M31 5/6; Progress M32 7/26; Mir 22 (Soyuz TM 24 73) 8/18–3/2; Progress M33 11/20
Mir crews		CDR: Dezhurov, ENG: Strekalov, Astronaut: Thagard (Mir 19); CDR: Solovyev ENG: Budarin; CDR: Ghidzhenko ENG: Avdyev ESA: Reiter	CDR: Onufrienko ENG: Usachev; CDR: Korzun ENG: Kalen CNES: Deshays
Mir astronauts		Norman Thagard NASA 1 3/14–7/7 (115 days)	Shannon Lucid NASA 2 3/22–9/26 (188 days); John Blaha NASA 3 9/16–1/22 (128 days)
U.S. flights & shuttle crews	STS-60 Discovery 2/3–2/11 CDR: Bolden PLT: Reightler MS: Chang-Diaz MS: J.Davis MS: Sega MS: Krikalev	STS-63 Discovery 2/3–2/11 (near Mir) CDR: Wetherbee PLT: Collins MS: Voss MS: Harris MS: Foale MS: Titov; STS-71 Atlantis 6/27–7/17 (docking 1) CDR: Gibson PLT: Precourt MS: E.Baker MS: Harbaugh MS: Dunbar MS: Thagard Cosmonaut: Solovyev Cosmonaut: Budarin Cosmonaut: Dezhurov Cosmonaut: Strekalov; STS-74 Atlantis 11/12–11/20 (docking 2) CDR: Cameron PLT: Halsell MS: J. Ross MS: McArthur MS: Hadfield	STS-76 Atlantis 3/22–3/31 (docking 3) CDR: Chilton PLT: Searfoss MS: Clifford MS: Godwin MS: Lucid MS: Sega; STS-79 Atlantis 9/16–9/26 (docking 4) CDR: Readdy PLT: Wilcutt MS: Akers MS: Apt MS: Walz MS: Blaha MS: Lucid
Events	Krikalev is 1st cosmonaut on shuttle	Shuttle approaches Mir; 1st U.S. Astronaut to launch on Soyuz 1st U.S. Astronaut on Mir; 1st Shuttle-Mir docking; Docking Module and cooperative solar array delivered	1st EVA during docked mission; Lucid sets record for women's longest time in space; 1st double Spacehab Module

TABLE 5.3

Shuttle-Mir timeline, 1994–98 [CONTINUED]

Year	1997	1998
Month	J F M A M J J A S O N D	J F M A M J J A S O N D
Russian flights/ progress resupply	Mir 23 (Soyuz TM 25 74) 2/10–8/14; Progress M34 4/6; Mir 24 (Soyuz TM 26 75) 8/5–2/19; Progress 235 7/5; Progress 237 10/5	Mir 25 (Soyuz TM 26 75) 1/29–8/25; Progress 236 12/20; Progress 240 3/15; Progress 238 5/15
Mir crews	CDR: Tsibliev ENG: Lazutkin DARA: Ewald; CDR: Solovyev ENG: Vinogradov	CDR: Musabayev ENG: Budarin CNES: Eyherts
Mir astronauts	Jerrry Linenger NASA 4 1/12–5/24 (132 days); Micheal Foale NASA 5 5/15–10/7 (144 days); David Wolf NASA 6 9/26–1/3 (128 days)	Andrew Thomas NASA 7 1/22–6/12 (140 days)
U.S. flights & shuttle crews	STS-81 Atlantis 1/12–1/22 (docking 5) CDR: M. Baker PLT: Jett MS: Grunsfeld MS: Ivins MS: Wisoff MS: Linenger↓ MS: Blaha↑; STS-84 Atlantis 5/15–5/24 (docking 6) CDR: Precourt PLT: Collins MS: Noriega MS: Lu MS: Foale↓ MS: Linenger↑ MS: Kondakova ESA: Clervoy; STS-86 Atlantis 9/25–10/6 (docking 7) CDR: Wetherbee PLT: Bloomfield MS: Lawrence MS: Parazynski MS: Foale↑ MS: Wolf↓ Cosmonaut: Titov CNES: Chretien	STS-89 Endeavour 1/22–1/31 (docking 8) CDR: Wilcutt PLT: Edwards MS: Dunbar MS: Anderson MS: Reilly MS: Wolf↑ MS: Thomas↓ Cosmonaut: Sharipov; STS-91 Discovery 6/2–6/12 (docking 9) CDR: Precourt PLT: Gorie MS: Lawrence MS: Chang-Diaz MS: Kavandi MS: Thomas↑ Cosmonaut: Ryumin
Events	Fire aboard Mir 1st Astronaut to perform EVA on Mir wearing Russian space suit.; Spektr collison Foale/ Solovyev EVA.; Titov 1st cosmonaut to perform EVA in U.S. space suit	Wolf Solovyev EVA; Phase 1 ends

SOURCE: "Graphic Timeline," in *Timeline of Shuttle-Mir*, National Aeronautics and Space Administration, December 4, 2003, http://spaceflight.nasa.gov/history/shuttle-mir/images/timeline.pdf (accessed February 4, 2006)

Throughout the spring the crew struggled to repair the ailing ship. In June another calamity struck. On June 27, 1997, the crew was conducting a docking test using a Progress supply ship. The cosmonauts did not trust the station's television images of the maneuver and tried to guess the distance by eyesight. The freighter slammed into the station and cracked its hull. The incident is described in detail in a 2003 book titled *Leaving Earth: Space Stations, Rival Superpowers, and the Quest for Interplanetary Travel,* by Robert Zimmerman.

According to Zimmerman, the *Mir* crew felt their ears pop as the station began to lose pressure and they could hear oxygen hissing out into outer space. The crash had breached the hull of the module called *Spektr.* Mission control ordered the crew to close the hatch to that module and seal off the breach. This was impossible, because previous crews had run electrical cables and wires through the doorway. The hatch could not be closed all the way. The crew frantically began cutting and unhooking the wiring. Finally they closed the door, isolating themselves in the base unit away from the leak.

The impact with the freighter knocked the station into an uncontrollable spin. Disconnection from *Spektr* had cut power to vital systems. The crew floated in darkness for nearly thirty hours. Finally, they used the rockets on the Soyuz lifeboats to nudge the station out of its spin and into proper position.

The mishaps aboard *Mir* could not be downplayed anymore. Politicians and the press in the United States called for NASA to stop sending American astronauts to the trouble-prone station. Despite the pressure, NASA and the White House felt it was important to complete the Phase 1 program. Shuttle flights continued to *Mir* throughout 1997 and into 1998.

AN INTERNATIONAL EFFORT

In his 1984 *State of the Union* speech, U.S. President Ronald Reagan directed NASA to develop a space station before the end of the decade. The project was expected to cost the United States only around $8 billion due to the participation of foreign governments. By 1988 Canada, Japan, and nine European countries had signed formal agreements with the United States to participate in the project. Reagan chose the name *Freedom* for the new station.

Freedom

Freedom was to include three separate components: a pressurized base unit in which the crew would live and work and two automated platforms that would conduct scientific experiments and record observations of Earth's climate. At that time, designers envisioned a station that could accommodate a crew of up to ten people.

The project was plagued immediately by financial and technological problems that continued to get worse. Development costs increased even as NASA's budget shrank. Congress demanded several redesigns to save money. NASA eliminated the two automated platforms and scaled back the base unit. Each redesign resulted in a smaller station with less usable space and less electrical power available to scientists. NASA's foreign partners became increasingly annoyed about the design changes.

Meanwhile the space shuttle program was enmeshed in its own difficulties. The space shuttle was crucial to the station program, because it was to be the only means by which American flight crews could reach the station. The catastrophic breakup of the space shuttle *Challenger* in 1986 grounded the entire shuttle fleet for more than two years. It also raised questions about the safety and quality of NASA's operations.

Even as the space station shrank in volume, its weight increased. This required shuttle design changes to accommodate the extra weight. It was also decided to include some kind of lifeboat capability in the station in case its crew had to leave in an emergency. By December 1990 the cost of *Freedom* was estimated at $38 billion (approximately $58.9 billion in 2006 dollars). This included the cost of shuttle launches required to build the station.

In 1991 a Congressional committee recommended that NASA cancel the *Freedom* program. A vote was held on the House floor to determine its fate, and an amendment was passed to continue the program. This was the first of nearly two dozen votes that would take place over the next decade in the U.S. Congress and U.S. Senate on the fate of the space station. Each time the program was allowed to continue. However, some of the votes were very close. In a June 1993 vote the station survived on the basis of a one-vote margin.

Station Alpha

When U.S. President Bill Clinton took office in 1993 he ordered a sweeping revision to cut costs in the space station program. By that time, NASA had already spent in excess of $11 billion on design costs alone. Not one piece of hardware had been launched into space yet. NASA designers presented President Clinton with several different options for station components and functions, and the president selected a plan called Design A or Design Alpha. The new station was unofficially named Station Alpha.

By this time, the Soviet Union had collapsed. The Clinton administration began talks with the new Russian government and welcomed Russia's eagerness to participate in the space station project. According to a 2001 NASA account of the history of the space station program, President Clinton saw Russian participation as a way to improve foreign relations and put pressure on the Russians to abide by newly signed ballistic missile agreements (http://www.hq.nasa.gov/office/pao/History/smith.htm). He also wanted to provide Russian scientists with jobs to keep them from selling valuable information to America's enemies. NASA estimated that Russian participation would cut $2 billion off the $19 billion cost of completing Project Alpha and speed it up by an entire year.

On September 2, 1993, the two countries signed the Joint Declaration on Cooperation in Space. By the end of the year, NASA and the Russian space agency Rosaviakosmos had ironed out a detailed work plan for what was now called the *International Space Station* (*ISS*).

The *ISS* Plan

It was decided that the *ISS* would comprise individual segments called modules that would dock together to form the station. Each module was to be constructed on the ground and then launched into space. The space shuttle was expected to carry the heaviest loads to the station. Russia agreed to transport supplies and propellants to the *ISS* aboard its unmanned Progress spacecraft. In addition, Russian Soyuz spacecraft were offered as the station's lifeboats.

A rotating schedule for the first four modules was developed in which Russia would build the first and third modules and the United States would build the second and fourth modules. NASA agreed to pay for one of the Russian modules.

The collaboration with Russia raised some concerns among U.S. politicians and scientists. They publicly expressed fears that the cash-strapped Russian government would not be able to fulfill its obligations to the program. NASA had given Rosaviakosmos responsibility for two of the most important modules in the station. Failure to deliver them would cripple the entire project. U.S. fears grew even greater during the late 1990s as problems surfaced on the Russian space station *Mir*.

On January 29, 1998, the U.S. government signed the *1998 Intergovernmental Agreement on Space Station Cooperation* with fourteen other countries. This document outlined the agreement among the partners for design, development, operation, and utilization of the *ISS*. The previous year the United States had signed a separate agreement with Brazil giving that country utilization rights in exchange for supplying *ISS* parts. Table 5.4 lists the nations that are *ISS* partners.

In early 1998 it was expected that at least forty spacecraft launches spread over five to seven years would be required to assemble more than 100 components into the *ISS*. At that time the station was designed for a crew of seven people. The space station was expected to be

TABLE 5.4

International Space Station partner nations

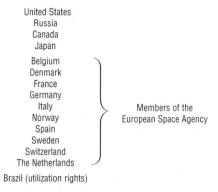

United States	
Russia	
Canada	
Japan	
Belgium	
Denmark	
France	
Germany	
Italy	Members of the
Norway	European Space Agency
Spain	
Sweden	
Switzerland	
The Netherlands	
Brazil (utilization rights)	

SOURCE: Created by Kim Weldon for Thomson Gale, 2004

FIGURE 5.2

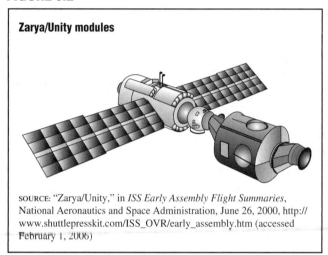

Zarya/Unity modules

SOURCE: "Zarya/Unity," in *ISS Early Assembly Flight Summaries*, National Aeronautics and Space Administration, June 26, 2000, http://www.shuttlepresskit.com/ISS_OVR/early_assembly.htm (accessed February 1, 2006)

completed by 2005 or 2006 and cost the U.S. $26 billion to complete.

ISS Assembly Begins

On November 20, 1998, a Russian Proton K rocket was launched, carrying the first module of the *ISS*. The module was named *Zarya*, which means "sunrise" in English. It is known by the Russian acronym FGB. The Russians built *Zarya*, but the Americans paid for it. It was to be a U.S. component of the station. The module provides control and cargo capabilities. *Zarya* was self-propelled and was designed to keep the station in orbit until the service module arrived. After that it would serve as a passageway, storage facility, docking port, and fuel tank. *Zarya* weighed 44,000 pounds at launch.

Two weeks later, an American shuttle, STS-88, carried the module *Unity* to the station. Figure 5.2 shows the two modules docked together. *Zarya* is on the left, and *Unity* on the right. *Unity* is an American-built module with six docking ports for attachment to other modules.

FIGURE 5.3

Service module attached to Zarya/Unity modules

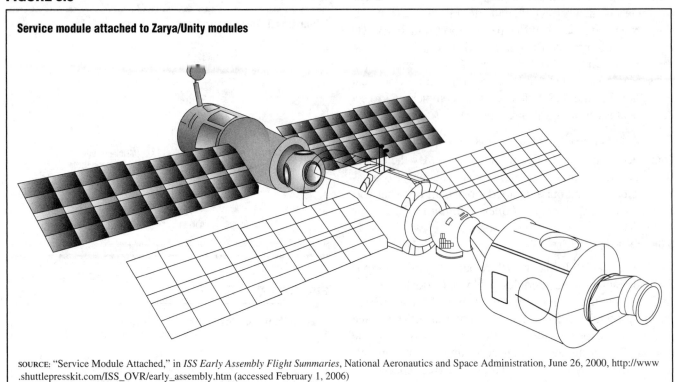

SOURCE: "Service Module Attached," in *ISS Early Assembly Flight Summaries*, National Aeronautics and Space Administration, June 26, 2000, http://www.shuttlepresskit.com/ISS_OVR/early_assembly.htm (accessed February 1, 2006)

Unity provides a node (link) between modules. It is basically a passageway with some internal storage space. *Unity* is eighteen feet long and fifteen feet in diameter and weighs 25,600 pounds.

Two more shuttle crews visited the *ISS* during 1999 and 2000 to outfit the modules with logistical equipment and supplies. On July 12, 2000, the Russians launched the service module *Zvezda* into space aboard a Proton K rocket. *Zvezda* is the Russian word for "star." This module was built and funded by the Russians and was to be the core of their segment.

It included functions for station control, navigation, communications, and life support systems (including the crew quarters). At that time, it was expected that many of these functions would be taken over by an American module to be added later. *Zvezda*'s design was based on the core module of the *Mir* space station. *Zvezda* is forty-three feet long with a wing span of ninety-eight feet and weighed 42,000 pounds at launch. It includes docking ports for Russian Soyuz and Progress spacecraft. Figure 5.3 shows the three modules hooked together. *Zvezda* is on the far left.

Throughout the summer of 2000 visiting Russian and American crews continued outfitting the modules and assembling the station. The station was also visited by an unmanned Russian Progress spacecraft carrying consumables (food and water), spare parts, and propellants.

On November 2, 2000, the first crew to actually inhabit the *ISS* arrived aboard a Soyuz spacecraft. This is called *ISS* Expedition 1. It included two Russian cosmonauts and one American astronaut. Table 5.5 lists the names of all Expedition crewmembers from Expedition 1 through Expedition 12. The Expedition 1 crew lived aboard the station until March 2001. They were visited by two shuttles, one of which brought the American lab module named *Destiny*. *Destiny* was to be the primary research laboratory for U.S. payloads. It included numerous racks that could support a variety of electrical and fluid systems during the performance of experiments. *Destiny* also contains the control center for the *ISS* robotic arm.

In March 2001 the Expedition 2 crew arrived at the station aboard shuttle STS-102. The Expedition 1 crew returned to Earth aboard the space shuttle, leaving their Soyuz spacecraft at the station to serve as a lifeboat. The Expedition 2 crew included two American astronauts and one Russian cosmonaut. It had been decided to swap out the Expedition crews every four to six months and to rotate back and forth between crews that were predominantly Russian and crews that were predominantly American.

The Expedition 2 crew brought a multi-purpose logistics module (MPLM) called *Leonardo*. MPLMs are reusable pressurized shipping containers designed to be temporarily attached to the *ISS* for unloading. They are transported back and forth on space shuttles and carry a

TABLE 5.5

International Space Station expedition crews

	Launch date	Landing date	Time in orbit	Crew members	Crew titles
Expedition 1	10/31/00	3/21/2001	140 days, 23 hr, 38 min	William Shepherd Yuri Gidzenko Sergei Krikalev	Commander Soyuz Commander Flight Engineer
Expedition 2	03/08/01	8/22/2001	167 days, 6 hr, 41 min	Yury Usachev Susan Helms James Voss	Commander Flight Engineer Flight Engineer
Expedition 3	08/10/01	12/17/2001	128 days, 20 hr, 45 min	Frank Culbertson Vladimir Dezhurov Mikhail Tyurin	Commander Soyuz Commander Flight Engineer
Expedition 4	12/05/01	6/19/2002	195 days, 19 hr, 39 min	Yury Onufrienko Dan Bursch Carl Walz	Commander Flight Engineer Flight Engineer
Expedition 5	06/05/02	12/7/2002	184 days, 22 hr, 14 min	Valery Korzun Peggy Whitson Sergei Treschev	Commander NASA ISS* Science Officer Flight Engineer
Expedition 6	11/23/02	5/3/2003	161 days, 1 hr, 17 min	Ken Bowersox Nikolai Budarin Don Pettit	Commander Flight Engineer NASA ISS Science Officer
Expedition 7	04/25/03	10/27/2003	184 days, 21 hr, 47 min	Yuri Malenchenko Ed Lu	Commander NASA ISS Science Officer
Expedition 8	10/18/03	4/29/2004	194 days, 18 hr, 35 min	Michael Foale Alexander Kaleri	Commander/NASA ISS Science Officer Flight Engineer
Expedition 9	04/18/04	10/19/2004	187 days, 21 hr, 17 min	Gennady Padalka Mike Fincke	Commander Flight Engineer/NASA ISS Science Officer
Expedition 10	10/13/04	4/24/2005	192 days, 19 hr, 2 min	Leroy Chiao Salizhan Sharipov	Commander/NASA ISS Science Officer Soyuz Commander/Flight Engineer
Expedition 11	04/14/05	10/10/2005	179 days, 23 min	Sergei Krikalev John Phillips	Commander Flight Engineer/NASA ISS Science Officer
Expedition 12	09/30/05	to be determined	to be determined	William McArthur Valery Tokarev	Commander/NASA ISS Science Officer Flight Engineer

*ISS = International Space Station

SOURCE: Adapted from "Mission Archives," in *Space Station Expeditions*, National Aeronautics and Space Administration, February 3, 2006, http://www.nasa .gov/mission_pages/station/expeditions/index.html (accessed January 31, 2006)

variety of cargo, equipment, and experiment racks. MPLM *Leonardo* was named by the Italian company that built it. The module was loaded with garbage (expired batteries, used filters, etc.) for its return trip to Earth.

The Expedition 2 crew also attached the station's remote manipulator system. This is a robotic arm that was nicknamed Canadarm2, because it was built by Canadian companies. The space shuttle fleet utilizes another robotic arm called Canadarm1.

Tito Comes Aboard

In April 2001 the Russians sent the first taxi to the station to replace the Soyuz lifeboat left by the Expedition 1 crew. The taxi crew included a U.S. millionaire named Dennis Tito (1940–). A year earlier Tito had paid $20 million for a visit to the *Mir* space station. He immediately began training in Russia. When the Russians decided to deorbit *Mir*, they rescheduled Tito for a trip to the *ISS*. They hoped he would be the first of many "space tourists" to pay to fly on *ISS* taxis. Rosaviakosmos desperately needed the money.

NASA and the European partners in the *ISS* were not happy with the decision. When Tito and two cosmonauts showed up at the Houston Space Center for training,

NASA refused to allow Tito entry. The cosmonauts responded by refusing to undergo training. The standoff resulted in a flurry of negotiations between NASA and Rosaviakosmos. The Americans finally agreed reluctantly to allow Tito to train in the United States. They continued to argue that Tito posed a safety risk to the station and banned him from the American segments of the *ISS*. NASA repeatedly asked Rosaviakosmos to postpone Tito's flight, but the Russians would not agree.

On April 28, 2001, a Soyuz blasted off with Tito and two cosmonauts aboard. Two days later it docked with the *ISS*. Tito spent his time aboard *ISS* in the Russian segment taking pictures of Earth and listening to opera. The crew stayed at the station for four days before returning to Earth aboard the old Soyuz lifeboat. After landing Tito told reporters, "I've finally had my dream."

The Price Goes Up

In April 2001 NASA announced that the U.S. cost to complete the *ISS* by 2006 was going to be $4 billion more than expected. The Clinton administration responded by setting up the ISS Management and Cost Evaluation Task Force to assess the project. In November 2001 the Task

Force released a forty-page report that called for serious downsizing of the *ISS*.

The report recommended that the *ISS* be configured for only a three-person crew. Previous plans had called for a seven-person crew. The Task Force noted that scientists were not happy with this proposal and feared it would "have a significant adverse impact on science." However, the report noted that the cutback was necessary to save money and suggested that some of the research planned for the *ISS* could take place aboard space shuttle flights instead.

The report recommended many management changes within NASA to save money and suggested that the agency prioritize its research goals for the station. One goal considered crucial was installation of a centrifuge. A centrifuge is a machine commonly used in research to separate different substances, to remove moisture, or to simulate certain gravitational effects. The report noted that NASA kept putting off centrifuge installation on the *ISS*, much to the disappointment of the scientific community.

The Task Force called for NASA to establish a specific "end state" for station construction that could be achieved within NASA's existing budget. At that point, station construction would be complete and a much cheaper operation stage could begin. The task force recommended that several planned *ISS* features be eliminated: a crew return vehicle; Node 3, a habituation module; and a propulsion module.

The report was issued only days before the United States election that ended with George W. Bush being declared the new president by the Supreme Court. The *ISS* project had now fallen under the terms of four U.S. presidents. The Bush administration agreed wholeheartedly with the findings of the Task Force. The end state recommended by the report came to be called the "core complete" point. Bush appointed a new NASA administrator, Sean O'Keefe (1956–), and charged him to achieve core complete as soon as possible. NASA estimated that construction could be completed by 2004.

America's international partners were not happy with the plan for a smaller crew size because it meant fewer chances for their personnel to visit the station. Many scientists were disappointed with the reduction in research potential afforded by the smaller station.

The Expeditions Continue

By the end of 2001 the station had been visited by Expedition 3 and Expedition 4 crews, which included three members each. These crews installed a new Russian docking and airlock module called *Pirs* ("Pier" in English) and began construction on the truss. The truss is a long girder-like structure that is perpendicular to the row of existing modules. The truss is designed to hold the solar panels that power the station and to hold any new modules constructed in the future.

Deliveries of food, water, and supplies to the *ISS* continued to occur every few months aboard automated Russian Progress spacecraft. These vehicles were loaded up with *ISS* trash and disposed of during reentry to Earth's atmosphere.

The Russian Soyuz lifeboat docked to the *ISS* was exchanged every six months by taxi crews. In April 2002 the taxi crew again included a space tourist. This time it was a South African Internet entrepreneur named Mark Shuttleworth (1973–). He paid approximately $20 million to visit the *ISS*.

After two years of negotiation, the *ISS* partners had worked out an agreement specifying who could visit the station. It was titled *Principles Regarding Processes and Criteria for Selection, Assignment, Training and Certification of ISS (Expedition and Visiting) Crewmembers*. The agreement listed very strict requirements regarding the personal character and communication skills of any visitors. It disqualified anyone found to have a drinking or drug problem, those with poor employment or military records, convicted criminals, people who had engaged in "notoriously disgraceful conduct," and anyone known to be affiliated with organizations that wished to "adversely affect the confidence of the public" in the space program. Visitors also had to speak English.

Shuttleworth passed the review process and launched aboard a Russian Soyuz rocket on April 25, 2002, with a Russian cosmonaut and an Italian flight engineer. Two days later they entered the station. Shuttleworth performed some simple scientific experiments while on board and conducted numerous interviews with school children. During his flight he said, "I have truly never seen anything as beautiful as the Earth from space." Shuttleworth and the taxi crew spent eight days aboard the *ISS*.

Throughout the remainder of 2002 the station was visited by the crews of Expedition 5 and Expedition 6 who continued construction of the *ISS* truss. A Soyuz taxi flight launched in October of that year included a visiting astronaut from the European Space Agency. Frank DeWinne (1961–) and two taxi cosmonauts spent eight days at the station before returning to Earth.

The last visitors of the year came aboard space shuttle flight STS-113. The orbiter *Endeavour* docked at the *ISS* in late November to deliver the Expedition 6 crew and a new truss segment. In early December the shuttle returned safely to Earth carrying the Expedition 5 crew. It was the sixteenth American shuttle flight to the *ISS*. It was to be the last for a long while.

TABLE 5.6

The International Space Station construction statistics as of February 2006

International Space Station:

Major elements:

Zarya:	Launched Nov. 20, 1998
Unity:	Attached Dec. 8, 1998
Zvezda:	Attached July 25, 2000
Z1 truss:	Attached Oct. 14, 2000
P6 integrated truss:	Attached Dec. 3, 2000
Destiny:	Attached Feb. 10, 2001
Canadarm2:	Attached April 22, 2001
Joint airlock:	Attached July 15, 2001
Pirs:	Attached Sept. 16, 2001
S0 Truss:	Attached April 11, 2002
S1 Truss:	Attached Oct. 10, 2002
P1 Truss:	Attached Nov. 26, 2002

Weight: 404,069 pounds (183,283 kg)

Habitable volume: 15,000 cubic feet (425 cubic meters)

Surface area (solar arrays): 9,600 square feet (892 square meters)

Dimensions:

Width:	240 feet (73 meters) across solar arrays
Length:	146 feet (44.5 meters) from Destiny Lab to Zvezda; 171 feet (52 meters) with a Progress resupply vessel docked
Height:	90 feet (27.5 meters)

SOURCE: "Vital Statistics," in *The ISS to Date (11/07/05)*, National Aeronautics and Space Administration, November 7, 2005, http://spaceflight.nasa.gov/station/isstodate.html (accessed December 28, 2005)

ISS Assembly Halts

On February 1, 2003, the space shuttle *Columbia* broke apart as it entered Earth's atmosphere over the western United States. The shuttle had been on a research mission and did not visit the *ISS*. The catastrophe killed the seven crewmembers and shook the U.S. space program to its core. An investigation revealed that the shuttle's thermal protection tiles were likely damaged by a foam strike shortly after launch. During reentry, hot gases seeped past the tiles into the orbiter structure. The resulting turbulence tore it apart.

The entire shuttle fleet was grounded. Flights planned to the *ISS* during 2003 to deliver truss segments and research facilities were cancelled. There was no other way to transport these heavy components to the *ISS*. The Russian Soyuz spacecraft can carry only around 5,000 pounds, compared to the approximately 36,000-pound capacity of a shuttle. Russia's automated Progress spacecraft can carry even less weight, only 1,000 pounds. *ISS* assembly came to a halt. As shown in Table 5.6, the last major element of the *ISS* was attached in November 2002.

THE COST OF WAITING

The shutdown of the shuttle program had a number of operational and cost effects on the *ISS*. In 2003 the General Accounting Office (now the Government

Accountability Office, an investigatory agency of the U.S. Congress) released a report titled *Space Station: Impact of the Grounding of the Shuttle Fleet*. The report noted that modules and other equipment already ready to fly to the *ISS* would have to be unpacked, undergo maintenance, be repacked, and retested prior to flight. Batteries had to be recharged due to prolonged storage. All of these problems resulted in unexpected costs in NASA's *ISS* program.

Grounding the shuttle also had negative effects on *ISS* research projects. NASA had planned to launch three major research facilities to the station during 2003. Onboard experiments must be conducted using existing facilities. However, some of this equipment needs to be replaced or repaired, particularly refrigeration and freezer units in the science section. These units have suffered some failures. NASA had planned to replace them during 2003 with the launch of a new and larger cold-temperature facility.

The GAO also found that shuttle delays affect the safety of the *ISS*. NASA had planned to transport a new on-orbit gyro to the station in March 2003 to replace a broken unit. The station includes four gyros that maintain the structure's orbital stability and permit navigational control. NASA scientists fear that problems could arise in the station's three remaining working gyros during a prolonged delay in shuttle flights. NASA had also planned to finish installing shielding on the Russian module *Zvezda* during 2003. *Zvezda* houses the *ISS* expedition crews. The module is supposed to be covered with twenty-three shielded panels to protect it from impacts by space debris. Only six panels have been installed so far. Every day that goes by without the additional shielding increases the risk that the module could be struck and damaged by debris.

Shuttle delays also affect America's *ISS* partners. The original cost-sharing plan was worked out in the 1998 *Agreement among the Government of Canada, Governments of Member States of the European Space Agency, the Government of Japan, the Government of the Russian Federation, and the Government of the United States of America Concerning Cooperation on the Civil International Space Station*. This plan calls for NASA to pay the entire cost for ground operations and common supplies for the station. NASA is then reimbursed by the partner countries for their shares depending on their level of participation. Partner countries also fund operations and maintenance for any elements they contribute to the *ISS*, any research activities they conduct, and a share of common operating expenses. The GAO concluded that these costs would have to be adjusted as the shuttle fleet remained grounded and planned activities were cancelled.

The GAO estimates that the United States spent $32 billion on the *ISS* between 1985 and 2002.

THE EXPEDITIONS ARE DOWNSIZED

The *ISS* partners decided to limit future station crews to only two people. This would make it easier for the Russians to assume all responsibility for resupplying the crew with food, water, and other necessities. In April 2003 the two-member Expedition 7 crew flew to the station aboard a Soyuz spacecraft. The crew included one American astronaut and one Russian cosmonaut. Unable to proceed with assembly, they were kept busy maintaining the station and performing limited scientific research. One of the largest drawbacks to *ISS* science is the presence of only two crewmembers. The number of new and continuing experiments that could be conducted had to be reduced so the crews could devote more time to station maintenance and operation.

In October 2003 the Expedition 8 crew arrived aboard a Soyuz to replace them. This crew also included one Russian and one American. In addition, an ESA astronaut from Spain came along and visited the station for several days. He returned to Earth aboard the Soyuz with the Expedition 7 crew.

The Expedition 8 crew were delivered supplies by a Progress spacecraft in late January 2004. They returned to Earth aboard a Soyuz rocket in April 2004 after the arrival of Expedition 9, which consisted of a Russian, an American, and an ESA astronaut from the Netherlands. The Expedition 10 crew launched in October 2004 aboard a Soyuz rocket. Also along was Russian Space Forces Test Cosmonaut Yuri Shargin (1960–), who returned to Earth with the Expedition 9 crew. Expedition 11 arrived at the *ISS* in April 2005. It included an ESA astronaut who caught a flight home with the Expedition 10 crew aboard a Soyuz rocket. In late September 2005 the Expedition 12 crew and space tourist Greg Olsen (1945–) left Earth for the *ISS*. Olson returned to Earth with the Expedition 11 crew. The American commander and Russian flight engineer of Expedition 12 were scheduled to remain at the space station for six months.

Table 5.7 lists all *ISS* missions flown as of February 2006. From the time that assembly began, the United States made seventeen flights to the *ISS*, and the Russians made thirty-four flights.

A NEW PLAN FOR THE *ISS*

In January 2004 U.S. President George W. Bush announced a new plan for America's space program. This plan calls for retirement of the space shuttle fleet by 2010. President Bush also wants to end *ISS* assembly as soon as the core-complete configuration is obtained and eliminate all *ISS* research projects that do not support the new plans for space travel. The core-complete *ISS* would support a crew of only three people and not include some of the modules, habitat enhancements, and scientific facilities and equipment originally planned for the space station.

The plan for a downsized *ISS* was criticized by many scientists and the agency's international partners. The small crew size was a major point of contention. According to the National Research Council of the National Academies and National Academy of Public Administration, at least 2.5 crew members per expedition are required to maintain and operate the *ISS* (*Factors Affecting the Utilization of the International Space Station for Research in the Biological and Physical Sciences [2003]*). Thus, a crew of three people would have very little time to conduct scientific experiments.

In November 2005 NASA administrator Michael Griffin (1949–) appeared before the U.S. Congressional Committee on Science to provide an update on NASA's plans for the future, including assembly of the *ISS*. Griffin stated that the agency planned to assemble enough infrastructure on the station to house a six-person crew and to allow "meaningful utilization" of the *ISS*. The assembly plan included modules and laboratories from international partners, with the exception of a Centrifuge Accommodation Module being developed by JAXA (the Japanese space agency) and the Russian Solar Power Module. Final details were to be worked out in 2006. Griffin projected that eighteen space shuttle flights could be undertaken to the *ISS* before the shuttle is retired in 2010. After that NASA hopes to solicit commercial spaceflight to handle delivery and return of *ISS* crews and cargo, rather than using government spacecraft.

ISS SCIENCE

The *ISS* was intended to be a world-class laboratory for conducting experiments under microgravity conditions. Three broad areas of research are conducted aboard the station: life sciences, biomedicine, and materials processing. The chief goal of life sciences and biomedicine research is to determine the effects on humans of long-duration space travel.

The Station's Glovebox

One of the difficulties of performing typical chemistry experiments in space is the microgravity condition. Liquids will not stay inside beakers or test tubes. The droplets float away. This could be extremely dangerous for the crew and the station's electronic components. To overcome this obstacle engineers at NASA and the European Space Agency developed an enclosed work space for the *ISS* called a microgravity science glovebox (MSG).

TABLE 5.7

International Space Station missions as of September 2005

Flight number	Launch date	Mission name	Spacecraft flying to ISS	Primary cargo	Purpose
1	11/20/98	1A/R	Proton K	Control module FGB (Zarya)	Assembly
2	12/04/98	2A	Shuttle/STS-88	Node1 (Unity), PMAs 1, 2	Assembly
3	05/27/99	2A.1	Shuttle/STS-96	Spacehab DM	Outfitting
4	05/19/00	2A.2a	Shuttle/STS-101	Spacehab DM	Outfitting
5	07/12/00	1R	Proton K	Service Module (Zvezda)	Assembly
6	08/06/00	1P	Progress M1-3	Consumables, spares, props	Logistics
7	09/08/00	2A.2b	Shuttle/STS-106	Spacehab DM	Outfitting
8	10/11/00	3A	Shuttle/STS-92	Z1 truss, 4 CMGs, PMA 3	Assembly
9	10/31/00	2R/1S	SoyuzTM-31	Expedition 1 crew	1st crew
10	11/15/00	2P	Progress M1-4	Consumables, spares, props	Logistics
11	11/30/00	4A	Shuttle/STS-97	P6 module, PV array	Assembly
12	02/07/01	5A	Shuttle/STS-98	U.S. Destiny Lab module, racks	Assembly
13	02/26/01	3P	Progress M-44	Consumables, spares, props	Logistics
14	03/08/01	5A.1	Shuttle/STS-102	Expedition 2 crew, MPLM Leonardo	2nd crew
15	04/19/01	6A	Shuttle/STS-100	SSRMS, MPLM Raffaello	Outfitting
16	04/28/01	2S	Soyuz TM-32	1st taxi (plus Tito)	New CRV
17	05/20/01	4P	Progress M1-6	Consumables, spares, props	Logistics
18	07/12/01	7A	Shuttle/STS-104	U.S. Airlock, HP O2/N2 gas	Assembly
19	08/10/01	7A.1	Shuttle/STS-105	Expedition 3 crew, MPLM Leonardo	3rd crew
20	08/21/01	5P	Progress M-245	Consumables, spares, props	Logistics
21	09/14/01	4R	"Progress 301"	Docking compartment 1	Assembly
22	10/21/01	3S	Soyuz TM-33	2nd taxi	New CRV
23	11/26/01	6P	Progress M-256	Consumables, spares, props	Logistics
24	12/05/01	UF-1	Shuttle/STS-108	Expedition 4 crew, MPLM Raffaello	4th crew
25	03/21/02	7P	Progress M1-8 (257)	Consumables	Logistics
26	04/08/02	8A	Shuttle/STS-110	S0 truss segment	Assembly
27	04/25/02	4S	Soyuz TM-34	3rd taxi (plus Shuttleworth)	New CRV
28	06/05/02	UF-2	Shuttle/STS-111	Expedition 5 crew, MBS, MPLM Leonardo	5th crew
29	06/26/02	8P	Progress M-24 (246)	Consumables, spares, props	Logistics
30	09/25/02	9P	Progress M1-9 (258)	Consumables, spares, props	Logistics
31	10/07/02	9A	Shuttle/STS-112	S1 truss segment	Assembly
32	10/30/02	5S	Soyuz TMA-1 (211)	4th taxi (plus Frank DeWinne)	New CRV
33	11/23/02	11A	Shuttle/STS-113	Expedition 6 crew, P1 truss segment	6th crew
34	02/02/03	10P	Progress M-47 (247)	Consumables, spares, props.	Logistics
35	04/26/03	6S	Soyuz TMA-2 (212)	Expedition 7 crew	7th crew
36	06/08/03	11P	Progress M1-10 (259)	Consumables, spares, props.	Logistics
37	08/28/03	12P	Progress M-48 (248)	Consumables, spares, props.	Logistics
38	10/18/03	7S	Soyuz TMA-3 (213)	Expedition 8 crew (plus Duque)	8th crew
39	01/29/04	13P	Progress M1-11 (260)	Consumables, spares, props.	Logistics
40	04/18/04	8S	Soyuz TMA-4 (214)	Expedition 9 crew (plus Kuipers)	9th crew
41	05/25/04	14P	Progress M-49 (249)	Consumables, spares, props.	Logistics
42	08/11/04	15P	Progress M-50 (250)	Consumables, spares, props.	Logistics
43	10/14/04	9S	Soyuz TMA-5 (215)	Expedition 10 crew (plus Shargin)	10th crew
44	12/23/04	16P	Progress M-51 (351)	Consumables, spares, props.	Logistics
45	02/28/05	17P	Progress M-52 (352)	Consumables, spares, props.	Logistics
46	04/14/05	10S	Soyuz TMA-6 (216)	Expedition 11 crew (plus Vittori)	11th crew
47	06/17/05	18P	Progress M-53 (353)	Consumables, spares, props.	Logistics
48	07/26/05	LF-1	Shuttle/STS-114 (RTF)	MPLM Raffaello	Logistics, utilization
49	09/08/05	19P	Progress M-54 (354)	Consumables, spares, props.	Logistics
50	09/30/05	11S	Soyuz TMA-7 (217)	Expedition 12 crew (plus Olsen)	12th crew
51	12/21/05	20P	Progress M-55 (355)	Consumables, spares, props.	Logistics

Notes:
Acronyms:
CMG	Control moment gyro
CRV	Crew return vehicle
DM	Double cargo module
HP	High pressure
MBS	Mobile remote services base system
MPLM	Multi purpose logistics module
PMA	Pressurized mating adapter
Props	Propellents
PV	Photo voltaic
SSRMS	Space station remote manipulator system

SOURCE: Adapted from "International Space Station ISS Assembly Progress," in *International Space Station*, National Aeronautics and Space Administration, June 1, 2005, http://www.hq.nasa.gov/osf/station/assembly/ISSProgress.html (accessed January 31, 2006)

The MSG includes a pair of built-in gloves that crewmembers can use to handle tools and equipment within the box. The MSG was installed in the American *Destiny* module by the Expedition 5 crew. Figure 5.4 shows a slice of the cylindrical *Destiny* module. The experimental racks are positioned around the outside of the circle. An astronaut stands at the MSG research station.

FIGURE 5.4

Space station Destiny laboratory and microgravity science glovebox

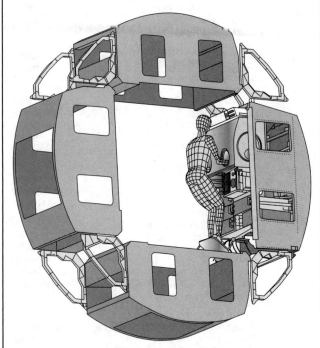

SOURCE: "This Cut-Away of the Cylindrical, Destiny Laboratory Module on the Space Station Shows How the New Microgravity Science Glovebox Fits Inside," in *NASA Fact Sheet: Microgravity Science Glovebox (MSG)*, National Aeronautics and Space Administration, Marshall Space Flight Center, November 27, 2004, http://www.nasa.gov/centers/marshall/news/background/facts/MSG .html (accessed January 31, 2006)

The MSG is used to handle chemicals or burning or molten specimens in experiments involving fluid physics, materials science, biotechnology, and combustion science.

Bone Loss

Scientists have known for some time that human bones in the legs and feet undergo deterioration during prolonged stays in space. This was first discovered in Soviet and Russian cosmonauts who spent many months aboard space stations. Scientists believe that the effect is due to lack of mechanical loading in microgravity. Mechanical loading refers to the weight of the upper body pressing down on the lower body as a person's body is pulled toward the ground by gravity on Earth. The type of bone loss and muscle deterioration experienced by space travelers is similar to that resulting from prolonged bed rest. It has long been known that using legs and feet keeps them healthy. In a spaceship people do not experience the force of gravity or the downward load of the upper body. Also, they rarely use muscles in their legs and feet to move around. They rely much more

on muscles in their arms and upper body to maneuver through hatches and accomplish tasks.

During the Russian *Mir* program, cosmonauts reported that the skin on the soles of their feet became very soft. They also lost muscle tone in their legs and feet due to lack of use. These factors caused them great difficulty walking when they returned to Earth. Scientists incorporated exercise regimens on the *ISS* to help prevent these problems. For example, stationary bicycles help crewmembers maintain foot muscle strength. However, the exercises have had little effect on bone loss.

Historical data show that humans experience a rate of bone loss in space of approximately 1% to 2% per month. This means a bone loss of 12% to 24% per year. Scientists know that the bone loss problem has to be resolved before humans can make interplanetary journeys. A trip from Earth to Mars could take as long as six months. Crewmembers have to be able to walk on the planet's surface when they get there.

One of the most important biomedical studies ongoing aboard the *ISS* is called Foot/Ground Reaction Forces during Spaceflight (FOOT). The experiment began with an astronaut in the Expedition 6 crew. He wore an instrumented suit called a lower extremity monitoring suit, or LEMS. An astronaut wearing a LEMS is depicted in Figure 5.5. The LEMS consists of a pair of Lycra pants equipped with numerous sensors that can measure the electrical activity of muscles, the angular motions of joints, and the force underneath the feet. Sensor data is recorded on a small wearable computer. The LEMS is also being worn by an Expedition 8 astronaut.

The LEMS is only one part of the FOOT study. The astronauts participating in the study are subjected to extensive bone scans and tests prior to launch and after landing. Scientists hope the FOOT results will help them to design new exercise equipment that can counteract bone loss in space.

ISS CREW TRAINING

The *ISS* Expedition crews undergo extensive training at facilities around the world. Figure 5.6 shows the locations of these training facilities, including:

- Johnson Space Center (JSC)—JSC is located in Houston, Texas. This is the primary training center for Expedition crews. It features laboratories, classrooms, and simulators to prepare crewmembers for living and working aboard the *ISS*.

- Kennedy Space Center (KSC)—KSC is located on Merritt Island, Florida, adjacent to the Cape Canaveral Air Force Station. This is the launch site for U.S. shuttle flights. Cargo bound for the *ISS* is packed and loaded here. Crewmembers familiarize themselves

FIGURE 5.5

The lower extremity monitoring suit

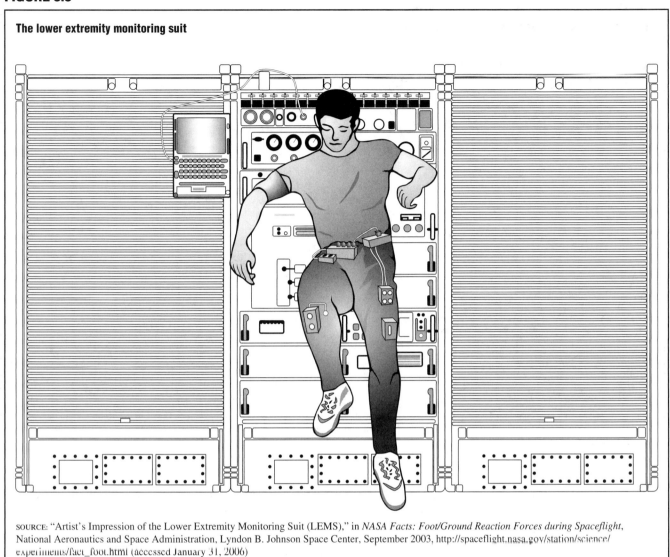

SOURCE: "Artist's Impression of the Lower Extremity Monitoring Suit (LEMS)," in *NASA Facts: Foot/Ground Reaction Forces during Spaceflight*, National Aeronautics and Space Administration, Lyndon B. Johnson Space Center, September 2003, http://spaceflight.nasa.gov/station/science/experiments/fact_foot.html (accessed January 31, 2006)

with *ISS* components and practice launch procedures at KSC.

- Gagarin Cosmonaut Training Center—This center is located in a Russian town called Zvezdny Gorodok (better known as Star City). The training facilities are similar to those at JSC, including classrooms, laboratories, and simulators. Expedition crewmembers learn about the Russian modules of the *ISS* and train for spacewalks in the Center's Hydrolab (a forty-foot-deep pool).

- Baikonur Cosmodrome—The Cosmodrome is located in Baikonur, Kazakhstan. This is the launch site for Russian Soyuz flights to the station. Expedition crews scheduled to fly aboard Soyuz spacecraft practice launch procedures at the Cosmodrome.

- Canadian Space Agency (CSA) Headquarters—The CSA headquarters is located in Saint-Hubert, Quebec. Canadian companies built the remote manipulator system (the robotic arm known as Canadarm2) used on the *ISS*. Expedition crewmembers undergo robotics training at this location to familiarize themselves with the Canadarm2.

Expedition crewmembers undergo approximately eighteen months of rigorous training. Training time is divided among various tasks. The single task receiving the most time is extravehicular activity (EVA) or spacewalking.

ISS assembly requires a lot of EVA. By the time assembly is completed, *ISS* crewmembers will have performed twice as much EVA as was performed during all spaceflight conducted between 1958 and 1998.

Space walking is not possible without a pressurized spacesuit. Figure 5.7 shows the many components that make up a typical American spacesuit. The suit is designed to protect space walkers from the wide temperature variations encountered outside the station. These temperatures

FIGURE 5.6

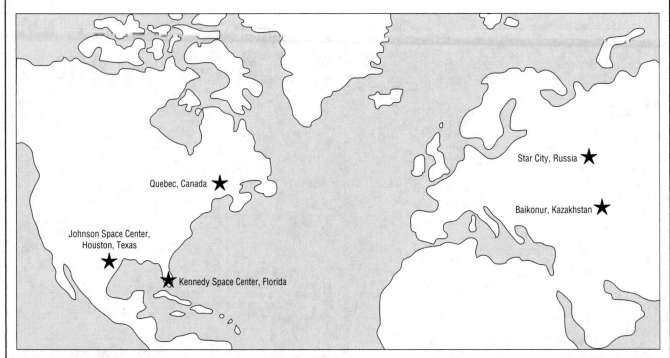

Training locations for space expedition crewmembers

Quebec, Canada ★

Star City, Russia ★

Baikonur, Kazakhstan ★

Johnson Space Center, Houston, Texas ★

★ Kennedy Space Center, Florida

SOURCE: "Expedition Crewmembers in Training Travel around the World to Prepare for Their Missions Long before They Begin Orbiting It," in *Behind the Scenes: Training Locations*, National Aeronautics and Space Administration, August 14, 2003, http://spaceflight.nasa.gov/shuttle/support/training/isstraining/locations.html (accessed January 31, 2006)

range from −250° Fahrenheit in the shade to 250° Fahrenheit in the sun. Temperature protection is provided by the liquid cooling and ventilation garment (LCVG). This is a tight-fitting garment that fits against a crewmember's body. It contains a network of tubing in which water circulates to maintain a constant body temperature.

Another important component of the spacesuit is the primary life support module (PLSM). The PLSM is worn like a backpack and includes oxygen tanks, water tanks, electrical power, ventilating fans, and scrubbers and filters to remove carbon dioxide. Fresh oxygen flows from the PLSM into the back of the helmet. Exhaled water vapor and carbon dioxide are carried through ductwork in the LCVG and returned to the PLSM.

Each space walker also wears a secondary oxygen pack that can provide up to thirty minutes of oxygen if something goes wrong with the PLSM.

During spacewalks crewmembers are usually tethered to the station structure to prevent them from floating away. American astronauts also wear mini jetpacks called SAFER units. SAFER stands for simplified aid for extra-vehicular activity rescue. An astronaut that becomes untethered for some reason could turn on the SAFER unit and use its jets to return to the station.

FIGURE 5.7

Spacesuit enhancements for the International Space Station

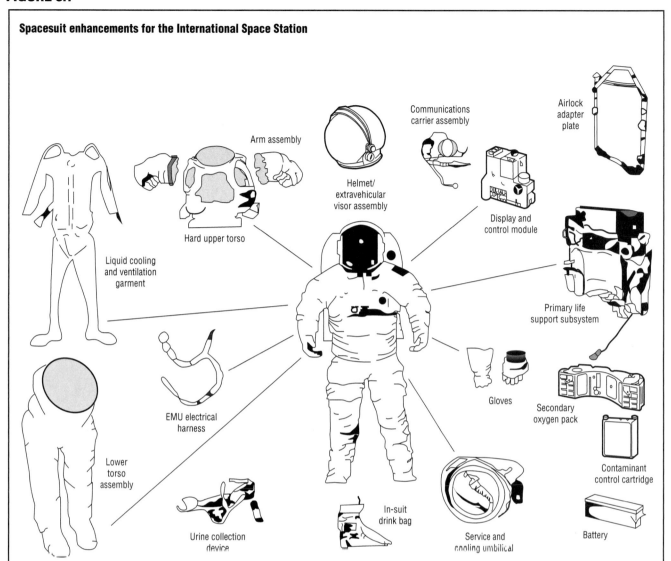

SOURCE: "Workclothes for Orbit: Spacesuit Enhancements for the International Space Station," in *NASA Facts: International Space Station Assembly: A Construction Site in Orbit*, National Aeronautics and Space Administration, Lyndon B. Johnson Space Center, June 1999, http://spaceflight.nasa.gov/spacenews/factsheets/pdfs/assembly.pdf (accessed January 31, 2006)

CHAPTER 6
ROBOTIC MISSIONS IN SUN-EARTH SPACE

The Universe is an explosive, energetic and continually changing place.

—NASA

Sending humans into space is expensive and risky. It takes great resources to protect them and sustain them every time they leave the planet. Losing a crewed spacecraft means loss of lives. This is a high price to pay to learn about the universe. This explains why robotic spacecraft are so vital to space science. Since the beginning of the space age, satellites and probes have been sent out to gather data about Earth's surroundings. The earliest ones were rather crude in their technology. People referred to them simply as unmanned spacecraft. Times changed, and technology improved significantly. Today these mechanized explorers are called robotic spacecraft.

Only a handful of robotic spacecraft are sent to other planets. The vast majority of them circle the Earth or the Sun. Spacecraft in Earth orbit serve commercial, military, and scientific purposes. Scientists rely on satellites to collect data about Earth's weather, climate, atmospheric conditions, sea levels, ocean circulation, and gravitational and electromagnetic fields. These satellites are not space explorers but Earth observers that reside in space.

Other satellites in Earth orbit look outward toward the cosmos. These space observatories carry high-powered telescopes that beam images back to Earth. They can detect the light and radiation of celestial objects hidden from human view. They peer into the deep, dark regions of space to explore what is out there. Scientists use the images sent back from such satellites to learn about the origins of stars and planets and unravel some of the mysteries of the universe.

Closer to Earth is our own star, the Sun. It emits radiation and heat and produces a flow of energetic particles called the solar wind that constantly blows against Earth. Gusts of solar wind can upset the planet's electromagnetic field. Every so often the Sun spits out globs of plasma and intense bursts of radiation. These too can have a profound effect on Earth. Investigating the Sun-Earth connection is a major goal of modern space science. Robotic spacecraft are put into Earth orbit or sent out into space to collect data about this vital connection.

For decades robotic spacecraft have been designed for one-way trips. Eventually their batteries give out and their radios fail. Then they quit reporting back to Earth. They wander around in space or are incinerated by reentry into Earth's atmosphere. During the 1990s scientists developed robotic spacecraft that can do more than report back. They can come back. These sampling ships are designed to grab samples of space particles and return safely to Earth. The first one returned in September 2005. It carried samples of solar wind collected a million miles from Earth. The spaceship's name was *Genesis*. It marked the beginning of a whole new way for humans to explore space without even leaving the planet.

NASA'S SCIENCE GOALS

The vast majority of all robotic spacecraft are operated by NASA, under the agency's space science enterprise.

This enterprise focuses on six areas of research:

- The origins of life and the universe
- The existence of life on other planets
- The formation of stars and galaxies
- The evolution of the solar system and the universe
- The mechanisms of life
- The relationship between matter and energy

NASA's Office of Space Science (OSS) oversees a number of programs and individual projects that utilize robotic missions to pursue these research goals.

NASA'S EXPLORER PROGRAM

The Explorer program includes robotic missions that conduct relatively low-cost scientific investigations with specific objectives in the fields of astronomy or space physics. The program dates back to the late 1950s. The first satellite launched into space by the United States was called *Explorer 1*. It was the first of a series of scientific satellites named Explorer.

In 1958 the *Explorer 1* satellite used temperature gauges and a Geiger counter to collect data in space. A Geiger counter is an instrument that detects the presence and intensity of radiation. *Explorer 1* detected radiation levels as expected during most of its orbit. However, at the highest altitudes the Geiger counter recorded no radiation. This was most puzzling. Scientists expected radiation levels to increase, not decrease, farther from Earth. Data from later Explorers and the Soviet Sputnik satellites revealed that *Explorer 1* had encountered radiation levels so high, its detector was overwhelmed.

These missions led to the discovery of the Van Allen radiation belts. These are two doughnut-shaped regions of high radioactivity that encircle Earth. The inner belt is centered approximately 2,000 miles above Earth and is roughly 4,000 miles thick. The outer belt lies between approximately 11,500 and 25,000 miles from Earth. All future spacecraft venturing very far from Earth had to be designed to withstand the intense radiation of the Van Allen belts.

The Explorer program continued over the next three decades with mission series given different names. These series included AE (Atmosphere Explorer), EPE (Electrostatic Particle Explorer), IMP (Interplanetary Monitoring Platform), and SOLRAD (Solar Radiation Satellites). There were also Explorer spacecraft with colorful names such as *Hawkeye*, *Injun*, and *Uhuru*. Uhuru means "freedom" in the African language Swahili.

During the 1990s NASA specified that Explorer missions be dedicated to three specific themes in space science:

- Astronomical search for origins and planetary systems

- Structure and evolution of the universe

- The Sun-Earth connection

NASA also established different classes of Explorer missions based on their total cost to the agency.

- University-Class Explorer (UNEX)/Student Explorer Demonstration Initiative Program (STEDI)—Total cost less than $15 million

- Small Explorer (SMEX)—Total cost less than $120 million

- Medium-Class Explorer (MIDEX)—Total cost less than $180 million

Another type of Explorer mission is called a Mission of Opportunity (MO). This is a mission operated by another office within NASA or by the space agency of another country in which an OSS investigation "hitches a ride" with the mission of the other agency or country. The total cost to NASA of an MO mission cannot exceed $35 million.

As of 2006 there have been nearly ninety Explorer program missions. Only four spacecraft failed. The other missions are considered successes. Operational Explorer missions as of February 2006 are listed in Table 6.1. Most of the spacecraft are small observatories in low-Earth orbit (LEO, 125–1,200 miles above Earth's surface). The Advanced Composition Explorer (ACE) and Wilkinson Microwave Anisotropy Probe (WMAP) are positioned at Lagrange points around Earth (designated L1 and L2 in Table 6.1). These are points at which the gravitational pulls of the Sun and Earth are relatively even, meaning that a spacecraft can stay "parked" between the two heavenly bodies. Point L1 is approximately one million miles from the Earth toward the Sun; Point L2 is approximately one million miles from Earth in the other direction (away from the Sun). IMAGE is in orbit just outside the outer Van Allen radiation belt, which extends to about seven Earth radii. (The radius of the Earth is approximately 3,960 miles.)

NASA'S DISCOVERY PROGRAM

The Discovery program was initiated in 1989 to develop missions that could be accomplished using small spacecraft at a low cost in a relatively short time to address focused scientific goals. The emphasis is on projects that can be managed by academic and/or research organizations.

The program allows scientists to conduct small space investigations that complement NASA's larger and more expensive interplanetary missions. Each Discovery mission must have a fast development time (less than thirty-six months) and relatively low cost (less than $299 million). NASA calls it the "faster, better, cheaper" approach to space science. Mission of Opportunity projects are also allowed, but cannot exceed a cost to NASA of $35 million. The Discovery program focuses on specific research objectives related to planets, moons, comets, and asteroids.

Approximately every two years NASA publishes an Announcement of Opportunity (AO) in which it invites organizations to submit proposals for a future Discovery mission. The organizations can be educational institutions, commercial enterprises, nonprofit organizations, NASA Centers (or the JPL), or other government agencies. The most recent AO was released in January 2006.

TABLE 6.1

Explorers Program: operational missions as of February 2006

Name	Long title	Launch date	Orbit location	Mission
ACE	Advanced Composition Explorer	08/25/1997	L1	Samples low-energy particles of solar origin and high-energy galactic particles
IMP-8	Interplanetary Monitoring Platform	10/26/1973	35 Earth radii	Measures magnetic fields, plasmas, and energetic charged particles
RXTE	Rossi X-ray Timing Explorer	12/30/1995	LEO	Observes black holes, neutron stars, X-ray pulsars and bursts of X-rays
Medium-Class Explorers (MIDEX)				
Swift	(Named after a small nimble bird)	11/20/2004	LEO	Multi-wavelength observatory dedicated to the study of gamma-ray burst
WMAP	Wilkinson Microwave Anisotropy Probe	06/30/2001	L2	Measures cosmic microwave background radiation over the full sky
FUSE	Far Ultraviolet Spectroscopic Explorer	06/24/1999	LEO	Uses high-resolution spectroscopy in the far-ultraviolet spectral region
IMAGE	Imager for Magnetopause-to-Aurora Global Exploration	03/25/2000	7.2 Earth radii	Images Earth's magnetosphere
Small Explorers (SMEX)				
FAST	Fast Auroral Snapshot Explorer	08/21/1996	215–2,500 miles	Investigates the plasma physics of the auroral phenomena at Earth's poles
SAMPEX	Solar Anomalous and Magnetospheric Particle Explorer	07/03/1992	LEO	Measures the composition of solar energetic particles and cosmic rays
SWAS	Submillimeter Wave Astronomy Satellite	12/05/1998	LEO	Surveys emissions of water, O_2, C, and CO in galactic star forming regions
TRACE	Transition Region and Coronal Explorer	04/02/1998	LEO	Images the solar corona and transition region
RHESSI	Reuven Ramaty High Energy Solar Spectroscopic Imager	02/05/2002	LEO	Explore basic physics of solar flares
GALEX	Galaxy Evolution Explorer	04/28/2003	LEO	Space telescope that observes galaxies in ultraviolet light
University-Class Explorers (UNEX)/Student Explorer Demonstration Initiative program (STEDI)				
SNOE	Student Nitric Oxide Explorer	02/26/1998	LEO	Measues effects of energy on nitric oxide in the Earth's upper atmosphere
CHIPS	Cosmic Hot Interstellar Plasma Spectrometer	01/12/2003	LEO	Conducts all-sky spectroscopy of the diffuse background
Missions of Opportunity (MO)				
HETE-2	High-Energy Transient Explorer	10/09/2000	LEO	Detects and localizes gamma-ray bursts
Suzaku	(Formerly called Astro-E2, Suzaku is a vermillion bird)	07/10/2005	LEO	Studies X-rays emitted by stars, galaxies, and black holes
Integral	International Gamma Ray Laboratory	10/17/2002	1.4–23.5 Earth radii	Gamma-ray observatory

SOURCE: Adapted from "Current Missions," in *Explorers Program,* National Aeronautics and Space Administration, Goddard Space Flight Center, 2005, http://explorers.gsfc.nasa.gov/missions.html (accessed December 28, 2005).

As of 2006 there were eleven missions listed under the Discovery program, as shown in Figure 6.1. Four of the missions (*NEAR, Lunar Prospector, CONTOUR,* and *NetLander*) ended during the late 1990s or early 2000s.

NEAR (*Near Earth Asteroid Rendezvous*) was the first mission launched under the Discovery program. On February 17, 1996, *NEAR* was launched into space to meet up with the asteroid 433 Eros. Asteroids are small celestial bodies that orbit larger ones. Most of the asteroids in the solar system are found in a massive asteroid belt that circles around the Sun between the orbits of Mars and Jupiter. 433 Eros is twenty-one miles long and has an elliptical orbit that carries it outside the Martian orbit and then in close to Earth orbit. On February 12, 2001, the *NEAR* spacecraft softly touched down on the asteroid. It was the first time in history that a spacecraft had landed on an asteroid. When it happened, 433 Eros was 196 million miles from Earth.

NEAR orbited 433 Eros for nearly a year before landing and returned dozens of high-resolution photographs of the asteroid. The spacecraft was built by the Johns Hopkins University Applied Physics Laboratory in Laurel, Maryland, which also managed the mission for

NASA. Before it landed, the spacecraft was renamed *NEAR Shoemaker* in honor of the late geologist Dr. Eugene M. Shoemaker (1928–97). The first Discovery mission was considered an overwhelming success.

Lunar Prospector launched on January 6, 1998, and assumed an orbit around Earth's moon to collect scientific data. In July 1999 the ship was purposely crashed into the lunar surface. *CONTOUR* launched on July 3, 2002 and was intended to encounter two comets in solar orbit. However, the spaceship was lost six weeks after launch. NetLander was a mission of opportunity in which NASA instruments were to fly aboard a French spaceship to Mars. However, the French mission was cancelled during the development phase.

Operational Discovery missions as of February 2006 were as follows:

- *Stardust*
- *Genesis*
- *ASPERA-3*
- *Deep Impact*
- *MESSENGER*

FIGURE 6.1

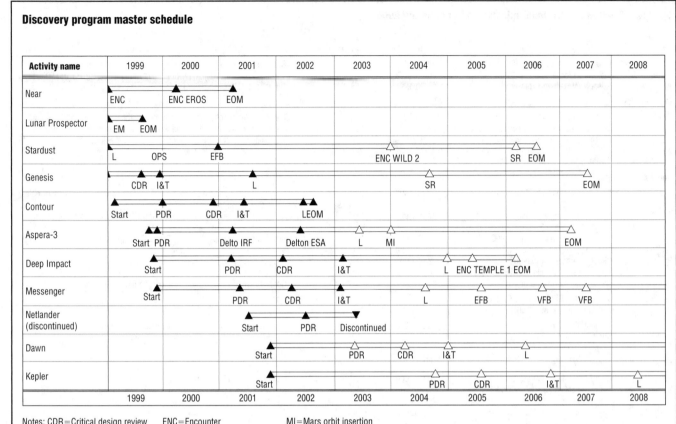

Discovery program master schedule

Activity name	1999	2000	2001	2002	2003	2004	2005	2006	2007	2008
Near	▲ ENC	▲ ENC EROS	▲ EOM							
Lunar Prospector	▲ EM ▲ EOM									
Stardust	▲ L	OPS	▲ EFB			△ ENC WILD 2		△ SR △ EOM		
Genesis		▲ CDR ▲ I&T	▲ L			△ SR			△ EOM	
Contour	▲ Start	▲ PDR	▲ CDR	▲ I&T	▲▲ LEOM					
Aspera-3		▲▲ Start PDR	▲ Delto IRF	▲ Delton ESA	△ △ L MI				△ EOM	
Deep Impact	▲ Start	▲ PDR	▲ CDR	▲ I&T		△ △ △ L ENC TEMPLE 1 EOM				
Messenger	▲ Start	▲ PDR	▲ CDR	▲ I&T		△ L	△ EFB	△ VFB	△ VFB	
Netlander (discontinued)			▲ Start	▲ PDR	▼ Discontinued					
Dawn			▲ Start		△ PDR	△ CDR	△ I&T	△ L		
Kepler			▲ Start			△ PDR	△ CDR	△ I&T		△ L
	1999	2000	2001	2002	2003	2004	2005	2006	2007	2008

Notes: CDR=Critical design review   ENC=Encounter   MI=Mars orbit insertion
DES=Design   EOM=End of mission   MOI=Mercury orbit insertion
DEV=Development   I&T=Begin integration & test   OPS=Operations
EFB=Earth fly by   L=Launch   PDR=Preliminary design review
EM=Extended mission   MFB=Mercury fly by   SR=Sample return
VFB=Venus fly by

SOURCE: "Discovery Program Master Schedule," in *Discovery Program: Program Description*, National Aeronautics and Space Administration, April 2, 2004, http://discovery.nasa.gov/images/missionschedule.gif (accessed January 31, 2006)

Stardust

Stardust was designed to collect particle samples from comet Wild 2 and return them safely to Earth. Comet Wild 2 is named after Swiss astronomer Paul Wild, who discovered it in 1978. (His name is pronounced "Vilt." Thus, comet Wild 2 is pronounced Vilt 2.)

Wild 2 is a relatively small comet with a nucleus measuring about three miles in diameter. A comet nucleus is a dense core of rock and frozen gas. The coma is the thick cloud of gases and sand that surrounds a comet nucleus. Comets follow heliocentric (sun-centered) orbits.

On February 7, 1999, *Stardust* was launched on its journey. The spacecraft carries a particle collector about the size of a tennis racket. (See Figure 6.2.) The collector is made of an ultra-lightweight material called aerogel that can capture and hold tiny comet dust particles. Following capture the collector folds down into the sample return capsule for its journey back to Earth. *Stardust* is also equipped with special detectors for analysis of comet dust and interstellar dust.

Stardust collected and analyzed interstellar dust in 2000 and 2002. On January 2, 2004, it successfully sampled particles from the coma of Wild 2. These particles are believed to be more than 4.5 billion years old.

As the spaceship approached Earth it released the sample return capsule. Shortly before 5:00 AM (Eastern Standard Time) on January 15, 2006, the capsule entered Earth's atmosphere. Parachutes were deployed, and the capsule landed safely in the Utah desert. The main spacecraft was put into orbit around the sun and will likely be used on a future mission.

A few days later NASA scientists confirmed that the mission had been a success. They reported finding "thousands of impacts" on the aerogel in the sample return capsule. *Stardust* was the first-ever comet sample return mission.

Genesis

In August 2001 *Genesis* was launched into outer space toward the L1 Lagrange point. It was a revolution-

FIGURE 6.2

The Stardust spacecraft

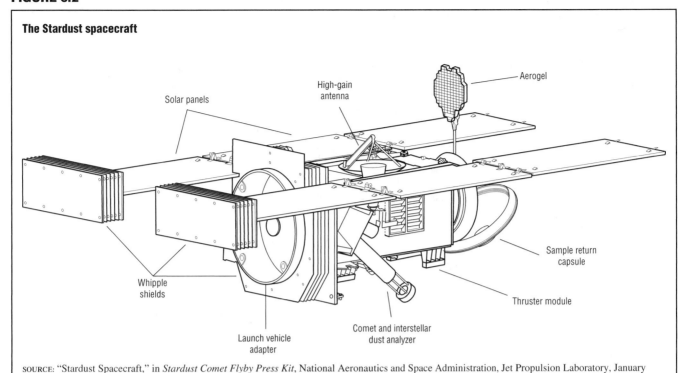

SOURCE: "Stardust Spacecraft," in *Stardust Comet Flyby Press Kit*, National Aeronautics and Space Administration, Jet Propulsion Laboratory, January 2004, http://stardust.jpl.nasa.gov/news/stardustflyby.pdf (accessed January 31, 2006)

ary spacecraft designed to collect charged particles emitted from the sun and return them safely to Earth. *Genesis* successfully reached its destination and collected samples. It assumed a tight orbit around L1 for 2.5 years and then headed back to Earth. After reentering Earth's atmosphere *Genesis* was supposed to deploy its parachutes for a slow descent toward the surface. A specially equipped helicopter was to snag the spaceship and carry it to land.

On September 8, 2004, *Genesis* began its descent to a Utah landing site. However, its parachutes did not open, and the spacecraft plummeted at high speed into the ground. The capsule containing the samples split open during the crash, exposing the sample medium to the outside atmosphere. After six months of recovery and analysis NASA announced that some of the samples had survived intact and would be made available for scientific study.

ASPERA-3

ASPERA-3 is a mission of opportunity in which NASA equipment flies aboard an ESA spacecraft—the Mars Express orbiter. Mars Express launched in June 2003 and assumed orbit around the Red Planet five months later. As of February 2006 it is still orbiting Mars and collecting data. The full name of the instrument is the *Analyzer of Space Plasma and Energetic Atoms*. It was designed to assess the interaction between the solar wind

and the Martian atmosphere. NASA funded two of the four sensors embedded in the instrument.

Deep Impact

Deep Impact was launched on January 12, 2005, for an encounter with the comet Tempel-1. The comet was discovered by French astronomer Ernst Tempel in 1867. It orbits the sun between Mars and Jupiter with a 5.5-year orbital period. *Deep Impact* included an impactor probe that was released and penetrated the comet surface on July 4, 2005. The probe relayed data to the flyby portion of the spaceship. NASA scientists are using the data to determine the physical and chemical makeup of the comet. In September 2005 NASA announced that preliminary analyses showed the comet included expected components, such as silicates (sand), and unexpected components, including clay, carbonates, iron-bearing compounds, and aromatic hydrocarbons.

As of February 2006 the *Deep Impact* flyby spacecraft remains in a flight trajectory among the inner planets (Mercury, Venus, Earth, and Mars). It is scheduled to make a flyby of Earth in late 2007. NASA expects that the spacecraft will be used for a future Discovery Program mission of opportunity.

MESSENGER

MESSENGER stands for *Mercury Surface, Space Environment, Geochemistry, and Ranging*. The spacecraft was launched on August 3, 2004, and will achieve orbit around Mercury in March 2011 for a one-year visit. Mercury is the planet closest to the Sun. During its journey *MESSENGER* will conduct flybys of Venus in October 2006 and June 2007. When it reaches Mercury the spacecraft will produce images of the entire surface of the planet and collect data on its composition, geology, atmosphere, and magnetosphere.

Future Discovery Missions

Future Discovery missions will explore asteroids in our solar system and look for planets in other solar systems. The *Dawn* mission is scheduled to launch in June 2006. Beginning in 2010 the spacecraft will assume an orbit around the asteroids Vesta and Ceres. These are the oldest and largest asteroids in our solar system. The *Kepler* mission will launch in June 2008 and assume an Earth-trailing heliocentric orbit (an orbit behind the Earth around the Sun). It will survey a region of the Milky Way looking for planets similar to Earth. The spacecraft will be pointed toward the constellations Cygnus and Lyra. Primary stars in this region include Vega, Deneb, and Albireo.

In addition to these primary missions NASA plans to conduct a mission of opportunity aboard an Indian spacecraft prior to 2010. *Chandrayaan-1* will be a lunar orbit spacecraft launched by the Indian Space Research Organization. In the Hindi language the term Chandrayaan means "voyage to the Moon." NASA expects to contribute two scientific instruments to the mission.

SPACE OBSERVATORIES

Celestial objects emit all kinds of radioactive waves that give clues about their shape, size, age, and location. Earth's atmosphere filters and blocks out most of this radiation, keeping it from reaching the planet's surface. This is one reason why biological life is possible on planet Earth. The atmospheric shield is good for human health but bad for observing the universe.

In order to clearly detect all radioactive waves traveling through space an observer would have to be above the shielding effects of Earth's atmosphere. The human eye can only detect one type of radioactive wave—visible light energy. These waves comprise only a small fraction of the radiation flying around the cosmos. This is why scientists invented instruments that can detect and measure waves that are invisible to humans. These instruments are put on satellites and launched into outer space to provide a clearer "picture" of the universe.

The largest and most powerful space telescopes in the world are NASA's Great Observatories. This is a family of observatories developed over many years. As of December 2005 NASA has put four Great Observatories into space:

- *Hubble Space Telescope*
- *Chandra X-Ray Observatory*
- *Compton Gamma Ray Observatory*
- *Spitzer Space Telescope*

These observatories were designed to observe the universe across a wide range of the light energies making up the electromagnetic spectrum.

The Electromagnetic Spectrum

Scientists use a scale to describe different types of light energies. This scale is called the electromagnetic spectrum. (See Figure 6.3.) It categorizes energy by wavelength and frequency. A radio wave is the largest type, measuring between one millimeter and one hundred kilometers in length. This is a range of roughly 0.04 inches to 328,000 feet. At the other end of the scale is the gamma ray. Its length is the size of subatomic particles. Near the middle of the scale is visible light. This is the only kind of energy that humans can see.

For most celestial objects there is a relationship between their temperature and the main type of radiation that they emit. Astronomers use the Kelvin scale to measure temperature. A temperature of 0° Kelvin is called absolute zero. This is equivalent to −460° Fahrenheit.

Most stars have temperatures between 3,500° and 25,000° Kelvin. Much of the energy they emit is visible. This explains why humans can see stars in the nighttime sky. However, stars emit radiation at other wavelengths across the spectrum, particularly at infrared.

Everything that has a temperature above absolute zero emits infrared waves. This makes infrared observation an extremely important astronomical tool. There are huge murky clouds of dust and gas floating around in the universe. These clouds, called nebulae, can hide visible light from human sight. Scientists poetically refer to this phenomenon as the "cosmic veil." An infrared telescope is said to lift the cosmic veil by allowing humans to see through the clouds.

Many of the most mysterious objects and explosions in the cosmos are associated with very high temperatures and releases of gamma-ray, x-ray, and ultraviolet radiation. Only space-based telescopes equipped with special instruments can detect these waves.

This is important because x-rays and gamma rays are associated with phenomena such as black holes, supernova remnants, and neutron stars. Table 6.2 defines these terms and describes other cosmic objects. According to NASA, ultraviolet observations are useful for studying the "youngest, hottest, and most massive" stars, in addition to quasars and white dwarf stars. Visible light is emitted by stars and emitted or reflected by cooler

FIGURE 6.3

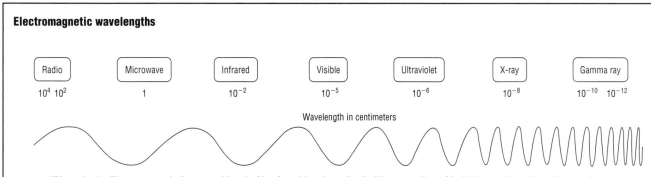

Electromagnetic wavelengths

| Radio | Microwave | Infrared | Visible | Ultraviolet | X-ray | Gamma ray |

$10^4 \ 10^2$ 1 10^{-2} 10^{-5} 10^{-6} 10^{-8} $10^{-10} \ 10^{-12}$

Wavelength in centimeters

SOURCE: "Waves in the Electromagnetic Spectrum Vary in Size from Very Long Radio Waves the Size of Buildings, to Very Short Gamma-Rays Smaller than the Size of the Nucleus of an Atom," in "Electromagnetic Waves Have Different Wavelengths," in *The Electromagnetic Spectrum*, National Aeronautics and Space Administration, Goddard Space Flight Center, 2005, http://imagers.gsfc.nasa.gov/ems/waves3.html (accessed February 4, 2006)

TABLE 6.2

Definitions

Term	Meaning
Black hole	Invisible celestial object believed created when a massive star collapses. Its gravitational field is so strong that light cannot escape from it.
Brown dwarf	Celestial object much smaller than a typical star that radiates energy, but does not experience nuclear fusion.
Dark matter	Hypothetical matter that provides the gravity needed to hold hot gases within a galaxy cluster.
Event horizon	The boundary of a black hole.
Galaxy cluster	Huge collection of galaxies bound together by gravity.
Nebulae	Large gas and dust clouds that populate a galaxy.
Neutron star	Hypothetical dense celestial object consisting primarily of closely packed neutrons. Believed to result from the collapse of a much larger star.
Nova	Star that suddenly begins emitting much more light than before and continues to do so for days or months before returning to its former state of illumination.
Pulsar	Celestial body emitting pulses of electromagnetic radiation at short relatively constant intervals
Quasar	Mysterious celestial object that resembles a star, but releases a tremendous amount of energy for its size.
Superflare	Massive explosion from a young star. A superflare is thousands of times stronger than a solar flare emitted by Earth's Sun.
Supernova	Catastrophic explosion of a star.
Supernova remnant	Bubble of multimillion degree gas released during a supernova.
White dwarf	Small dying star that has used up all its fuel and is fading.

SOURCE: Created by Kim Weldon for Thomson Gale, 2004

objects in the universe, including planets, nebulae, and brown dwarf stars. Radio waves are associated with cold clouds of molecules and radiation left over from celestial events that happened billions of years ago.

Telescopes as Time Machines

It takes a long time for energy waves to travel through space. Although there are many different wavelengths of radiation, they all travel at a speed called the speed of light. The speed of light through a vacuum is 186,280 miles per second. That seems very fast. However, most celestial objects are so incredibly far away from Earth that it takes their energy a long time to reach us despite traveling at such a high speed.

Astronomers use a unit of measure called the light year to describe long distances in space. During one year light travels 5.9 trillion miles, so a celestial object that is 5.9 trillion miles from Earth is said to be one light year away. If a spaceship could move as fast as the speed of light, it would reach that object in one year. Earth's Sun is eight light minutes away. This means that it takes eight minutes for an energy wave from the Sun to reach the Earth. The sunlight visible on Earth at any given instant actually left the Sun eight minutes before.

When a telescope captures an image of an object in the universe, the image shows what that object looked like in the past. The farther away a telescope can see, the farther back into time it looks. NASA's Great Observatories capture images of celestial objects that are very far from Earth. This means that the radiation reaching the observatories left its source a long time ago. This is why NASA refers to its observatories as time machines. They allow humans to look back into time and learn about the origins of the universe.

Most astronomers believe that the universe started ten to twenty billion years ago with a massive explosion called the Big Bang. The energy released during this explosion would have been tremendous. Since then the universe has been cooling and expanding. Scientists believe that most radiation left over from the Big Bang now travels throughout space as radio waves and microwaves. This is called cosmic background radiation.

Ground-Based Telescopes

Telescopes were first developed in Europe around the beginning of the 1600s. Glass lenses had been used to make eyeglasses for several centuries. The first telescopes included a series of lenses within a long slender tube that magnified distant objects. In 1610 the Italian scientist Galileo Galilei (1564–1642) popularized the telescope when he published *Sidereus Nuncius* (*Starry Messenger*). The book described Galileo's astronomical observations, including the discovery of four moons around Jupiter.

Like the human eye, optical telescopes can discern only visible light. During the 1800s scientists first developed detectors for infrared radiation coming from outer space. Because the water in Earth's atmosphere blocks out most infrared radiation, these detectors were placed atop mountains where the air was thinner. Later they were sent up in high-altitude balloons and even airplanes.

During the 1930s an American engineer named Karl Jansky (1905–50) built an instrument capable of detecting radio waves. His radio telescope picked up waves generated by thunderstorms and from another unknown source in outer space. Over time scientists constructed larger and more powerful radio telescopes to pick up the radio waves beaming toward Earth.

RADIO ARRAYS. In the 1980s dozens of radio telescopes in New Mexico were linked together to enhance their capability. This was called the Very Large Array. In 1993 the National Science Foundation created a more powerful system called the Very Long Baseline Array. It includes ten eighty-two-foot radio antennas located across the United States, from Hawaii in the west to the Virgin Islands in the east. The telescopes are connected by a computer network and provide the best radio wave images in the world.

ARECIBO OBSERVATORY. The largest single-dish radio telescope on Earth is the Arecibo Observatory in Puerto Rico. The observatory was built in the 1960s and is operated by Cornell University for the National Science Foundation. NASA also provides funding support. On November 16, 1974, scientists used the observatory to send a radio message out into the galaxy. The message was coded in binary, meaning that a series of zeroes and ones were transmitted by shifting frequencies. The total broadcast took less than three minutes. It was a pictorial message.

The message shows the numbers one through ten, the atomic numbers of five chemical elements, the chemical formula of DNA, information about the human form and Earth's population, a stick figure person, the location of Earth in relation to the Sun and the other planets, a representation of the Arecibo telescope, and information about its size.

MAUNA KEA OBSERVATORIES. Mauna Kea is a 14,000-foot-high extinct volcano in Hawaii. A number of observatories have been erected atop Mauna Kea, because the very thin atmosphere allows detection of near-infrared radiation. This is a narrow band of infrared radiation that lies closest to visible light in the electromagnetic spectrum.

The Keck Observatory is operated at Mauna Kea by NASA in conjunction with the California Institute of Technology and the University of California. The Subaru Telescope is a project of the National Institutes of Natural Science of Japan. The Gemini Observatory is an international collaboration between the United States, United Kingdom, Canada, Chile, Australia, Argentina, and Brazil. Other universities and institutions around the world also operate telescopes atop Mauna Kea.

Early Space Telescopes

In 1923 German scientist Hermann Oberth (1894–1989) proposed a space telescope in *Die Rakete zu den Planetenraumen* (*The Rocket into Planetary Space*). Two decades later an American physicist named Lyman Spitzer (1914–97) more fully outlined the scientific benefits of putting telescopes in space. At the time space travel was not even possible.

In April 1968 NASA launched its first space telescope, the *Orbital Astronomical Observatory* (*OAO-1*). It was small and sensitive to ultraviolet light. This launch was first in a series of many satellites put into Earth orbit to detect energy sources in space. Most of these projects included NASA and international partners. In 1975 the European Space Agency (ESA) mission *COS-B* provided the first complete gamma-ray map of the galaxy.

Spitzer and other astronomers continued to urge the National Academy of Sciences and NASA to develop more powerful space telescopes for the scientific community. During the 1970s development began on a project called the *Large Space Telescope* (*LST*). Following the U.S. Moon landings NASA's budget was severely cut. This forced scientists to downsize the *LST* project several times. In 1975 the ESA joined the project and agreed to fund a percentage of *LST* costs in exchange for a guaranteed amount of telescope time for its scientists. Two years later the U.S. Congress authorized funding for construction and assembly of the *LST*.

At the same time that the *LST* was under construction, the space shuttle was also undergoing development. *LST* planners decided to utilize shuttle crews to deploy and service the telescope and ultimately return it to Earth. This decision would turn out to be a fateful one. By 1985 the observatory was finished and given a new name, *Hubble Space Telescope*.

The *Hubble Space Telescope*

The *Hubble Space Telescope* (*HST*) was the first of NASA's Great Observatories. It is named after the American astronomer Dr. Edwin Hubble (1889–1953). *HST* detects three types of light radiation: ultraviolet, visible, and near-infrared. It is the only one of the Great Observatories that captures images in visible light.

FIGURE 6.4

The configuration of the Hubble space telescope

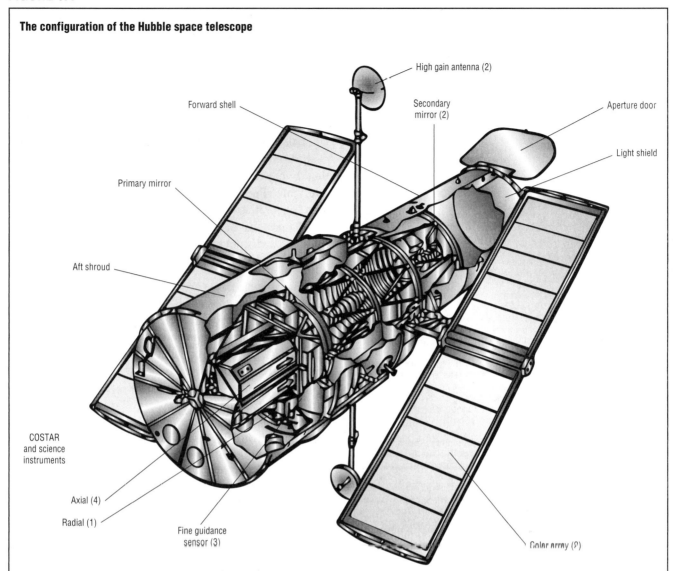

SOURCE: "Figure 1-2. HST Overall Configuration," in *Hubble Space Telescope: Servicing Mission 3B Media Reference Guide*, National Aeronautics and Space Administration, Goddard Space Flight Center, 2001, http://hubble.nasa.gov/a_pdf/news/sm3b_composite.pdf (accessed January 31, 2006)

Figure 6.4 shows the general layout of the *Hubble* spacecraft. It can hold eight science instruments in addition to the primary mirror and secondary mirrors. These instruments are powered by sunlight captured by the satellite's two solar arrays. Gyroscopes and flywheels are used to point the telescope and keep it stable.

Table 6.3 provides design, cost, and operational data about the *HST*. NASA's Goddard Space Flight Center (GSFC) in Greenbelt, Maryland, is responsible for oversight of all *HST* operations and for servicing the satellite. The Space Telescope Science Institute (STSI) in Baltimore, Maryland, selects targets for the observatory based on proposals submitted by astronomers. The STSI also analyzes the astronomical data generated by the observatory. The STSI is operated by the Association of Universities for Research in Astronomy, Inc., an international consortium of dozens of universities.

Hubble transmits raw data to one of NASA's Tracking and Data Relay Satellites (TDRSs). There are six of these satellites in low-Earth orbit. The TDRS relays the data via radio frequency communication links to a complex in White Sands, New Mexico. From there the data are relayed to GSFC and then to STSI.

The *HST* was originally supposed to go into space in 1986. The explosion of the space shuttle *Challenger* that year delayed *Hubble*'s launch for four years. On April 24, 1990, *HST* was carried into orbit by the space shuttle *Discovery* on mission STS-31. One day later the astronauts deployed the satellite approximately 380 miles

TABLE 6.3

The Hubble space telescope at a glance

Hubble's name:	NASA named the world's first space-based optical telescope after American astronomer Edwin P. Hubble (1889–1953). Dr. Hubble confirmed an "expanding" universe, which provided the foundation for the Big Bang theory.
Launch:	April 24, 1990, from space shuttle Discovery (STS-31)
Deployment:	April 25, 1990
Mission duration:	Up to 20 years
Servicing mission 1:	December 1993
Servicing mission 2:	February 1997
Servicing mission 3A:	December 1999
Servicing mission 3B:	February 2002
Length:	43.5 ft (13.2 m)
Weight:	24,500 lb (11,110 kg)
Maximum diameter:	14 ft (4.2 m)
Cost at launch:	$1.5 billion
Orbit:	At an altitude of 380 statute miles (612 km), inclined 28.5 degrees to the equator (low-Earth orbit)
Time to complete one orbit:	97 minutes
Speed:	17,500 mph (28,000 kph)
Hubble can't observe:	The Sun or Mercury, which is too close to the Sun
Sensitivity to light:	Ultraviolet through near infrared (110–2,500 nanometers)
First image:	May 20, 1990: Star Cluster NGC 3532
Data stats:	Each day the telescope generates enough data—3 to 4 gigabytes—to fill six CD-ROMs. The orbiting observatory's observations have amounted to more than 7 terabytes of data. Hubble's digital archive delivers 10 to 20 gigabytes of data a day to astronomers all over the world.
Power mechanism:	Two 25-foot solar panels
Power usage:	3,000 watts. In an average orbit, Hubble uses about the same amount of energy as 30 household light bulbs.
Pointing accuracy:	In order to take images of distant, faint objects, Hubble must be extremely steady and accurate. The telescope is able to lock onto a target without deviating more than 7/1000th of an arcsecond, or about the width of a human hair seen at a distance of 1 mile. Pointing the Hubble Space Telescope and locking onto distant celestial targets is like holding a laser light steady on a dime that is 200 miles away.
Primary mirror	
Diameter:	94.5 in (2.4 m)
Weight:	1,825 lb (828 kg)
Secondary mirror	
Diameter:	12 in (0.3 m)
Weight:	27.4 lb (12.3 kg)
Power storage	
Batteries:	6 nickel-hydrogen (NiH)
Storage capacity:	Equal to 20 car batteries

SOURCE: "Hubble at a Glance," in *About Hubble*, Space Telescope Science Institute, Undated, http://hubblesite.org/newscenter/news_media_resources/reference_center/about_hubble/glance.php (accessed January 31, 2006)

from Earth. It assumed an orbit that is nearly a perfect circle.

When scientists first started using *Hubble* they discovered that its images were fuzzy due to a defect in the telescope's optical mirrors. The defect had existed before the satellite was launched into space. Media publicity about the problem caused an uproar and brought harsh criticism of NASA. In December 1993 astronauts aboard the space shuttle *Endeavour* (STS-61) were sent to intercept the *HST* and install corrective devices. The mission was a success.

This was the first of four servicing missions conducted by shuttle crews between 1993 and 2002. During these missions astronauts performed repairs and maintenance, installed new components to enhance *HST*'s performance, and reboosted the satellite to a higher altitude. Like all objects in low-Earth orbit the *HST* experiences some drag and gradually loses altitude. The satellite does not have its own propulsion system.

HUBBLE **NEEDS SERVICING.** *Hubble* was originally designed for a twenty-year lifetime. Planners assumed that five shuttle servicing missions would be required over this time period to keep the observatory in orbit and functioning properly. At the end of 2002 four of these missions had been performed. A fifth servicing mission was anticipated in 2004 followed by satellite retrieval in 2010.

The loss of the space shuttle *Columbia* in February 2003 altered the fate of the *HST*. The shuttle fleet was grounded while NASA examined the cause of the accident and developed ways to make shuttle flight safer. In January 2004 NASA Administrator Sean O'Keefe announced his decision to cancel the fifth servicing mission to *HST*. A shuttle flight to the observatory was considered too risky, because *HST* orbits far from the *International Space Station* (*ISS*). The *ISS* is the only safe haven available to shuttle astronauts in the event of an emergency.

The decision caused an uproar in the scientific community. Without a reboost the *HST* will gradually fall out of orbit and be destroyed by reentry to Earth's atmosphere. This is anticipated to occur before 2010. Even before that time the observatory will likely cease to be useful due to failures in its scientific instruments, computer, gyroscopes, batteries, data relay system, or solar arrays. Many scientists are disappointed that such a valuable resource may be lost before the end of its useful life.

In December 2004 the National Academy of Sciences (NAS) issued a report examining the options for extending the life of the *HST*. The NAS recommended that NASA perform a servicing mission as soon as possible after the space shuttle successfully returns to flight. It also recommended that a robotic spacecraft be developed to safely deorbit the *HST* once it reaches the end of its usefulness. Reentry does not always completely incinerate space objects. *HST* reentry over a populated area could pose a danger to people on the ground. A robotic "tow truck" could guide the satellite to reentry over a desolate area of the Pacific Ocean.

In April 2005 a new NASA Administrator, Mike Griffin, indicated the agency is rethinking its decision to cancel the servicing mission. He instructed NASA engineers to begin preparing for such a mission in case the funding and opportunity for it become available.

However, there was no money included for an *HST* servicing mission in NASA's 2006 budget request. In July 2005 the space shuttle returned to flight with the successful "test flight" of the orbiter *Discovery*. Unfortunately photographs revealed that a chunk of foam was torn from the shuttle's external tank during lift-off. A foam impact doomed the shuttle *Columbia* in 2003. The next return-to-flight mission was postponed until summer 2006 to allow engineers to reexamine the problem.

In August 2005 scientists began operating the *HST* with only two gyroscopes, the minimum necessary to accurately point the telescope at its target. At that time only three of the *HST*'s six original gyroscopes were still operational. One of the three was shut down to hold in reserve as a backup. In addition, NASA reported that the satellite's batteries and outside heat shielding are deteriorating.

During its mission *Hubble* has captured more than 700,000 dramatic images of celestial objects throughout the universe, including nebulae, galaxies, stars, and even some bodies believed to be planets. In late 2005 NASA began using the telescope to search Earth's moon for oxygen-bearing minerals. Scientists hope to find a potential source of oxygen for use by future astronauts visiting the lunar surface. The *HST* provided the first-ever high-resolution ultraviolet images of the Moon.

Compton Gamma-Ray Observatory

The *Compton Gamma-Ray Observatory* (*CGRO*) was the second of NASA's Great Observatories. It was designed to detect high-energy gamma rays, the most powerful type of energy in the electromagnetic spectrum. The observatory was named after the late American physicist Dr. Arthur Compton (1892–1962).

On April 5, 1991, *CGRO* was launched into space aboard the space shuttle *Atlantis* as part of mission STS-37. The satellite was deployed at an orbit altitude of 280 miles above Earth. It was equipped with four highly sensitive detecting instruments and a pair of solar arrays. (See Figure 6.5.) The satellite also had a small propulsion system and three gyroscopes that were used to point the telescope and maintain flight stability. The propulsion system was not powerful enough to boost *CGRO* to higher altitudes.

The original plan was for a five-year mission. The observatory proved to be much more durable than expected and remained in space for nine years.

In early 2000 NASA learned that one of *CGRO*'s gyroscopes had failed. Even though the observatory instruments were still in working order, NASA decided to purposely deorbit the satellite. Scientists feared that left to reenter on its own, *CGRO* could rain pieces down on populated areas. This was of particular concern

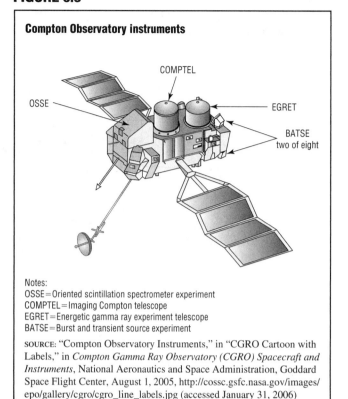

FIGURE 6.5

Compton Observatory instruments

Notes:
OSSE=Oriented scintillation spectrometer experiment
COMPTEL=Imaging Compton telescope
EGRET=Energetic gamma ray experiment telescope
BATSE=Burst and transient source experiment

SOURCE: "Compton Observatory Instruments," in "CGRO Cartoon with Labels," in *Compton Gamma Ray Observatory (CGRO) Spacecraft and Instruments*, National Aeronautics and Space Administration, Goddard Space Flight Center, August 1, 2005, http://cossc.gsfc.nasa.gov/images/epo/gallery/cgro/cgro_line_labels.jpg (accessed January 31, 2006)

because the satellite was unusually massive at seventeen tons and also included a fair amount of titanium in its components. Titanium is not easily disintegrated during atmospheric reentry.

On June 4, 2000, *CGRO*'s propulsion system was used to guide the satellite to a safe reentry over a deserted area of the Pacific Ocean.

The *CGRO* program included participation by scientists from Germany, the Netherlands, the European Space Agency, and the United Kingdom. During its mission the observatory investigated high-energy phenomena associated with black holes, novas, supernovas, quasars, pulsars, solar flares, cosmic rays, and gamma-ray bursts. Gamma-ray bursts are sudden short flashes of gamma rays that do not seem to occur at predictable locations. *CGRO* recorded more than 2,500 gamma-ray bursts, far more than had ever been detected before. It also mapped hundreds of gamma-ray sources.

Chandra X-Ray Observatory

The *Chandra X-Ray Observatory* (*CXRO*) is the third of NASA's Great Observatories and the world's most powerful x-ray telescope. It was named after the late Indian-American scientist Dr. Subrahmanyan Chandrasekhar (1910–95). The *CXRO* can detect x-ray sources that are billions of light years away from Earth.

FIGURE 6.6

The Chandra X-Ray Observatory

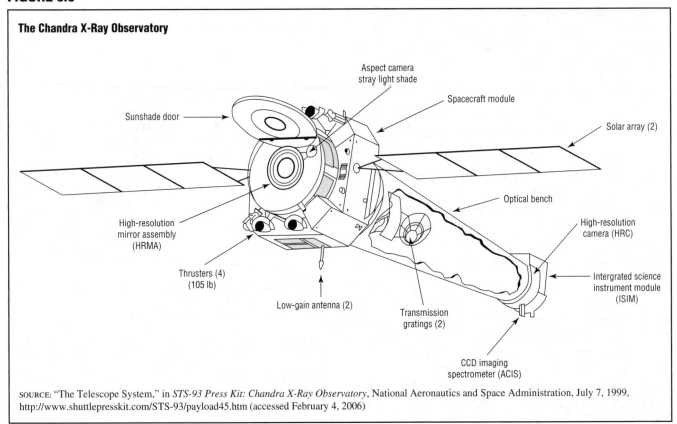

SOURCE: "The Telescope System," in *STS-93 Press Kit: Chandra X-Ray Observatory*, National Aeronautics and Space Administration, July 7, 1999, http://www.shuttlepresskit.com/STS-93/payload45.htm (accessed February 4, 2006)

On July 22, 1999, the space shuttle *Columbia* blasted off from Cape Canaveral, Florida. Mission STS-93 carried the *CXRO* as its primary payload. The *CXRO* spacecraft was equipped with a rocket called an Inertial Upper Stage (IUS). The shuttle crew released the satellite and then moved a safe distance away. The IUS was fired for four minutes to move the satellite into a temporary orbit high above the Earth.

At that point, the *CXRO*'s twin solar wings were unfolded and began converting sunlight into electrical power. Figure 6.6 shows a diagram of the spacecraft. Important design information is included in Table 6.4. Solar power is used to run *CXRO*'s equipment and charge its batteries. The spacecraft's own propulsion system was used to boost it to its final orbit.

Chandra circles the Earth once every sixty-four hours. The satellite's orbit is highly elliptical. At its closest point the observatory is about 6,200 miles away from Earth. At its farthest point it is about 87,000 miles away. This is nearly one-third of the distance to the Moon. *Chandra*'s flight path carries it through the Van Allen radiation belts that encircle Earth. The satellite's sensitive electronic instruments are turned off while the spacecraft is in or near this highly radioactive area of space.

Scientists around the world are using *CXRO* images to create an x-ray map of the universe. They hope to use this map to learn more about black holes, supernovas, superflares, quasars, extremely hot gases at the center of galaxy clusters, and the mysterious substance known as dark matter.

In February 2004 researchers at the Max Planck Institute for Extraterrestrial Physics in Germany announced that *CXRO* had recorded an event called a stellar tidal disruption (STD). This is when a star wanders too close to a massive black hole and gets ripped apart. STDs are predicted by astronomical theory, but had never been observed before.

CXRO detected a tremendous burst of x-ray energy coming from the center of a galaxy called RXJ1242-11. This galaxy is about 700 million light years from Earth. Scientists believe that the energy burst was due to superheating of gases from the star as it was being devoured by the black hole.

The *CXRO* was designed for a five-year mission. In August 2005 NASA celebrated the observatory's sixth anniversary. It is expected to keep operating at least another four years. During its sixth year the *CXRO* was focused on targets closer to home, including Jupiter and Saturn. Scientists observed electromagnetic activities at Jupiter's poles and learned that Saturn reflects solar x-ray flares from the sun.

The *CXRO* is managed by the Marshall Space Flight Center for NASA's Office of Space Science. NASA's Jet

TABLE 6.4

The Chandra Observatory at a glance

Mission duration

Chandra science mission	Approx. 5 yrs
Orbital activation & checkout period	Approx. 2 mos

Orbital data

Inclination	28.5 degrees
Altitude at apogee	86,992 sm
Altitude at perigee	6,214 sm
Orbital period	64 hrs
Observing time per orbital period	Up to 55 hrs

Dimensions

Length—(Sun shade open)	45.3'
Length—(Sun shade closed)	38.7'
Width—(solar arrays deployed)	64.0'
Width—(solar arrays stowed)	14.0'

High resolution mirror assembly

Configuration	4 sets of nested, grazing incidence araboloid/hyperboloid mirror pairs
Mirror weight	2,093 lbs
Focal length	33 ft
Outer diameter	4 ft
Length	33.5 in
Material	Zerodur
Coating	600 angstroms of iridium

Science instruments

Charged coupled imaging spectrometer (ACIS)
High resolution camera (HRC)
High energy transmission grating (HETG)
Low energy transmission grating (LETG)

SOURCE: Adapted from "Chandra at a Glance," in *STS-93 Press Kit: Chandra X-Ray Observatory*, National Aeronautics and Space Administration, July 7, 1999, http://www.shuttlepresskit.com/STS-93/payload45.htm (accessed February 4, 2006)

Propulsion Laboratory in Pasadena, California, provides communications and data links. The Smithsonian Astrophysical Observatory in Cambridge, Maryland, controls science and flight operations. NASA's High-Energy Astrophysics Science Archive (HEASARC) collects and maintains all astronomy data obtained from the *CXRO* and NASA's smaller gamma-ray, x-ray, and extreme ultraviolet observatories.

Spitzer Space Telescope

The *Spitzer Space Telescope* is the fourth of NASA's Great Observatories and detects infrared radiation. When it was originally conceived during the 1970s the observatory was called the *Shuttle Infrared Telescope Facility* (*SIRTF*). Its planners hoped to carry the telescope into space as a science payload on numerous space shuttle missions. In 1983 NASA decided the telescope needed to be in continuous orbit and able to fly by itself. The *SIRTF* was renamed the *Space Infrared Telescope Facility*.

During the 1990s NASA's science programs suffered harsh budget cuts. The *SIRTF* was redesigned several times and downsized from a large $2 billion telescope with numerous capabilities to a modest telescope costing less than half a billion dollars.

On August 25, 2003, the telescope was launched into orbit atop a Delta rocket. It was the only Great Observatory not carried into space by a space shuttle. Figure 6.7 shows the components of the two-stage launch vehicle. The telescope was housed inside a protective case called a fairing that fell away when the satellite reached orbit. Figure 6.8 shows details of the observatory's structure. The dark circle at the top of the observatory is actually a door that opens to expose the telescope to space.

In December 2003 NASA chose a new name for the *SIRTF* from thousands of names suggested during a naming contest. More than 7,000 people from around the world entered the contest. Some of the most frequently proposed names were Red Eye, Sagan (after scientist Carl Sagan), Herschel (after scientist William Herschel, who in 1800 discovered infrared radiation), and Roddenberry (after science-fiction writer Gene Roddenberry, creator of the *Star Trek* television show). In the end NASA chose the name Spitzer to honor the late American physicist. It was Spitzer's groundbreaking work of the 1940s that inspired astronomers to build space telescopes.

The *Spitzer Space Telescope* is unique among the Great Observatories because it orbits the Sun instead of the Earth. It actually follows along behind Earth as the planet travels around the Sun. This is called an Earth-trailing heliocentric (sun-centered) orbit.

This orbit was chosen for several reasons. It allows the spacecraft to avoid interference from Earth's infrared-absorbing atmosphere. The orbit also provides a colder environment for the telescope than would an Earth orbit. Space infrared telescopes must be maintained at a very cold temperature to operate properly. Everything that has a temperature emits infrared radiation. The spacecraft must be kept as cold as possible to prevent heat within it from interfering with its own detection equipment. The relatively cold Earth-trailing heliocentric orbit was selected to minimize the amount of liquid helium required onboard to keep the telescope cool. Scientists believe that *Spitzer*'s liquid helium supply will last as long as five years.

The *Spitzer Space Telescope* can detect infrared energy with wavelengths of three to 180 microns. A micron is one-millionth of a meter. Detection of infrared waves at these lengths allows the telescope to image celestial objects hidden by dense clouds of dust and gas. During its first few months of operation *Spitzer* provided images of galaxies and nebulae up to three billion light years away. In January 2004 NASA released colorful *Spitzer* images of massive newborn stars at the core of the Tarantula Nebula. This nebula is a known star-forming region 170,000 light years from Earth.

In early 2005 the telescope captured the first-ever infrared images of two extrasolar (or exosolar) planets.

FIGURE 6.7

Delta II Rocket launches Spitzer space telescope

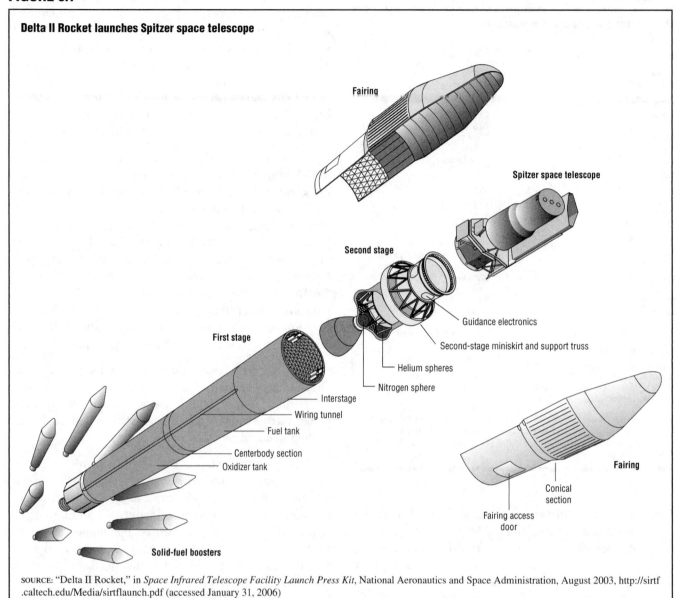

Fairing

Spitzer space telescope

Second stage

Guidance electronics

Second-stage miniskirt and support truss

Helium spheres

Nitrogen sphere

First stage

Interstage

Wiring tunnel

Fuel tank

Centerbody section

Oxidizer tank

Fairing

Conical section

Fairing access door

Solid-fuel boosters

SOURCE: "Delta II Rocket," in *Space Infrared Telescope Facility Launch Press Kit*, National Aeronautics and Space Administration, August 2003, http://sirtf .caltech.edu/Media/sirtflaunch.pdf (accessed January 31, 2006)

Extrasolar and exosolar mean "outside our solar system." The two large gas planets, known as HD 209458b and TrES-1, orbit very close to their respective suns. This makes them impossible to image directly with visible light telescopes. HD 209458b circles a yellow sun 150 light-years from Earth in the constellation Pegasus. TrES-1 is located 500 light-years from Earth in the constellation Lyra.

In October 2005 NASA released infrared images of the spiral galaxy Messier 31, commonly called Andromeda. The galaxy is about 2.5 million light-years from Earth. *Spitzer* recorded nearly 11,000 separate snapshots of the galaxy that were pieced together to form a dramatic mosaic. The images reveal old giant stars at Andromeda's center and a system of arms spiraling out to a prominent ring of stars.

Humans cannot see infrared radiation. The energy data that *Spitzer* collects are transformed into visible pictures by assigning different colors to different energy levels. The resulting pictures are called false-color images. Scientists either choose colors to highlight a specific area of scientific interest or to make a celestial object appear realistic to the human eye. For example, coloring the highest energy areas of an image with white or red indicates brightness and heat to a human looking at the picture. Likewise low-energy regions are shaded with darker colors to indicate relative coolness.

The *Spitzer Observatory* is managed by NASA's Jet Propulsion Laboratory, a division of the California Institute of Technology in Pasadena, California. The university also includes the Spitzer Science Center, which conducts science operations for the telescope.

FIGURE 6.8

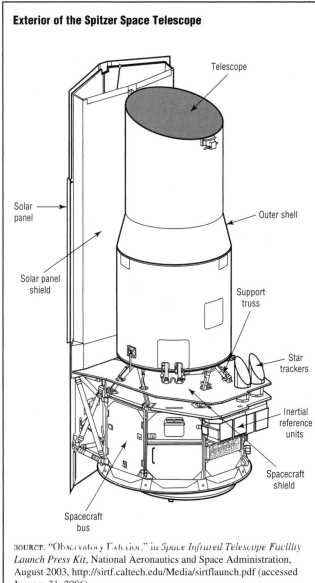

Exterior of the Spitzer Space Telescope

Telescope

Solar panel

Solar panel shield

Outer shell

Support truss

Star trackers

Inertial reference units

Spacecraft shield

Spacecraft bus

SOURCE: "Observatory Exterior," in *Space Infrared Telescope Facility Launch Press Kit*, National Aeronautics and Space Administration, August 2003, http://sirtf.caltech.edu/Media/sirtflaunch.pdf (accessed January 31, 2006)

The Next Great Observatory

As of 2006 NASA's next Great Observatory is under development and scheduled for launch in 2013. It is called the *James Webb Space Telescope* (*JWST*) in honor of NASA's director of flight operations during the Apollo program. The expected cost of the *JWST* mission is $825 million.

JWST will detect infrared radiation from its orbit location at L2 approximately one million miles from Earth in the direction away from the Sun.

When it is stationed at the L2 point, the *JWST* will make a small circle called a halo orbit around the L2 point. This point will not only provide the *JWST* with a relatively cold location in space, but make it easier to shield the telescope from the massive infrared radiation

coming from the Sun, Earth, and Moon. *JWST* will look outward into space.

Other Space Observatories

Besides the Great Observatories NASA operates a number of smaller space observatories in Sun-Earth orbits. These spacecraft were funded and built under NASA's Explorer program and are listed in Table 6.1. The projects are conducted with the help of scientists from academic institutions and/or foreign space agencies.

The European Space Agency operates an observatory of its own called the *XMM-Newton*, named after scientist Isaac Newton (1642–1727). This powerful telescope is similar to one of NASA's Great Observatories. The *XMM-Newton* is the largest science satellite ever built in Europe and includes three x-ray telescopes. It was launched into space on December 10, 1999, by an Ariane rocket. *XMM-Newton* is expected to have a ten-year lifetime.

In September 2005 NASA summarized the major findings of the observatory in "XMM-Newton Highlights" (http://heasarc.gsfc.nasa.gov/FTP/xmm/doc/XMMHandout.pdf). At that time the satellite had made nearly 5,000 observations of phenomena. Scientists analyzing *XMM-Newton* images discovered the largest and most distant object in the known universe, a cluster of galaxies located nine billion light-years from Earth.

THE SUN-EARTH CONNECTION

One of NASA's primary research goals is to learn about the Sun-Earth connection. The Sun bathes the Earth with life-sustaining warmth. It also emits radiation, charged particles, and other substances that can disrupt electrical systems and electronic equipment vital to modern societies. Since the beginning of the space age scientists have put satellites in Earth orbit to monitor the Sun and its effects on Earth. During the 1990s NASA began sending spacecraft even farther out into space to investigate solar phenomena that directly impact our planet.

Lagrange points are places found in the space around a large body (like the Earth) orbiting another large body (like the Sun). At a Lagrange point, the gravity of the Earth and the gravity of the Sun are balanced, so that a satellite can remain at one of these points and not be pulled away by the gravity of either of the larger bodies. The concept was discovered by the Italian-born French mathematician Joseph-Louis Lagrange (1736–1813).

There are five Lagrange points in the Sun-Earth system. Because the Earth constantly orbits the Sun, the Lagrange points constantly move also. However, they remain a fixed distance from the Earth and the Sun. The L1 point is roughly one million miles from the Earth, between the Earth and the Sun. The L2 point is approximately the same distance in the opposite direction, outside the Earth's orbit. Point L3 is located on the opposite side of the Sun from the Earth at a distance equal to that of the distance from the center of the Sun to the center of the Earth. The L1, L2, and L3 points lie on a centerline that bisects (cuts through the middle of) the Earth and the Sun. The L4 and L5 points are located at the end points of two equilateral (equal-sided) triangles whose shared base is the segment of centerline that falls between the center of the Earth and the center of the Sun.

Any object at the L3 point is always hidden from Earth view by the Sun. Point L3 has been an intriguing location to science fiction writers who fantasize that an unseen mirror-image of Earth could be there. The idea was popularized in the 1951 movie *The Man from Planet X.*

Solar Wind

Solar wind is a constant flow of charged particles from the Sun into space. It moves away from the Sun at a speed of 200 to 400 miles per second and flows across the solar system reaching just past Pluto. The region of outer space exposed to the solar wind is called the heliosphere. Helios was the mythical Greek god of the Sun. His Roman counterpart was named Sol.

Astronomers see visible proof of the solar wind when they look at comets. Comet tails are blown outward by the solar wind and always point away from the sun. The solar wind also contributes to the magnificent auroras seen on Earth. An aurora is a magnetic phenomenon that causes colorful streaks of bright light to appear in the upper atmosphere near Earth's two polar regions. Aurora is a Latin word meaning dawn. It was also the name of the mythical Roman goddess of dawn.

The aurora centered on Earth's northern hemisphere is called *aurora borealis* or the northern lights. The southern lights are known as *aurora australis*. Usually auroras are only visible to people in the northern and southern regions of the Earth. For example, people in Canada, Alaska, and northern states are far more likely to see auroras than people in the southern United States.

The solar wind constantly pushes and shapes Earth's magnetosphere. This is the region of space around the Earth dominated by the planet's magnetic field. (See Figure 6.9.) The magnetosphere helps protect Earth from dangerous electromagnetic radiation moving through space. The outer boundary of Earth's magnetosphere is called the bow shock.

The speed, composition, density, and magnetic field strength of the solar wind are not constant but vary depending on conditions on the Sun. A "gust" in the solar wind can energize Earth's magnetosphere, produc-

cing beautiful auroras, but wreaking havoc on sensitive electronics.

Sunspots

Sunspots look like dark blemishes on the surface of the Sun. They are actually areas of plasma that are slightly cooler than their surroundings due to magnetic activity. Sunspots can be enormous in size, even bigger than planet Earth. Sunspots can appear, change size, and disappear. Each typically lasts from a few hours to a few months. Most sunspots are visible only through telescopes.

Sunspots were first identified in western literature by the Greek philosopher and scientist Theophrastus (c. 372–287 BC). However, some historians believe that Chinese astronomers knew about sunspots centuries before this. After the invention of the telescope in the 1600s many scientists reported seeing dark spots when they looked toward the Sun. Galileo argued that these spots were clouds moving across the Sun's surface. However, most astronomers of the time believed they were seeing a small unnamed planet moving near the Sun.

During the late 1800s European astronomers searched for this mysterious planet thought to orbit between Mercury and the Sun. It was called Vulcan. In 1826 an amateur German astronomer named Heinrich Schwabe (1789–1875) was looking for the planet Vulcan. He began keeping detailed records of the number of sunspots he observed every day. After nearly two decades of doing this he noticed a cyclical pattern to the data. By the early 1900s scientists knew that the number of sunspots peaks approximately every eleven years and then drops off dramatically. (See Figure 6.10.) This is called the solar cycle. The period of peak sunspot activity is called the solar maximum.

In 1952 the International Council of Scientific Unions proposed that scientists around the world cooperate in conducting extensive earth science investigations between July 1957 and December 1958. This was the next expected solar maximum. Scientists hoped to collect large amounts of data on the relationship between sunspots and solar activity and the resulting effects on Earth's magnetic fields. This research remained a priority throughout the space age. The most recent solar maximum occurred in 2000–01. The next peak is expected in 2011–12.

Solar Flares

Solar flares are sudden energy releases that burst out from the Sun near sunspots. They are most common during the solar maximums. Solar flares emit electromagnetic radiation, and can include energy particles and bulk plasma. They can last from minutes to hours.

Solar flares that erupt in Earth's direction can shower the planet with energetic particles within thirty minutes.

FIGURE 6.9

The Earth and the magnetosphere

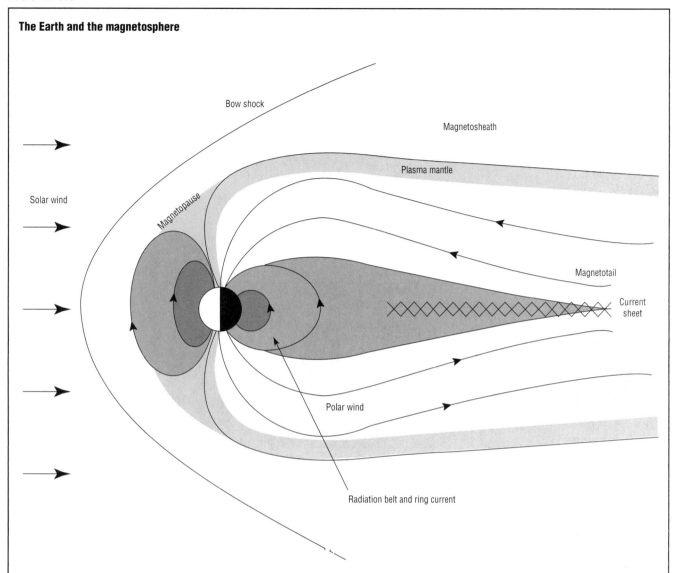

SOURCE: Adapted from Roger P. Briggs and Robert J. Carlisle, "Figure 4-12. A 'Side View' of the Earth and Magnetosphere Showing Some of the Important Regions," in *Solar Physics and Terrestrial Effects: A Curriculum Guide for Teachers, Grades 7–12*, U.S. Department of Commerce, National Oceanic and Atmospheric Administration, Space Environment Center, June 1996, http://www.sel.noaa.gov/Curric_7-12/Chapter_4.pdf (accessed December 28, 2005)

Solar flares are associated with auroras and magnetic storms on Earth and can cause radio interference if large amounts of x-ray radiation are released.

Scientists classify solar flares by the intensity of their x-ray radiation and its effects on Earth. These classifications are:

- C Class—Small flares with few consequences to Earth systems.

- M Class—Medium-sized flares resulting in brief radio blackouts in Earth's polar regions. May cause some minor radiation storms.

- X Class—Large flares that can cause worldwide radio blackouts and long-lasting radiation storms on Earth.

Auroras typically appear on Earth about two days after a solar flare occurs. On July 14, 2001, a powerful solar flare occurred that resulted in the northern lights appearing as far south as Florida. In November 2003 the largest solar flare ever recorded burst out from the Sun. Luckily it was pointed away from Earth and did not cause any major problems.

Solar Prominences

Solar prominences are giant clouds of dense plasma suspended in the Sun's corona (outermost atmosphere). They are usually calm, but occasionally erupt and snake out from the sun along magnetic field lines. Prominences can break away from the Sun and hurtle through space carrying large amounts of solar material.

FIGURE 6.10

The sunspot cycle, 1750–September 2005

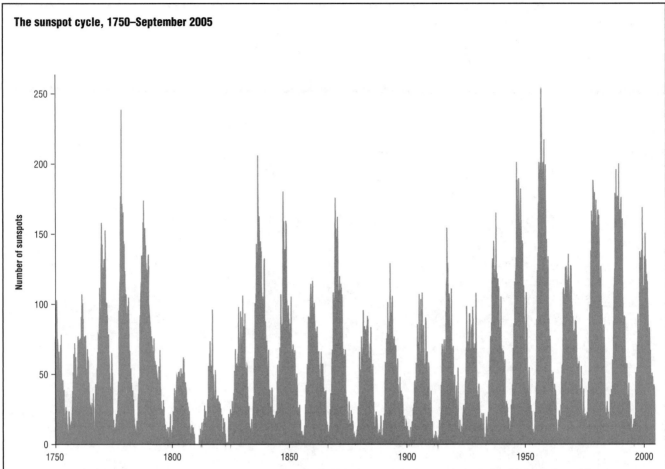

SOURCE: Adapted from "Monthly Averages (Updated Monthly) of the Sunspot Numbers," in *The Sunspot Cycle*, National Aeronautics and Space Administration, Marshall Space Flight Center, October 25, 2005, http://science.nasa.gov/ssl/pad/solar/greenwch/spot_num.txt (accessed December 28, 2005)

Coronal Mass Ejections

Coronal mass ejections (CMEs) occur when billions of tons of particles are slung away from the sun by broken magnetic field lines. CMEs move away from the Sun into space at a tremendous speed (millions of miles per hour). These ejections happen frequently and are most common during the solar maximum, when they can occur several times a day. CMEs are known informally as solar storms.

When a CME blasts toward Earth the planet's magnetosphere is hit within two to four days by a cloud of electrically charged particles. This cloud distorts the entire magnetosphere, as shown in Figure 6.11. The resulting disturbance is called a geomagnetic storm. These storms can harm satellites and disrupt telecommunications and electric power generation around the world.

Space Weather

The term space weather was created by scientists to refer to the overall effects of solar activities on the space around Earth (or geospace). Space weather encompasses all the solar phenomena (solar wind, sunspots, flares, prominences, and coronal mass ejections) and the resulting conditions in Earth's magnetosphere and upper atmosphere.

The National Oceanic and Atmospheric Administration (NOAA) monitors space weather and issues warnings about geomagnetic storms and solar radiation storms (increases in the number of energy particles in geospace). The NOAA's Space Environment Center (SEC) is located in Boulder, Colorado. It serves as the national and worldwide warning center for space weather disturbances. The Space Weather Operations branch is jointly operated by NOAA and the U.S. Air Force.

The SEC uses a scale system to characterize space weather storms according to severity, effects, and frequency. Geomagnetic storms are ranked from level G1 (least severe) to G5 (most severe). Solar radiation storms are similarly ranked from level S1 to level S5. Minor storms occur far more frequently than major storms. The SEC also ranks radio blackouts from level R1 to level R5. These blackouts are caused by x-ray emissions from the Sun that disturb Earth's ionosphere.

FIGURE 6.11

Geomagnetic storm effects

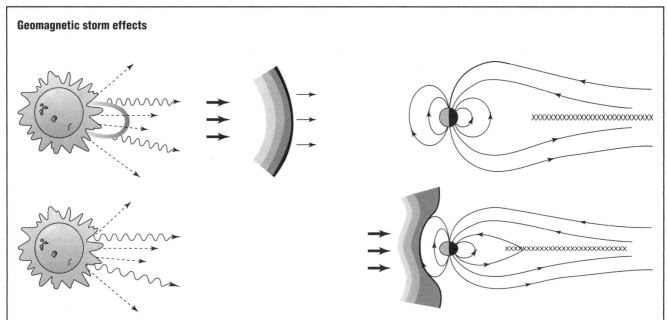

SOURCE: Norm Cohen and Kenneth Davies, "Figure 5. An Ejection from the Sun Travels to Earth and Distorts Earth's Magnetic Field, Resulting in Geomagnetic Activity," in *Space Environment Topics: Radio Wave Propagation*, U.S. Department of Commerce, National Oceanic and Atmospheric Administration, Space Environment Laboratory, 1994, http://www.sec.noaa.gov/info/Radio.pdf (accessed January 31, 2006)

Monitoring of space weather is important. A major solar event like a coronal mass ejection can cause a tremendous inflow of electrical energy into Earth's atmosphere. This is disruptive to electrical power grids and telecommunication systems on the ground and dangerous for Earth-orbiting satellites. NASA also worries about possible harmful exposure of astronauts to excessive radiation levels during solar storms.

Ongoing Solar Exploration Missions

NASA and other space agencies operate a number of missions dedicated to studying the Sun-Earth connection. Table 6.5 describes major missions ongoing as of February 2006. Most of these spacecraft are stationed in Earth orbit and study the effects of solar activity on the magnetosphere or upper atmosphere. Some of the spacecraft are in orbits far from Earth or on other paths designed for optimal study of the Sun-Earth connection. NASA spacecraft support several science programs, including Explorer, Discovery, and ISTP.

INTERNATIONAL SOLAR-TERRESTRIAL PHYSICS (ISTP) PROGRAM. During the 1990s NASA teamed with the European and Japanese space agencies in a project known as the International Solar-Terrestrial Physics (ISTP) Science Initiative. The purpose of the ISTP project is to combine international resources to conduct long-term investigations of the Sun-Earth space environment. The program includes both ground-based studies and space missions. The space activities fall into two categories: terrestrial (Earth-directed) and solar (Sun-directed).

Terrestrial space missions rely on satellites in Earth orbit that gather data about the sun's effects on the planet. This effort is supported by two other U.S. federal agencies—the National Oceanic and Atmospheric Administration (NOAA) and the Department of Energy's Los Alamos National Laboratory (LANL).

As of 2006 ISTP's solar space missions are ongoing. They are conducted by spacecraft called *Cluster*, *Geotail*, *Polar*, *SOHO*, and *Wind*. All but *SOHO* and *Wind* are kept in Earth orbit. *SOHO* and *Wind* are positioned a million miles from Earth to get a better look at the Sun.

SOHO and *ACE* (an Explorer mission) are in small orbits around the L1 Lagrange point. *ACE* collects samples of the solar wind and other particles emitted from the Sun. The spacecraft is equipped with a sophisticated communications system that allows it to transmit data very quickly. Many solar storms reach the L1 point an hour or more before hitting Earth. *ACE* is able to give scientists advance warning of impending storms that could disrupt Earth's magnetic field. The spacecraft has its own propulsion system and enough propellant to last until approximately 2019.

SOHO is the largest and most sophisticated solar observatory ever developed. It includes twelve instruments that collect data about the Sun's inner and outer workings and the solar wind. *SOHO* observes the Sun in visible light and four different wavelengths of extreme-ultraviolet radiation. This is ultraviolet radiation that lies

TABLE 6.5

Ongoing solar missions

Name	Primary mission sponsors	NASA science program	Launch date	Location in space	Mission
ACE (Advanced Composition Explorer)	NASA, California Institute of Technology	Explorer	8/25/1997	L1 orbit	Samples the solar wind, particles emitted during solar flares, and high-energy galactic particles. Monitors space weather and provides warning of impending geomagnetic storms.
Cluster	ESA[a]	ISTP[b]	7/16/00 and 8/9/00	Earth orbit	Includes 2 pair of two satellites that study the solar wind and its interaction with Earth's magnetospheric plasma.
FAST (Fast Auroral Snapshot Explorer)	NASA, University of California at Berkeley	Explorer	8/21/1996	Earth orbit	Studies Earth's aurora. Also supports ISTP[b].
Geotail	NASA, JAXA[c]	ISTP[b]	7/24/1992	Earth orbit	Studying the dynamics of the Earth's magnetotail.
IMAGE (Imager for Magnetopause-to-Aurora Global Exploration)	NASA, Southwest Research Institute	Explorer	3/25/2000	Earth orbit	Satellite imaging Earth's magnetosphere. Also supports ISTP[b].
RHESSI (Reuven Ramaty High Energy Solar Spectroscopic Imager)	NASA, University of California at Berkeley	Explorer	2/5/2002	Earth orbit	Explores the basic physics of particle acceleration and explosive energy release in solar flares. Named after the late NASA physicist Reuven Ramaty.
SAMPEX (Solar Anomalous and Magnetospheric Particle Explorer)	NASA, University of Maryland, German Universities	Explorer	7/3/1992	Earth orbit	Investigating the composition of local interstellar matter and solar material and the transport of magnetospheric charged particles into Earth's atmosphere. Also supports ISTP[b].
SOHO (Solar and Heliospheric Observatory)	NASA, ESA[a]	ISTP[b]	12/2/1995	L1 orbit	Solar observatory studying the Sun's internal structure, heating of its extensive outer atmosphere, and the origin of the solar wind.
Stardust	NASA	Explorer	2/7/1999	Comet encounter	Sampling ship sent to intercept comet Wild 2 as it orbits the sun and collect particle samples. Expected to return to Earth in January 2006.
TIMED (Thermosphere, Ionosphere, Mesosphere Energetics and Dynamics)	NASA, John Hopkins University Applied Physics Laboratory	Solar Terrestrial Probes	12/7/2001	Earth orbit	Studying solar influences on Earth's mesosphere and lower thermosphere/ionosphere.
TRACE (Transition Region and Coronal Explorer)	NASA, Lockheed Martin Solar & Astrophysics Laboratory	Explorer	4/1/1998	Earth orbit	Studies the connection between the Sun's magnetic fields and heating of its corona.
Ulysses	NASA, ESA[a]	Sun-Earth Connection	10/6/1990	Solar orbit	An observatory in polar solar orbit that studies the solar wind throughout the heliosphere.
Wind	NASA	ISTP[b]	11/1/1994	Traveling a path with multiple loops around L1, Earth, and L2	Investigating solar phenomena and their effects on Earth's magnetosphere. Provides almost continuous monitoring of the solar wind.

[a]European Space Agency
[b]International Solar Terrestrial Physics
[c]Japan Aerospace Exploration Agency

SOURCE: Created by Kim Weldon for Thomson Gale, 2004

closest to x-rays in the electromagnetic spectrum. *SOHO* sensors can detect waves of pressure under the Sun's surface, a process called helioseismology.

SOHO has captured hundreds of images of solar events, such as solar flares and coronal mass ejections, and has discovered more than 400 comets. The spacecraft is operated by NASA and the ESA. Dozens of universities and research institutions worldwide participate in *SOHO* as part of the ISTP initiative. *SOHO* is expected to remain in orbit at least until 2013.

Wind is an observational spacecraft that carries an array of scientific instruments. Following its launch in 1994 *Wind* was designed to follow a complicated path that winds back and forth between the L1 and L2 points. The path is a figure-eight shape with numerous loops around Earth in between. *Wind* collects data on plasma, energetic particles, and Earth's magnetic field. The

spacecraft provides nearly continuous monitoring of the solar wind approaching Earth.

Ulysses is a multi-functional spacecraft on a far-flying mission. It travels around the Sun on an orbital path that reaches out near Jupiter. As shown in Figure 6.12 the Earth and Jupiter orbit the sun in the same plane. This is called the ecliptic plane and appears as a circle within the diagram. All of the planets except Pluto orbit the Sun in the ecliptic plane. *Ulysses* follows a different path that takes it high above and far below the ecliptic plane as the spacecraft circles the Sun.

This provides *Ulysses* with exposure to vast regions of the heliosphere. The spacecraft investigates the magnetic field and solar wind at various heliospheric altitudes. It can also detect bursts of radio and plasma waves, solar x-rays, and solar and galactic cosmic rays. During its Jupiter flybys the spacecraft studies the magnetic fields surrounding

FIGURE 6.12

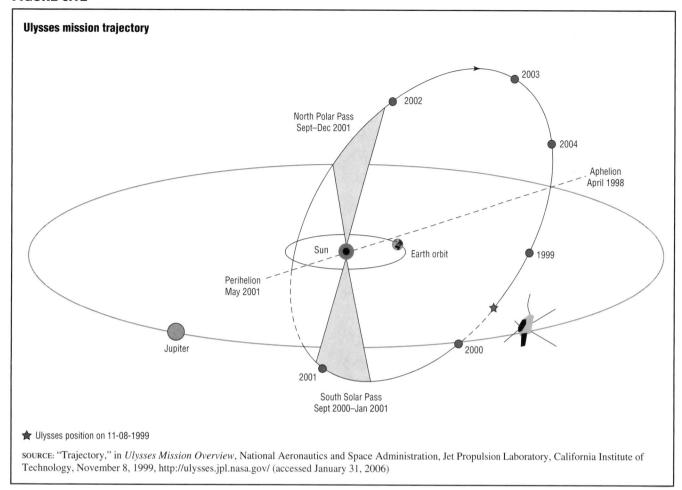

Ulysses mission trajectory

North Polar Pass
Sept–Dec 2001

2002

2003

2004

Aphelion
April 1998

Sun

Earth orbit

Perihelion
May 2001

1999

Jupiter

2000

2001

South Solar Pass
Sept 2000–Jan 2001

★ Ulysses position on 11-08-1999

SOURCE: "Trajectory," in *Ulysses Mission Overview*, National Aeronautics and Space Administration, Jet Propulsion Laboratory, California Institute of Technology, November 8, 1999, http://ulysses.jpl.nasa.gov/ (accessed January 31, 2006)

that planet. *Ulysses* is also unique because it flies over the polar regions of the Sun. This provides the observatory with a high-latitude view of the Sun's outer atmosphere.

Ulysses is a joint mission operated by NASA and the ESA. The spacecraft is named after a hero of Greek legend who went on a long sea journey and had many adventures along the way. The story is contained in a book called *Odyssey*, written by Homer, an epic poet who lived in the ninth or eighth century BC.

Future Solar Missions

Between 2006 and 2010 NASA plans to launch a number of solar missions as part of its "Living with a Star" program to study the solar maximum of 2011–12.

CHAPTER 7
MARS

Mars moves through our skies in its stately dance, distant and enigmatic, a world awaiting exploration.

—Astronomer Carl Sagan, 1967

Mars has been a mystery to humans for thousands of years. Although we know much about it now, there is still much more to learn. Mars is the fourth planet from the Sun, and the planet most like Earth in the solar system. It is named after the mythical god of war whom the Romans called Mars and the Greeks called Ares. Mars is also known as the Red Planet, because it looks reddish from Earth. Mars is a dusty, cold world. The average temperature is −64° Fahrenheit. Rays of ultraviolet radiation beat down on the surface continuously. The atmosphere is nearly all carbon dioxide.

People on Earth have always been fascinated with the idea of life on Mars. Ancient people could see Mars as a pale reddish light in the nighttime sky. They believed that it was stained with the blood of fallen warriors. Once telescopes were invented people had a better view of the planet, but many still thought it was inhabited. Patterns of straight lines could be seen on the surface. To some these were evidence of water canals dug into the ground by hard-working Martians. The notion lingered for decades in the public imagination.

At the dawn of the Space Age, humans sent robotic probes to Mars to settle the question once and for all. These probes found a frozen wasteland of fine powdery dust. Neither canals nor Martians could be located. There was some water vapor in the atmosphere and some frozen water at the planet's poles. Where there is water, there is potential for life similar to that found on Earth. Scientists continue to send probes to search for water and life.

In January 2004 President George W. Bush proposed that astronauts travel to Mars and explore the planet. It will be expensive and difficult. It takes six months to fly to Mars. The United States will need new rockets and spacecraft and some clever ways to keep astronauts healthy and happy on such a long journey. These are great challenges, but the idea is tantalizing—humans standing on another planet. Finally, there would be some life on Mars.

EARLY TELESCOPIC VIEWS OF MARS

The Italian astronomer Galileo Galilei (1564–1642) was probably the first to see Mars through a telescope. He noticed that sometimes it appeared larger than at other times. He believed that its distance from Earth was changing over time. During the seventeenth century Johannes Kepler (1571–1630) studied Mars's movement for years. His observations helped him to develop the laws of planetary motion for which he would become famous.

As telescopes improved, astronomers reported seeing dark and light patches on Mars that also varied in size over time. Some people thought these were patches of vegetation changing in response to the changing seasons. Others believed that they represented contrasting areas of land and sea.

In 1659 Dutch astronomer Christiaan Huygens (1629–95) recorded his Mars observations and noticed an odd-shaped feature that came to be called the hour-glass sea. Huygens kept an eye on the location of the sea over time and figured out that the Martian day lasts about twenty-four hours. The same conclusion was reached independently by the French astronomer Giovanni Cassini (1625–1712).

During the 1700s astronomers performed more detailed observations of the dark and light patches on Mars, particularly the whitish spots at the north and south poles. They could see that the spots changed in size over time, but did not guess that these were caps of ice. It was commonly believed that Mars was inhabited by some kind of beings. In 1774 the English astronomer Frederic

William Herschel (1738–1822) speculated that Martians lived on a world much like Earth, with oceans on the surface and clouds flying overhead.

GIOVANNI SCHIAPARELLI

During the late 1800s Mars became the topic of a debate that would go on for decades. The controversy was sparked by the observations of an Italian astronomer named Giovanni Schiaparelli (1835–1910). He created some of the first maps of the planet and assigned names to prominent features. Schiaparelli's naming system relied on place names taken from the Bible and ancient mythology.

Schiaparelli said he saw straight lines on the Martian surface and called them *canali*. In Italian *canali* could mean either channels or canals. Many people interpreted the word to mean that there were artificial canals on Mars. The Suez Canal had recently been constructed in Egypt to connect the Mediterranean Sea to the Red Sea. Obviously if there were artificial canals on Mars, they had been built by Martians.

Some other Mars observers also reported seeing the canals and claimed they connected dark and light patches on the planet. This reinforced the mistaken idea that the patches were areas of land and water.

ASAPH HALL

In August 1877 the American astronomer Asaph Hall (1829–1907) discovered that Mars has two moons. Centuries before him, Johannes Kepler guessed that Mars had two moons, but this had never been verified.

Hall reported that the moons were very small and orbited close to the planet's surface. This had made them impossible to see before. Hall made the discovery using a powerful new telescope recently installed at the U.S. Naval Observatory in Washington, D.C. He named the moons Phobos (meaning fear) and Deimos (meaning flight or panic). These are two characters mentioned in ancient Greek mythology as being servants to the god Mars.

PERCIVAL LOWELL

Percival Lowell (1855–1916) was a mathematician and amateur astronomer who greatly popularized the idea that Mars was inhabited by intelligent beings.

In 1894 he founded the Lowell Observatory in Flagstaff, Arizona. Perched at an altitude of 7,000 feet, the observatory provided one of the best views yet of the cosmos, including Mars. For fifteen years Lowell studied the Red Planet and wrote about his observations. He was convinced that there were artificial canals on Mars, because he could see hundreds of straight lines on the surface that intersected large patches of contrasting colors. Lowell argued that the canals must have been built to move water from the melting polar cap toward desert regions near the planet's equator.

Lowell publicized his theories in numerous articles and three popular books: *Mars* (1895), *Mars and Its Canals* (1906), and *Mars as the Abode of Life* (1908). In 1901 Lowell constructed a globe of Mars showing large geological features crisscrossed by a network of lines that intersected at certain points around the world. Lowell called these intersections oases, because he imagined they were fertile green areas amidst the desert.

In a series of articles published during 1895 in the *Atlantic Monthly*, Lowell explained his theories in detail. He believed that Mars had a thin air-based atmosphere containing lots of water vapor and that fresh water flowed from the polar ice caps through irrigation canals built by the highly intelligent Martians.

INHABITED OR NOT?

Lowell's ideas were not shared by most astronomers of the time. In 1908 the distinguished journal *Scientific American* noted that "Lowell is practically alone in the astronomical world in believing that he has proven that Mars is inhabited."

Scientists pointed out that artificial canals would have to be hundreds of miles wide to be visible from Earth. Furthermore, Mars was believed to be extremely cold, because of its great distance from the Sun. This made it even more unlikely that open flowing water was present on the planet's surface. Agnes Clerk (1842–1907) was a science writer trained in astronomy who wrote well-respected books about astronomical discoveries. She called Lowell's idea "hopelessly unworkable."

Lowell did have his supporters. The French astronomer Camille Flammarion (1842–1925) also believed that the lines on Mars were artificial canals built by an advanced civilization. Flammarion insisted that the reddish appearance of Mars was due to the growth of red vegetation on the planet.

In 1907 natural scientist Alfred Russel Wallace (1823–1913) wrote the book *Is Mars Habitable?*, which examined Lowell's claims one by one and attacked them with scientific data and reasoning. The book is considered a pioneering work in the field of exobiology (the investigation of possible life outside the Earth).

Wallace argued that Mars was a frozen desert and that the polar caps were probably frozen carbon dioxide, instead of water ice. Wallace ended the book with the following definitive statement: "Mars, therefore, is not only uninhabited by intelligent beings such as Mr. Lowell postulates, but is absolutely UNINHABITABLE."

Many astronomers of the time admitted seeing fine lines on the Martian surface. Most believed that these lines were either natural geological features or an optical illusion. Astronomer W. H. Pickering (1858–1938) believed the lines to be cracks in Mars's volcanic crust. He speculated that hot gases and water escaped through the cracks and supported vegetative growth. This explained the appearance of different-colored splotches on the planet. The general public and science fiction writers much preferred Lowell's explanation.

MARS IN SCIENCE FICTION

Around the turn of the nineteenth century Mars became a popular topic of science fiction. Before that time there is little mention of the Red Planet. One notable exception is a fanciful story published in 1726 by the Irish writer Jonathan Swift (1667–1745). *Gulliver's Travels* mentions the discovery of two Martian moons by astronomers living on the fictional island of Laputa. Oddly enough, Mars does have two moons, but they were not discovered until 151 years after the book was written.

After Schiaparelli and Lowell popularized the idea of intelligent life on Mars, science fiction writers gleefully embraced the notion. In 1898 H. G. Wells (1866–1946) portrayed Martians as lethal invaders in *War of the Worlds*. The insect-like creatures come to Earth looking for water and leave destruction in their path. They are finally wiped out by a common Earth germ to which they do not have immunity. The story was famously adapted for radio in 1938 and for film more than once, most recently in 2005.

Beginning in 1910 Edgar Rice Burroughs (1875–1950) wrote a series of adventure books in which an Earth man battles and romances his way around Mars. In his books the planet is called Barsoom by its exotic inhabitants. They come in various shapes, sizes, and colors. Some of them are green.

In 1924 the motion picture *Aelita: Queen of Mars* debuted in the Soviet Union. It features an engineer who takes a spaceship to Mars and falls in love with the planet's beautiful queen. At the end of the film he wakes up and discovers the journey was just a dream.

Three decades later Martians became popular characters in American media. Ray Bradbury published a series of stories called *The Martian Chronicles* in which well-intentioned humans travel to Mars and accidentally spread deadly Earth germs among the Martian population. It was an interesting twist on the theme introduced by H. G. Wells a half century before.

Evil invaders from Mars were common villains in low-budget horror movies of the 1950s. Historians now believe that these sinister creatures symbolized the threat that Americans felt from the Soviet Union during the Cold War. During the early 1960s the television show *My Favorite Martian* featured a friendly and wise Martian who crash lands on Earth and befriends a newspaper reporter.

SCIENTIFIC FACTS ABOUT MARS

Mars is a small planet. Its diameter is about half that of Earth. Mars is twice as large as Earth's moon.

A Martian day lasts twenty-four hours and thirty-nine minutes and is called a sol. It takes Mars 687 days to travel around the Sun. The planet has different seasons throughout its orbit, because it is tilted, just like Earth. During a Martian winter, at the poles the temperature can drop to −200° Fahrenheit. At the equator during the summer, the temperature can reach 80° Fahrenheit during the day.

The force of gravity is much weaker on Mars than it is on Earth. An astronaut standing on Mars would feel only 38% as much gravity as on Earth.

Martian Geology and Atmosphere

Mars is called a terrestrial planet, because it is composed of rocky material, like Mercury, Venus, and Earth. Mars has some of the same geological features as Earth, including volcanoes, valleys, ridges, plains, and canyons.

Most Martian features have two-word names. One of the words is a geological term, and is usually from Latin or Greek, for example *Mons* for mountain, *Planitia* for plains, and *Vallis* for valley. The other word comes from the classical naming system begun by Schiaparelli during the 1800s or from later astronomers. Beginning in 1919, the International Astronomical Union (IAU) became the official designator of names for celestial objects and the features upon them. Only the IAU has that authority.

There are two particularly prominent features on Mars. One is a volcano called Olympus Mons that is about seventeen miles high. This is three times higher than Mount Everest on Earth. In English Olympus Mons means Mount Olympus. This was the home of the gods in ancient Greek mythology. The other notable geological feature on Mars is a canyon called Valles Marineris (Mariner Valleys in English). This enormous canyon is 2,500 miles long by sixty miles wide and up to six miles deep in places. It was named after the Mariner spacecraft that photographed it during the 1960s.

The surface of Mars is covered with a fine powdery dust with a pale reddish tint. This is due to the presence of oxidized iron minerals (like rust) on the planet's surface. The Martian atmosphere is thin and contains more than 95% carbon dioxide. There is a tiny amount of oxygen, but not enough for humans to breathe. It is windy on Mars. Strong winds sometimes engulf the planet in dust storms that turn the atmosphere a hazy yellowish-brown color. The wind also blows clouds around the sky.

The poles of Mars are covered by solid carbon dioxide (dry ice) layered with dust and water ice. These polar caps change in size as the seasons change. Sometimes during the summer the uppermost dry ice evaporates away, only to reform again when the weather turns cold.

Martian Moons

The two Martian moons Phobos and Deimos are not round spheres like Earth's Moon. They are shaped like lopsided potatoes. Phobos is approximately 5,800 miles from Mars, while Deimos is nearly 15,000 miles away.

The Martian moons are very small compared to other moons in the solar system. Many scientists believe that Phobos and Deimos are actually asteroids that wandered too close to Mars and were captured by its gravity. There is a large asteroid belt located between the orbits of Mars and Jupiter. This could be where Phobos and Deimos originated.

Mars in Orbit and Opposition

Because their orbital paths are different, the Earth and Mars each take a different amount of time to complete an orbit around the Sun. This means that Mars and Earth are constantly changing position in relation to each other. At their most distant point the two planets are 233 million miles apart. At their closest point they are less than thirty-five million miles apart. This explains why in some years Mars looks closer to Earth than in others. During August 2003 Mars was only 34.7 million miles from Earth. It will not be that close again until the year 2287.

About every twenty-six months the Sun, Earth, and Mars line up in a row with Earth lying directly in the middle. This configuration is called Mars in opposition. It means that Mars is closer to Earth than usual and easier to observe. Most of the historic discoveries about Mars occurred when the planet was in opposition. This was particularly true for the 1877 opposition associated with the findings of Schiaparelli and Hall. Scientists now know that Mars was only thirty-five million miles from Earth during that opposition.

The most recent Mars opposition occurred in November 2005. The next one will be in December 2007. Oppositions are the best times to send spacecraft to Mars.

MISSIONS TO MARS

After the Moon the planet Mars was the destination of choice during the early days of space travel. The Soviet Union was particularly eager to reach the Red Planet before the United States. A historical log of all Mars missions is presented in Table 7.1.

Many Failures

Historically, spacecraft have had a difficult time making it to Mars in working order and staying that way. As shown in Table 7.1, more than half of the missions intended for Mars have failed for one reason or another. Some were plagued by launch problems, while others suffered malfunctions during flight, descent, or landing.

Mars missions undertaken during the 1960s by the former Soviet Union were particularly trouble-prone. All six of them failed. Although the next decade showed some improvement, little usable data were obtained from the spacecraft that reached their destination. The one attempt to reach Mars by the Russian space agency, in 1996, failed when the spacecraft was unable to leave Earth orbit.

In contrast to the Soviet Union's Mars attempts, most NASA Mars missions conducted during the 1960s achieved their objectives. There was also notable success over the next decade with the Viking spacecraft. There was a long lull after that in NASA's Mars exploration program.

During the 1990s NASA launched five separate missions to Mars. Their names were *Mars Observer* (1992), *Mars Global Surveyor* (1996), *Mars Pathfinder* (1996), *Mars Climate Orbiter* (1998), and *Mars Polar Lander* (1999). Only two of the missions were successful (*Mars Global Surveyor* and *Mars Pathfinder*). The other spacecraft were lost on arrival.

NASA lost contact with the *Mars Observer* just before it was to go into orbit around Mars. It is believed that some kind of fuel explosion destroyed the spacecraft as it began its maneuvering sequence. The *Observer* carried a highly sophisticated gamma-ray spectrometer designed to map the Martian surface composition from orbit. Failure of the mission resulted in a loss estimated at $1 billion. This was by far the most expensive of NASA's failed Mars missions.

In June 1999 the *Mars Climate Orbiter* was more than sixty miles off-course when it ran into the Martian atmosphere and was destroyed. The loss of the $85 million spacecraft was particularly embarrassing for NASA, because it was due to human error. An investigation revealed that flight controllers had made mistakes doing unit conversions between metric units and English system units. This resulted in erroneous steering commands being sent to the spacecraft. Outside investigators complained that the problem was larger than some mathematical errors. They blamed overconfidence and poor oversight by NASA management during the mission.

NASA's embarrassment deepened a few months later when the *Mars Polar Lander* was lost. The loss was attributed to a software problem that caused the spacecraft to think it had touched down on the surface even though it had not. The computer apparently shut down the engines during descent and let the spacecraft plummet

TABLE 7.1

Historical log of Mars expeditions, 1960–2005

Mission	Country	Launch date	Purpose	Results
[Unnamed]	USSR	10/10/1960	Mars flyby	Did not reach Earth orbit
[Unnamed]	USSR	10/14/1960	Mars flyby	Did not reach Earth orbit
[Unnamed]	USSR	10/24/1962	Mars flyby	Achieved Earth orbit only
Mars 1	USSR	11/1/1962	Mars flyby	Radio failed at 65.9 million miles (106 million km)
[Unnamed]	USSR	11/4/1962	Mars flyby	Achieved Earth orbit only
Mariner 3	U.S.	11/5/1964	Mars flyby	Shroud failed to jettison
Mariner 4	U.S.	11/28/1964	First successful Mars flyby 7/14/65	Returned 21 photos
Zond 2	USSR	11/30/1964	Mars flyby	Passed Mars but radio failed, returned no planetary data
Mariner 6	U.S.	2/24/1969	Mars flyby 7/31/69	Returned 75 photos
Mariner 7	U.S.	3/27/1969	Mars flyby 8/5/69	Returned 126 photos
Mariner 8	U.S.	5/8/1971	Mars orbiter	Failed during launch
Kosmos 419	USSR	5/10/1971	Mars lander	Achieved Earth orbit only
Mars 2	USSR	5/19/1971	Mars orbiter/lander arrived 11/27/71	No useful data, lander destroyed
Mars 3	USSR	5/28/1971	Mars orbiter/lander, arrived 12/3/71	Some data and few photos
Mariner 9	U.S.	5/30/1971	Mars orbiter, in orbit 11/13/71 to 10/27/72	Returned 7,329 photos
Mars 4	USSR	7/21/1973	Failed Mars orbiter	Flew past Mars 2/10/74
Mars 5	USSR	7/25/1973	Mars orbiter, arrived 2/12/74	Lasted a few days
Mars 6	USSR	8/5/1973	Mars orbiter/lander, arrived 3/12/74	Little data return
Mars 7	USSR	8/9/1973	Mars orbiter/lander, arrived 3/9/74	Little data return
Viking 1	U.S.	8/20/1975	Mars orbiter/lander, orbit 6/19/76–1980, lander 7/20/76–1982	Combined, the Viking orbiters and landers returned 50,000+ photos
Viking 2	U.S.	9/9/1975	Mars orbiter/lander, orbit 8/7/76–1987, lander 9/3/76–1980	Combined, the Viking orbiters and landers returned 50,000+ photos
Phobos 1	USSR	7/7/1988	Mars/Phobos orbiter/lander	Lost 8/88 en route to Mars
Phobos 2	USSR	7/12/1988	Mars/Phobos orbiter/lander	Lost 3/89 near Phobos
Mars Observer	U.S.	9/25/1992	Orbiter	Lost just before Mars arrival 8/21/93
Mars Global Surveyor	U.S.	11/7/1996	Orbiter, arrived 9/12/97	Currently conducting prime mission of science mapping
Mars 96	Russia	11/16/1996	Orbiter and landers	Launch vehicle failed
Mars Pathfinder	U.S.	12/4/1996	Mars lander and rover, landed 7/4/97	Last transmission 9/27/97
Nozomi (Planet-B)	Japan	7/4/1998	Mars orbiter	Could not achieve Martian orbit due to propulsion problem
Mars Climate Orbiter	U.S.	12/11/1998	Orbiter	Lost on arrival at Mars 9/23/99
Mars Polar Lander/ Deep Space 2	U.S.	1/3/1999	Lander/descent probes to explore Martian South Pole	Lost on arrival 12/3/99
Mars Odyssey	U.S.	4/7/2001	Orbiter	Currently conducting prime mission of science mapping
Mars Express	Europe	6/2/2003	Orbiter and Beagle 2 Lander	Orbiter currently collecting planetary data. Beagle 2 lost during descent.
Mars Exploration	U.S.	6/10/03 (Spirit) and 7/7/03 (Opportunity)	Two rovers: Spirit and Opportunity	Rovers landed in January 2004. Currently exploring planet surface.
Mars Reconnaissance Orbiter	U.S.	8/12/2005	Orbiter	Scheduled to arrive in March 2006

SOURCE: Adapted from "Historical Log," in *NASA's Mars Exploration Program*, National Aeronautics and Space Administration, Jet Propulsion Laboratory, California Institute of Technology, October 13, 2005, http://marsprogram.jpl.nasa.gov/missions/log/ (accessed January 31, 2006)

at high speed into the ground, where it was destroyed. The cost of the failed spacecraft was estimated at $120 million.

THE MARINER PROGRAM

The Mariner program included a series of spacecraft launched by NASA between 1962 and 1973 to explore the inner solar system (Mercury, Venus, and Mars). These were relatively low-cost missions conducted with small spacecraft launched atop Atlas-type rockets. Each spacecraft weighed between 400 and 2,200 pounds. They were designed to operate for several years and collect specific scientific data about Earth's nearest planetary neighbors.

Six of the Mariner spacecraft were scheduled for Martian missions. Two of these missions failed. In 1964 *Mariner 3* malfunctioned after take-off and never made it to Mars. In 1971 *Mariner 8* failed during launch. This left four successful Mars Mariner missions: *Mariner 4, 6, 7,* and *9*.

Mariner 4

In July 1965 *Mariner 4* was launched into a solar orbit and achieved the first successful flyby of Mars. A planetary flyby mission is one in which a spacecraft is put on a trajectory that takes it near enough to a planet for detailed observation, but not close enough to be pulled in by the planet's gravity.

During its flyby, *Mariner 4* took nearly two dozen photos, the first close-ups ever obtained of Mars. They showed a world pock-marked with craters, probably from meteor strikes.

Mariner 6 and 7

In 1969 *Mariner 6* and *Mariner 7* conducted a dual mission to Mars. Both spacecraft flew by the planet, and together sent back more than 200 photos. These photos revealed that the features once thought to be canals were not canals after all. Instead, it appears that a number of small features or shadows on Mars only looked like they were aligned when viewed through telescopes from Earth. The illusion was perpetuated by a human tendency to see order in a random collection of shapes. The mystery of the *canali* had finally been solved.

Mariner 9

Mariner 9 turned out to be the most fruitful of the Mariner missions. In November 1971 the spacecraft went into orbit around Mars after a five-and-a-half-month flight from Earth. It was the first artificial satellite ever to be placed in orbit around the planet.

When it first arrived *Mariner 9* found the entire planet engulfed in a massive dust storm. The spacecraft remained in orbit for nearly a year and returned more than 7,000 photos of the planet's surface. For the first time scientists got a good look at Mars's surface features, such as volcanoes and valleys. *Mariner 9* showed geological features that looked like dry flood channels. It also captured the first close-up photos of the Martian moons Phobos and Deimos.

Scientists learned from Mariner data that Mars had virtually no magnetic field and was bombarded with ultraviolet radiation. Earth's extensive magnetic field (or magnetosphere) helps protect our planet from dangerous electromagnetic radiation traveling through space. Scientists knew that lack of such protection on Mars would make it exceedingly difficult for life to exist on the planet.

THE VIKING MISSION

Within only four years NASA went from orbiting Mars to a landing on the planet. In 1976 the Viking mission was the first American spacecraft to land safely on Mars. For the mission NASA built two identical spacecraft, each containing an orbiter and lander. The two spacecraft entered orbit around Mars and released their landers to descend to the planet's surface.

The spacecraft launched only weeks apart during late summer 1975. It took them nearly a year to reach Mars. On July 20, 1976, the *Viking 1* lander set down on the western slope of Chryse Planitia (the Plains of Gold). On September 3, 1976, the *Viking 2* lander set down at Utopia Planitia (the Plains of Utopia).

The landers provided NASA with constant weather reports. They detected nitrogen in the atmosphere. Scientists reported that a thin layer of water frost formed on the ground during the winter near the *Viking 2* lander. Temperatures varied between −184° Fahrenheit in the winter to 7° Fahrenheit in the summer at the lander locations.

The orbiters mapped 97% of the Martian surface and observed more than a dozen dust storms. Scientists examined Viking images and decided that some geologic features on Mars could have been carved out millions of years ago by flowing water. The *Viking 2* orbiter continued functioning until 1978. The *Viking 1* orbiter lasted another two years. The *Viking 1* lander continued to make transmissions to Earth until 1982.

The landers were unique because they were powered by generators that created electricity from heat released during the natural decay of plutonium, a radioactive element. This method of power generation was selected because NASA feared that sunlight on the planet would not be consistent enough to provide solar power.

The Viking orbiters carried high-resolution cameras and were able to map atmospheric water vapor and surface heat from orbit. The landers included cameras and a variety of scientific instruments designed to investigate seismology, magnetic properties, meteorology, atmospheric conditions, and soil properties. They also tested for the presence of living microorganisms in the soil, but found no clear evidence of them. They did learn that the surface of Mars contains iron-rich clay. The Viking images revealed that Mars has a light yellowish-brown atmosphere due to the presence of airborne dust. In other words, the Red Planet is actually more the color of butterscotch.

Many scientists associated with the Viking project concluded that Mars is "self-sterilizing." This means that the natural planetary conditions are such that living organisms cannot form. The high radiation levels and the unique soil chemistry are actually destructive to life. The Martian soil was found to be extremely dry and oxidizing. Oxidizing agents destroy organic chemicals considered necessary for life to form. The self-sterilizing theory was not universally accepted, however, and remains controversial.

ALH84001

During the 1980s NASA was busy running the space shuttle program. There was no money to send spacecraft to Mars. Luckily, a piece of Mars turned up on Earth. On December 27, 1984, a meteorite hunter found a four-pound rock on the Allan Hills ice field in Antarctica (the South Pole). The rock was grayish-green and covered with pits and gouges. It was given the designation ALH84001.

The National Science Foundation (NSF) conducts annual searches in Antarctica looking for meteorites (rocks that have traveled through space to Earth). Each possible candidate is collected and assigned a tracking code. The letters in the tracking code represent the location of the find (Allan Hills). The first two numbers stand for the year of the find (1984). The last digits indicate the order in which rocks were processed by the NSF that year. ALH84001 was recognized immediately as a significant find, so it was the first rock investigated during that sampling year. In fact, the person who found it wrote "Yowza-Yowza" across the sample notes.

Scientists also got excited when they examined the rock, because they found gas trapped within it that matched the known atmosphere of Mars. They concluded that the rock formed on Mars 4.5 billion years ago. About sixteen million years ago an asteroid probably slammed into the planet and sent the rock hurtling through space. Scientists believe that the rock arrived on Earth 13,000 years ago.

The rock contains a small amount of carbonate (a carbon-containing compound). Some scientists believe that the carbonate formed inside the rock in the presence of liquid water. This would mean that liquid water existed on Mars billions of years ago. The exact origin of the carbonate is still under debate.

Rocks determined to be meteorites are kept in special laboratories at NASA or the Smithsonian Institution. In all there have been thirty-six known Martian meteorites found on Earth since 1815. ALH84001 is the oldest meteorite in the collection.

MARS GLOBAL SURVEYOR

More than twenty years passed between the launch of the highly productive Viking missions and another successful mission to Mars. In November 1996 the *Mars Global Surveyor* took off from Cape Canaveral, Florida, atop a Delta II rocket. The spacecraft arrived near the planet ten months later. To save on fuel the *Global Surveyor* was put into its final Martian orbit very slowly through a process called aerobraking.

During aerobraking a spacecraft is repeatedly skimmed through the thin upper atmosphere surrounding a planet. Each skim reduces the speed of the craft due to frictional drag. Aerobraking eliminates the need for extra fuel to do a retro-burn to slow down a spacecraft.

The *Global Surveyor* was put through a very long series of gentle skims for a year and a half. Generally aerobraking does not take this long. Flight controllers were extremely careful with the *Surveyor*, because one of its solar panels did not fully deploy during flight. Scientists were afraid that aggressive skimming might put too much stress on the panel.

In March 1999 the spacecraft began its mapping mission. This continued for one Martian year (687 days). The most significant finding during mapping was images of gullies and other flow features that scientists believe may have been formed by flowing water. *Surveyor* also captured close-up photographs of Mars's moon Phobos. The images reveal that the moon is covered with at least three feet of powdery material.

In April 2002 *Global Surveyor* began performing data relay and imaging services for other NASA spacecraft carrying out missions at Mars. As of 2006 *Surveyor* was still in orbit and functioning properly.

MARS PATHFINDER

Mars Pathfinder was a mission conducted as part of NASA's Discovery program. This was the agency's "faster, better, cheaper" approach to space science. The mission was developed in only three years and cost $265 million. On December 4, 1996, the spacecraft launched atop a Delta II rocket from the Cape Canaveral Air Station in Florida. *Pathfinder* traveled for seven months before entering into the gravitational influence of Mars.

On July 4, 1997, the spacecraft was ordered to begin its descent to the planet's surface. A giant parachute released to slow its fall. A landing craft separated from the spacecraft shell and began to drop. Eight seconds before hitting the ground the lander's air bags deployed around it like a cocoon to cushion its impact on the surface. The lander ball bounced and rolled for several minutes before coming to a stop more than half a mile from where it first impacted. It was in a rocky flood plain known as Ares Vallis (Valley of Ares).

After the successful landing NASA renamed the lander the Carl Sagan Memorial Station, in memory of the late astronomer Carl Sagan (1934–96). He died while *Pathfinder* was en route to Mars. The lander unfolded three hinged solar panels onto the ground, as shown in Figure 7.1. It released a small six-wheeled rover named *Sojourner* that began exploring the nearby area. The name resulted from a NASA contest in which schoolchildren proposed names of historical heroines for the mission. The winning entry suggested Sojourner Truth, an African-American woman who crusaded for civil rights during the 1800s.

For two and a half months the *Sojourner* rover collected data about Martian soil, radiation levels, and rocks. The robotic machine weighed twenty-three pounds and

FIGURE 7.1

The Pathfinder spacecraft

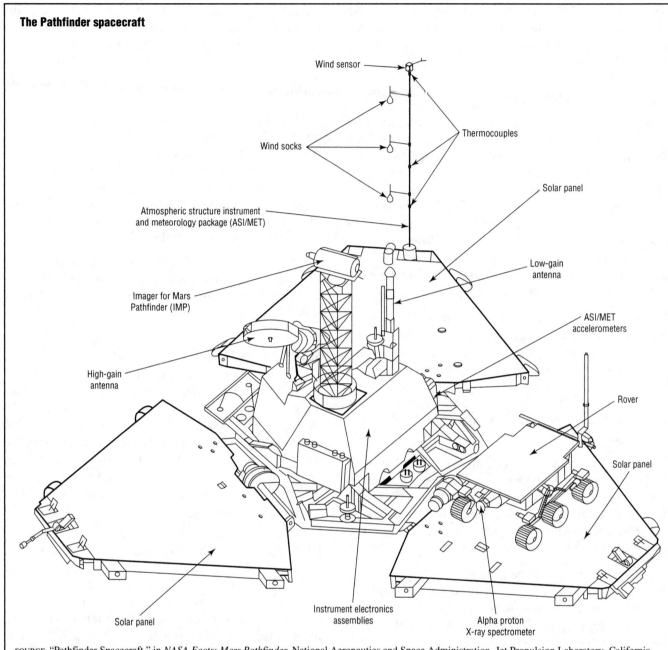

Wind sensor

Thermocouples

Wind socks

Solar panel

Atmospheric structure instrument
and meteorology package (ASI/MET)

Low-gain
antenna

Imager for Mars
Pathfinder (IMP)

ASI/MET
accelerometers

High-gain
antenna

Rover

Solar panel

Solar panel

Instrument electronics
assemblies

Alpha proton
X-ray spectrometer

SOURCE: "Pathfinder Spacecraft," in *NASA Facts: Mars Pathfinder*, National Aeronautics and Space Administration, Jet Propulsion Laboratory, California Institute of Technology, May 1999, http://www.jpl.nasa.gov/news/fact_sheets.cfm (accessed January 14, 2004)

could move at a top speed of two feet per minute. It was powered by a flat solar panel that rested atop its frame. (See Figure 7.2.)

Meanwhile the lander collected images and relayed data back to Earth. It also measured the amount of dust and water vapor in the atmosphere. The lander's forty-inch mast held little wind socks at different heights to determine variations in wind speed near the planet's surface. Magnets were mounted along the lander to collect dust particles for analysis. Scientists learned that airborne Martian dust is very magnetic and may contain the mineral maghemite, a form of iron oxide.

The *Pathfinder* returned more than 17,000 images and performed fifteen chemical analyses. Scientists studying this data concluded that Mars might have been warm and wet sometime in the past with a thicker, wetter atmosphere. In late September 1997, *Pathfinder* sent its last message home.

2001 MARS ODYSSEY

In 1895 Percival Lowell said "If Mars be capable of supporting life, there must be water upon his surface; for to all forms of life water is as vital a matter as air. To all organisms water is absolutely essential. On the question

FIGURE 7.2

The Sojourner rover

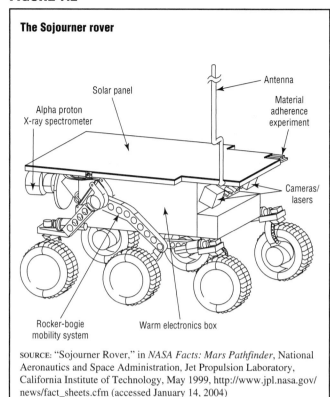

SOURCE: "Sojourner Rover," in *NASA Facts: Mars Pathfinder*, National Aeronautics and Space Administration, Jet Propulsion Laboratory, California Institute of Technology, May 1999, http://www.jpl.nasa.gov/news/fact_sheets.cfm (accessed January 14, 2004)

of habitability, therefore, it becomes all-important to know whether there be water on Mars."

A century later this same issue drove NASA to conduct its most extensive program of missions to the Red Planet, the Mars Exploration Program. This is a long-term program in which robotic explorers are used to investigate Mars in support of four science objectives:

- Determining whether life ever existed on Mars
- Characterizing the climate of Mars
- Characterizing the geology of Mars
- Preparing for future human exploration of Mars

The overall motto of the Mars Exploration Program is "Follow the Water." In other words, the mission scientists hope that discovery of liquid water on Mars will lead them to any microscopic life forms that exist on the planet or were ever present.

The *2001 Mars Odyssey* mission falls under NASA's Mars Exploration Program. The mission was named after the hit 1968 movie *2001: A Space Odyssey*, based on a short story by science fiction author Arthur C. Clarke.

On April 7, 2001, *Odyssey* was launched toward Mars atop a Delta II rocket. The spacecraft reached Mars six months later. To conserve fuel *Odyssey* was placed in Martian orbit via aerobraking. The spacecraft skimmed against the upper edge of the Martian atmosphere hun-

dreds of times over a three-month period to slow itself down.

In February 2002 *Odyssey* reached its final orbit and began mapping the planet's surface. The mission was intended to last for at least one Mars year. In early January 2004 *Odyssey* completed one Mars year in service. As of 2006 the spacecraft was still operational and functioning well nearly 275 million miles from Earth.

A schematic of the spacecraft is shown in Figure 7.3. It includes three scientific instruments: a thermal imaging system, a gamma ray spectrometer, and the Mars Radiation Environment Experiment (MARIE).

The thermal imaging system collects surface images in the infrared portion of the electromagnetic spectrum. Everything that has a temperature greater than absolute zero emits infrared radiation. Scientists use *Odyssey*'s images to identify and map minerals in the surface soils and rocks. This work is being coordinated with the mineral mapping being performed by the *Mars Global Surveyor*.

Odyssey's gamma ray spectrometer can detect the presence of various chemical elements on the planet's surface. This is particularly useful for finding water ice buried beneath the surface and for detecting salty minerals. *Odyssey* data indicate the presence of large amounts of water ice just beneath the surface in the polar regions. The MARIE instrument collects radiation data that will be useful to planning any future Mars expeditions by humans.

Odyssey's telecommunications system is performing a dual role. It transmits to NASA data collected by the spacecraft itself and also data collected by other NASA spacecraft conducting Mars missions.

THE PERIHELIC OPPOSITION OF 2003

Scientists all over the world knew that 2003 was going to be a good year to go to Mars, because Mars would be in opposition to Earth. On August 28, 2003, the Sun, Earth, and Mars were going to line up in a row. This happens every twenty-six months.

The opposition of 2003 was special, because it was going to occur while Mars was at its closest point to the Sun. This configuration is known as a perihelic opposition. When Mars is in perihelic opposition it is also much closer to Earth than usual. This means that less fuel and flight time are required to send a spacecraft from Earth to Mars near the time of a perihelic opposition.

Perihelic oppositions only happen every fifteen to seventeen years. During the late 1990s, the Japanese Space Agency, European Space Agency, and NASA began planning Mars missions to coincide with the perihelic opposition of 2003. The Japanese mission Nozomi

FIGURE 7.3

Scientific instruments on the Mars Odyssey Orbiter

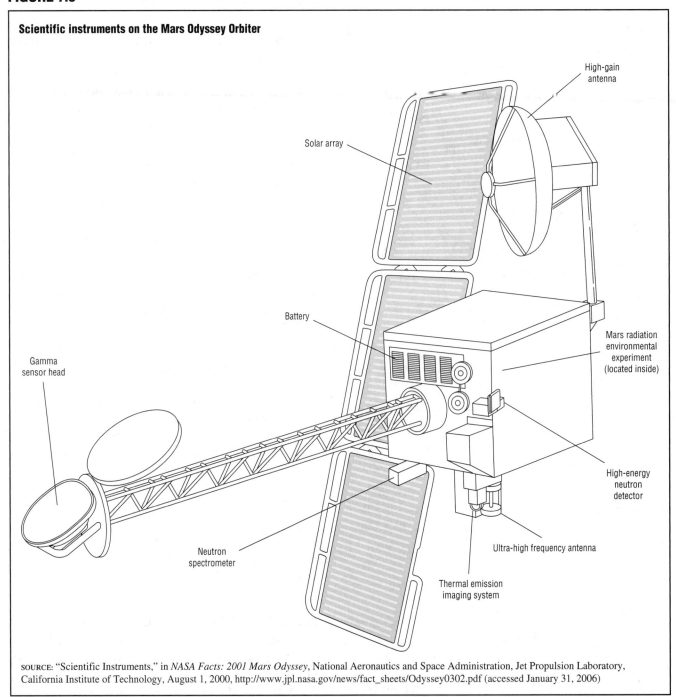

SOURCE: "Scientific Instruments," in *NASA Facts: 2001 Mars Odyssey*, National Aeronautics and Space Administration, Jet Propulsion Laboratory, California Institute of Technology, August 1, 2000, http://www.jpl.nasa.gov/news/fact_sheets/Odyssey0302.pdf (accessed January 31, 2006)

(which means hope) suffered radiation damage during its flight and never made it to Mars.

MARS EXPRESS

The *Mars Express* mission is the first mission to Mars by the European Space Agency (ESA). It was timed to put the spacecraft in flight near the time of Mars's perihelic opposition.

In June 2003 the spacecraft was launched toward Mars from the Baikonur launch pad in Kazakhstan.

A Russian Soyuz-Fregat rocket was used as the launch vehicle. The spacecraft included two parts—an orbiter and a lander named *Beagle 2*. The lander name was chosen in honor of the ship on which Charles Darwin (1809–92) traveled during the 1830s while exploring South America and the Pacific region on Earth.

In late November 2003 the *Mars Express* reached the planet's vicinity and prepared to go into orbit. On December 19, 2003, the *Beagle 2* was released from the orbiter. Six days later the lander entered the Martian atmosphere on its way to a landing site at Isidis Planitia

(Plains of Isis). Isis was the Egyptian goddess of heaven and fertility. The ESA lost contact with *Beagle 2* as it descended toward the planet. Repeated attempts to reestablish contact were made over the next few months, but were not successful.

The *Mars Express* orbiter achieved orbit to begin its mission of collecting planetary data. As of 2006 it was still operating. The orbiter carries seven instruments designed to investigate the Martian atmosphere and geological structure and to search for subsurface water. One of the instruments (called *ASPERA-3*) was supplied by NASA.

MARS EXPLORATION ROVERS

Another Mars mission began in 2003 with the launch of NASA's twin Mars Exploration Rovers (MERs). Each spacecraft carried a lander to Mars. Inside each lander was a rover about the size of a golf cart, designed to explore the Martian surface.

The rovers were named *Spirit* and *Opportunity*. The names were the winning entries in a naming contest NASA held in 2002. The winning entry came from a third-grade girl living in Scottsdale, Arizona. She was born in Russia and adopted by an American family. She chose the names to honor her feelings about America.

NASA adopted seven specific objectives for the MER missions:

- Find and sample rocks and soils that could reveal evidence of past water on the planet

- Characterize the composition of rocks, soils, and minerals near the landing sites

- Look for evidence of geological processes (such as erosion or volcanic activity) that could have shaped the Martian surface

- Use the rovers to verify data reported by the orbiters regarding Martian geology

- Probe for minerals containing iron or water or minerals known to form in water

- Analyze rocks and soils to characterize their mineral content and morphology (form and structure)

- Seek out clues about the geological history of the planet to determine whether watery conditions could have supported life

The Launches

Spirit launched first on June 10, 2003. *Opportunity* launched several weeks later on July 7, 2003. The launch dates were chosen to put the spacecraft in flight near the time of Mars's perihelic opposition.

Both spacecraft were launched atop Delta II rockets from Cape Canaveral Air Force Station in Florida.

Figure 7.4 shows a drawing of a Rover spacecraft being released by its rocket to make the journey to Mars.

The Flight Trajectories

The flight trajectories for the MERs were chosen to take advantage of the perihelic opposition configuration of the Sun, Earth, and Mars.

Figure 7.5 shows the flight trajectory for the *Spirit* spacecraft. The Sun is pictured at the middle of the diagram. Earth's orbit is the innermost circle, while Mars's orbit is the outer circle. Both planets travel in a counterclockwise direction around the Sun. The figure shows the position of the Earth and Mars at the time of *Spirit*'s launch and its arrival at Mars.

The Earth's path around the Sun is almost a perfect circle. Mars's path is more elliptical. The Sun does not sit at the center of the ellipse, but is offset to one side. The Earth travels around the Sun in 365.25 days. It takes Mars nearly twice that long to make an orbit (687 days). This is why Mars did not move as far around the Sun as the Earth during *Spirit*'s journey.

Figure 7.6 shows the flight trajectory for the *Opportunity* spacecraft. Both spacecraft were subjected to occasional flight maneuvers along the way to keep them on their path to intercept Mars in January 2004.

Notice that the MER missions took place when the paths of Mars and the Earth were relatively close to each other. The spacecraft would have taken longer and had to travel farther and use more fuel if scientists had timed them to occur when the orbital paths of Mars and Earth were farther apart.

Landing on Mars

Figure 7.7 shows the various parts of the spacecraft that traveled to Mars. Each rover was nestled inside a landing vehicle protected by an aeroshell connected to the cruise stage of the spacecraft. The cruise stage contained fuel tanks, solar panels, and the propulsion system for trajectory corrections during flight. The aeroshell included two parts, a back shell and a heat shield. The back shell carried a deceleration instrument to ensure that the parachute was deployed at the right altitude above the Martian surface. It also had some small rockets to stabilize the spacecraft as it fell. The heat shield protected the lander/rover package from the heat generated by entering the Martian atmosphere.

The stages of entry, descent, and landing are shown in Figure 7.8. At twenty-one minutes before landing (L-21 min) the cruise stage separated from the rest of the spacecraft. Fifteen minutes later the spacecraft entered the atmosphere about seventy-four miles above the surface. The parachute deployed at an altitude of five miles when the craft was traveling nearly 300 miles per hour. Seconds

FIGURE 7.4

Mars Exploration Rover spacecraft released by Delta II rocket

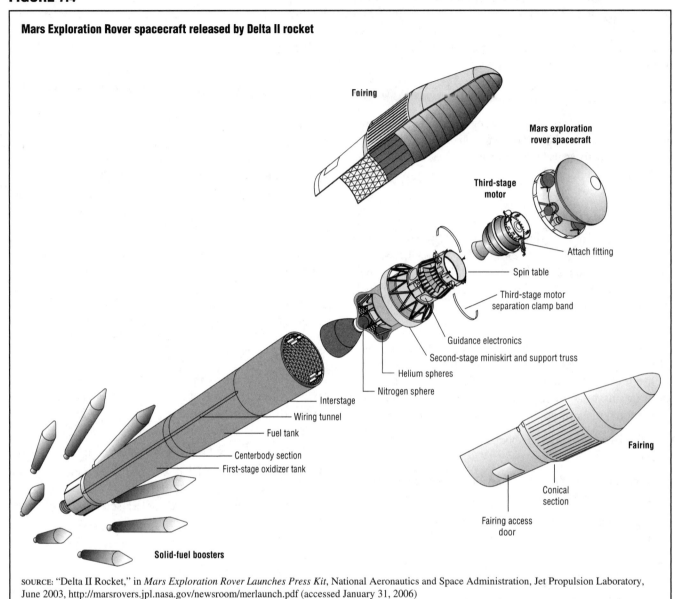

Fairing

Mars exploration
rover spacecraft

Third-stage
motor

Attach fitting

Spin table

Third-stage motor
separation clamp band

Guidance electronics

Second-stage miniskirt and support truss

Helium spheres

Nitrogen sphere

Interstage

Wiring tunnel

Fuel tank

Centerbody section

First-stage oxidizer tank

Solid-fuel boosters

Fairing

Conical
section

Fairing access
door

SOURCE: "Delta II Rocket," in *Mars Exploration Rover Launches Press Kit*, National Aeronautics and Space Administration, Jet Propulsion Laboratory, June 2003, http://marsrovers.jpl.nasa.gov/newsroom/merlaunch.pdf (accessed January 31, 2006)

later the heat shield was jettisoned away. Eight seconds before hitting the ground the spacecraft deployed its air bags to cushion its impact with the ground. Retro-rockets were fired to slow its descent. Three seconds later the parachute line was cut. The spacecraft ball bounced and rolled until it finally came to a stop. About an hour after landing the airbags were deflated and retracted so the lander could open its petal and release the rover.

On January 4, 2004, the *Spirit* MER landed on Mars. It was just after 8:30 PM at the mission control center in California. The landing site was in a crater named Gusev Crater in honor of the Russian astronomer Matvei Gusev (1826–66). The crater is about 100 miles in diameter and lies at the end of a long valley known as Ma'adim Vallis. This translates as Mars Valley, because Ma'adim is the Hebrew word for Mars. Major valleys on the Red Planet are named for the word Mars in different Earth languages.

On January 25, 2004, *Opportunity* set down near Mars's equator in an area called Meridiani Planum. Planum means plateau or high plains. The Meridiani Planum is considered the site of zero longitude on Mars. This is the longitude arbitrarily selected by astrogeologists to be the prime meridian for the rest of the planet. *Opportunity*'s landing site was nearly half way around Mars from Gusev Crater.

Both landing sites were chosen for their very flat terrain. Gusev Crater is of interest to scientists, because they believe it could be a dried-up lakebed. The Meridiani Planum is thought to contain a layer of hematite beneath the surface. Hematite is a gray iron ore mineral similar to red rust that on Earth usually only forms in a wet environment. Both landing sites were considered prime locations to look for evidence of ancient water.

FIGURE 7.5

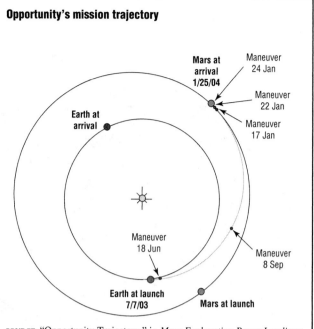

Spirit's mission trajectory

Mars at arrival 1/4/04
Maneuver 3 Jan
Maneuver 1 Jan
Maneuver 27 Dec
Maneuver 14 Nov
Earth at arrival
Maneuver 20 Jun
Earth at launch 6/10/03
Maneuver 1 Aug
Mars at launch

SOURCE: "Spirit Trajectory," in *Mars Exploration Rover Landings Press Kit*, National Aeronautics and Space Administration, Jet Propulsion Laboratory, January 2004, http://marsrovers.jpl.nasa.gov/newsroom/merlandings.pdf (accessed January 31, 2006)

FIGURE 7.6

Opportunity's mission trajectory

Mars at arrival 1/25/04
Maneuver 24 Jan
Maneuver 22 Jan
Maneuver 17 Jan
Earth at arrival
Maneuver 18 Jun
Maneuver 8 Sep
Earth at launch 7/7/03
Mars at launch

SOURCE: "Opportunity Trajectory," in *Mars Exploration Rover Landings Press Kit*, National Aeronautics and Space Administration, Jet Propulsion Laboratory, January 2004, http://marsrovers.jpl.nasa.gov/newsroom/merlandings.pdf (accessed January 31, 2006)

Roving *Spirit* and *Opportunity*

The components of an MER are labeled in Figure 7.9. The rovers are just over five feet long. The panoramic cameras sit about five feet above the ground atop a mast.

Each rover weighed about 380 pounds on Earth and carried a package of science instruments called an Athena science payload. Each payload includes two survey instruments, three instruments for close-up investigation of rocks, and a tool for scraping off the outer layer of rocks. The rovers were designed to move at a top speed of two inches per second. An average speed of 0.4 inches per second was expected when a rover was traveling over rougher terrain.

The rovers were designed to operate independently of their landers. Each rover carries its own telecommunications equipment, camera, and computer. The electronic equipment received power from batteries that were repeatedly recharged by solar arrays. It was late summer on Mars when the rovers began their mission. Scientists expected that power generation would taper off after about ninety sols (or ninety-two Earth days) and eventually stop as the arrays became too dust-coated to harness solar power. However, scientists were pleasantly surprised when dust devils kept sweeping by the rovers and blowing the dust off the arrays. These periodic cleanings have allowed the rovers to keep operating for much longer than expected.

The rovers completed their prime missions in April 2004. Since that time they have investigated dozens of additional sites. In May 2005 the *Opportunity* rover became stuck in a small sand dune when its wheels sank into soft sand and could not gain traction. Scientists worked for nearly five weeks to maneuver the rover back onto more solid ground. In September 2005 NASA reported that the *Spirit* rover had reached the summit of a Martian hill nearly 350 feet higher than where the rover landed. The hill was tentatively named "Husband Hill" in honor of Rick Husband, the commander of the doomed space shuttle *Columbia*. Scientists used the panoramic pictures captured from this vantage point to map out future exploration routes for the rover.

As of 2006 the rovers were still operating after a full Martian year (687 Earth days) in service. Each rover had traveled more than three miles from its original landing site. NASA scientists believe the rovers will operate indefinitely as long as their solar arrays continue to be cleaned by Mars's dust devils. Thus far, the planet has not experienced a global-wide dust storm during the MER missions. However, such storms do occur on Mars and could render the rovers inoperable.

The Name Game

Only the IAU has the authority to assign official names to planetary features. Major features, such as mountains, valleys, and large craters, have already been

FIGURE 7.7

Mars Exploration Rover flight system

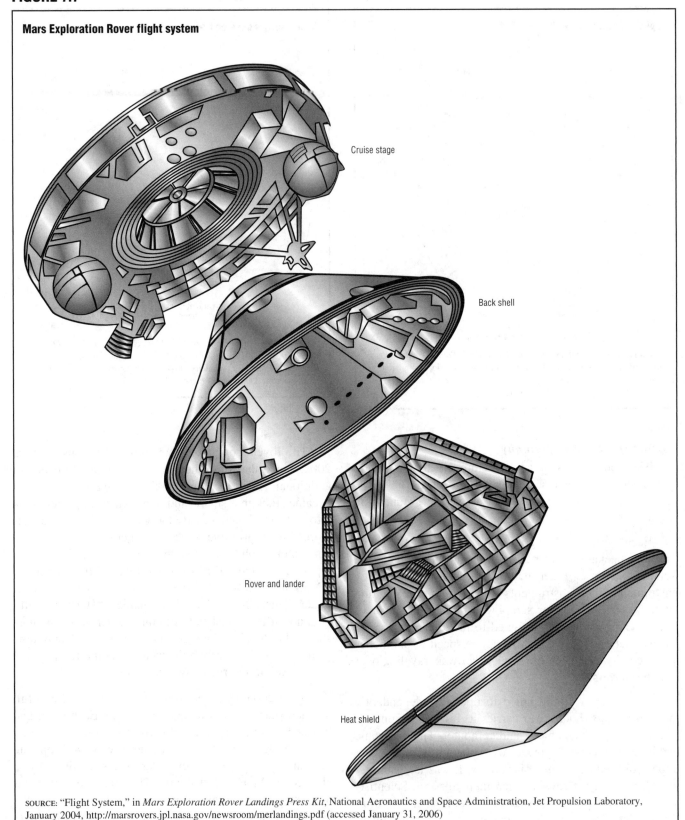

Cruise stage

Back shell

Rover and lander

Heat shield

SOURCE: "Flight System," in *Mars Exploration Rover Landings Press Kit*, National Aeronautics and Space Administration, Jet Propulsion Laboratory, January 2004, http://marsrovers.jpl.nasa.gov/newsroom/merlandings.pdf (accessed January 31, 2006)

named. The IAU naming process can take many months and even years to accomplish. NASA scientists handling images from the MER rovers had to quickly assign temporary working names to the many new smaller features being revealed. The evolution of this process is described in the article "Naming Mars: You're in Charge" (*Astrobiology Magazine*, NASA, Ames Research Center, June, 20, 2004).

FIGURE 7.8

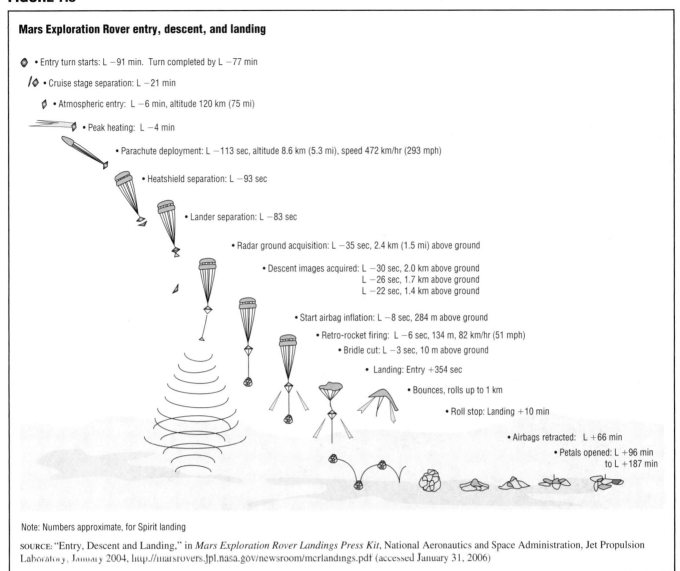

Mars Exploration Rover entry, descent, and landing

• Entry turn starts: L −91 min. Turn completed by L −77 min

• Cruise stage separation: L −21 min

• Atmospheric entry: L −6 min, altitude 120 km (75 mi)

• Peak heating: L −4 min

• Parachute deployment: L −113 sec, altitude 8.6 km (5.3 mi), speed 472 km/hr (293 mph)

• Heatshield separation: L −93 sec

• Lander separation: L −83 sec

• Radar ground acquisition: L −35 sec, 2.4 km (1.5 mi) above ground

• Descent images acquired: L −30 sec, 2.0 km above ground
L −26 sec, 1.7 km above ground
L −22 sec, 1.4 km above ground

• Start airbag inflation: L −8 sec, 284 m above ground

• Retro-rocket firing: L −6 sec, 134 m, 82 km/hr (51 mph)

• Bridle cut: L −3 sec, 10 m above ground

• Landing: Entry +354 sec

• Bounces, rolls up to 1 km

• Roll stop: Landing +10 min

• Airbags retracted: L +66 min

• Petals opened: L +96 min
to L +187 min

Note: Numbers approximate, for Spirit landing

SOURCE: "Entry, Descent and Landing," in *Mars Exploration Rover Landings Press Kit*, National Aeronautics and Space Administration, Jet Propulsion Laboratory, January 2004, http://marsrovers.jpl.nasa.gov/newsroom/mcrlandings.pdf (accessed January 31, 2006)

Most of the names are picked arbitrarily by whatever scientist first views an incoming image. Features are named after people, places, sailing ships, or other things the scientist fancies. The *Opportunity* rover landed within a tiny crater dubbed "Eagle Crater" in honor of the *Apollo 11* spacecraft that carried the first men to Earth's moon. When the *Spirit* rover landed in January 2004 it captured images of seven hilltops about two miles in the distance. Scientists dubbed them the "Columbia Hills" in honor of the seven shuttle *Columbia* astronauts that perished during 2003. Each hill was named after one of the astronauts. NASA hopes that the IAU will choose to make these names official.

Water and Blueberries

On March 2, 2004, NASA scientists announced that the *Opportunity* rover had uncovered strong evidence that its landing area Meridiani Planum had been "soaking wet" in the past.

The claim was based on examination of the chemical composition and structure of rocks found in an outcrop in the area. The rocks contained minerals, such as sulfate salts, known to form in watery areas on Earth. The rocks also had niches in which crystals appear to have grown in the past. These empty niches are called vugs, and are a strong indicator that the rocks sat in water for some time. Finally, there are round particles embedded in the rock that are about the size of ball bearings. Scientists have nicknamed them blueberries. The iron-rich composition of the "blueberries" and the way they are embedded in the rocks hints that water acted against the rocks in the past.

In November 2005 NASA published a series of reports in *Earth and Planetary Science Letters* detailing the latest findings from the MER rover *Opportunity*. Scientists believe that ancient conditions in the Meridiani Planum region were "strongly acidic, oxidizing, and sometimes wet." These harsh conditions are considered

FIGURE 7.9

Mars Exploration Rover

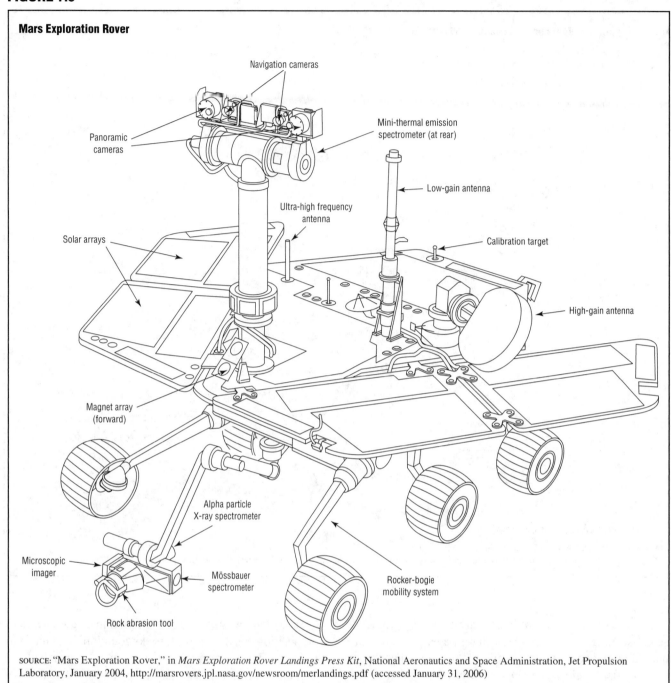

Navigation cameras

Panoramic cameras

Mini-thermal emission spectrometer (at rear)

Low-gain antenna

Ultra-high frequency antenna

Solar arrays

Calibration target

High-gain antenna

Magnet array (forward)

Alpha particle X-ray spectrometer

Microscopic imager

Mössbauer spectrometer

Rocker-bogie mobility system

Rock abrasion tool

SOURCE: "Mars Exploration Rover," in *Mars Exploration Rover Landings Press Kit*, National Aeronautics and Space Administration, Jet Propulsion Laboratory, January 2004, http://marsrovers.jpl.nasa.gov/newsroom/merlandings.pdf (accessed January 31, 2006)

unlikely to have allowed Martian life to develop at that time in the planet's history.

Mission Costs

The total cost of the MER missions has been estimated at $825 million. Each spacecraft cost about $325 million to develop, build, and equip with scientific instruments. Another $100 million was spent launching the spacecraft. About $75 million was devoted to operations and science costs.

MARS RECONNAISSANCE ORBITER

On August 12, 2005, NASA launched the *Mars Reconnaissance Orbiter* (*MRO*) toward the Red Planet.

The spacecraft was approximately twenty-one by forty-five feet in size and weighed more than two tons. A powerful Atlas V two-stage rocket was used to hoist the heavy *MRO* into space. Figure 7.10 shows some of the main features of the orbiter.

The *MRO* includes sophisticated radar, mineralogy, and atmospheric probes designed to learn about the atmosphere, terrain, and subsurface of the planet. It also carries a high-resolution camera to provide very detailed images of the Martian surface. NASA calls the spacecraft its "eyes in the sky." The *MRO* entered Mars orbit on March 10, 2006, and, following several months of aero-

FIGURE 7.10

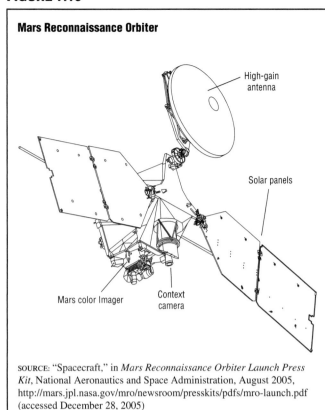

Mars Reconnaissance Orbiter

High-gain antenna

Solar panels

Context camera

Mars color Imager

SOURCE: "Spacecraft," in *Mars Reconnaissance Orbiter Launch Press Kit*, National Aeronautics and Space Administration, August 2005, http://mars.jpl.nasa.gov/mro/newsroom/presskits/pdfs/mro-launch.pdf (accessed December 28, 2005)

braking, will take up an orbiting position 190 miles from the surface of the planet. The *MRO* will also act as a communications relay satellite for future Mars missions. The orbiter is expected to operate at least through 2010. The total price of the mission is approximately $720 million.

THE FUTURE OF MARS EXPLORATION

NASA plans to launch robotic spacecraft to Mars during the oppositions of 2007 and 2009. In 2008 a spacecraft named *Phoenix Mars Scout* is scheduled to land on the Martian surface to investigate the water ice near the north polar region. It will also search for organic molecules in the soil. In 2009 a rover named the *Mars Science Laboratory* will be sent to conduct detailed chemical analysis of the Martian surface. NASA's long-term strategy calls for a robotic sample return mission sometime in the 2010s, possibly in cooperation with the ESA.

The ESA is considering launching a robotic Mars mission in 2011. It is expected to include a rover equipped with a drill and probes capable of studying the planet's atmosphere, rocks, and soil.

Human Missions to Mars

Human exploration missions will probably not be possible until the 2030s. There are several major obstacles that must be overcome to make these missions feasible. Most of the problems lie within bioastronautics (the field of biology concerned with the effects of space travel on humans).

Scientists worry that radiation exposure poses a major health risk to astronauts traveling in deep space (beyond Earth's magnetosphere). A solar flare while they are in flight or on Mars could be particularly hazardous. Mars has no magnetosphere of its own. Its atmosphere is very thin, with little shielding effect. The radiation levels are two to three times higher than around Earth. Special protective clothing and materials will have to be developed to protect the astronauts from the radiation hazards.

Bone loss due to long-term weightlessness is also a major concern. A trip to Mars takes about six months with current propulsion technology. This would mean a minimum of one year in space for the astronauts. They might have to spend a long time on the planet. It is considered likely that the astronauts would make their flights to and from Mars near the times of Mars oppositions. These occur twenty-six months apart. Thus, it is possible that an entire Mars mission could last around two years. Scientists know that humans lose 1% to 2% of their bone mass per month while in space. This bone loss would pose a serious health threat to the astronauts during such a long mission.

The psychological pressures of long space missions have not been well studied. A trip to Mars would require astronauts to live and work in tight quarters and under very stressful conditions for a year or two. The psychological strain could prove to be a major problem during such a long journey.

Another obstacle facing astronauts on a Mars mission would be access to medical care. On such a long flight the astronauts would have to have doctors aboard and some means of performing remote diagnosis and treatment of any medical problems that arose.

CHAPTER 8
THE FAR PLANETS

The farther we penetrate the unknown, the vaster and more marvelous it becomes.

—Charles A. Lindbergh, 1974

The far planets are Jupiter, Saturn, Uranus, Neptune, and Pluto. (See Figure 8.1.) They lie far from the Sun, in the coldest and darkest part of the solar system.

In ancient times people noticed that some lights in the sky followed odd paths around the heavens. The Greeks called them *asteres planetos*, or wandering stars. Later they would be called planets. The ancients could see only two of the far planets in the nighttime sky—Jupiter and Saturn.

Jupiter was named for the mythical Roman god of light and sky. He was the supreme god also known as Jove or *dies pater* (shining father). His counterpart in Greek mythology was named Zeus. Saturn was named after the god of agriculture, who was also Jupiter's father. His Greek counterpart was called Kronos.

Following the invention of the telescope, three more of the far planets were discovered—Uranus, Neptune, and Pluto. Uranus was named for the father of the god Saturn. Neptune was the god of the sea and Jupiter's brother in Roman mythology. Pluto was named after the Greek god of the underworld.

When the Space Age began humans sent robotic spacecraft to investigate the far planets. They returned images of strange and marvelous worlds composed of gas and slush instead of rock. Many new moons were revealed. Some of these moons are covered with ice and have atmospheres. There could be liquid water beneath that ice teeming with life. This possibility is particularly appealing to space scientists and to all people who wonder if we are alone in the universe.

THREE CENTURIES OF DISCOVERY

It took three centuries for humans to uncover the far planets in our solar system. In the 1600s the telescope opened up new opportunities for observation. People learned that Jupiter and Saturn had moons and that Saturn had rings. The telescope also showed that wandering stars were not stars at all, because they did not generate their own light, but reflected light from the Sun.

No new planets were discovered during the 1600s. The far planets were still too distant and fuzzy to be recognized for what they were. Uranus was discovered in the late 1700s. Another century passed before the discovery of Neptune. Pluto, discovered in 1930, was the last planet found in the solar system.

Astronomers categorize planets based on geology and composition. Mercury, Venus, Earth, and Mars are called the terrestrial planets, because they are made of rock and metal. They have solid surfaces on which space craft could land. Jupiter, Saturn, Uranus, and Neptune are called the gas giants. Some scientists think they may have solid cores, but the exterior of these planets consists of huge clouds of gas. These planets are also known as the Jovian planets (after Jove or Jupiter). All of them have ring systems.

Pluto is in a class by itself. It is a small ice world. Astronomers argue whether it is even a planet at all. Some believe that it is a dormant comet. Nevertheless, Pluto has been classified as a planet for decades now, and will likely continue to be so.

JUPITER

Jupiter is the fifth planet from the Sun and the largest planet in this solar system. It takes the planet nearly twelve Earth years to make one orbit around the Sun. Jupiter is eleven times larger than Earth. The planet is bright enough to be seen with the naked eye and appears yellowish from Earth. Jupiter is similar in composition to a small star and has an incredibly powerful magnetic field that stretches out millions of miles. The poles experience

FIGURE 8.1

Our solar system

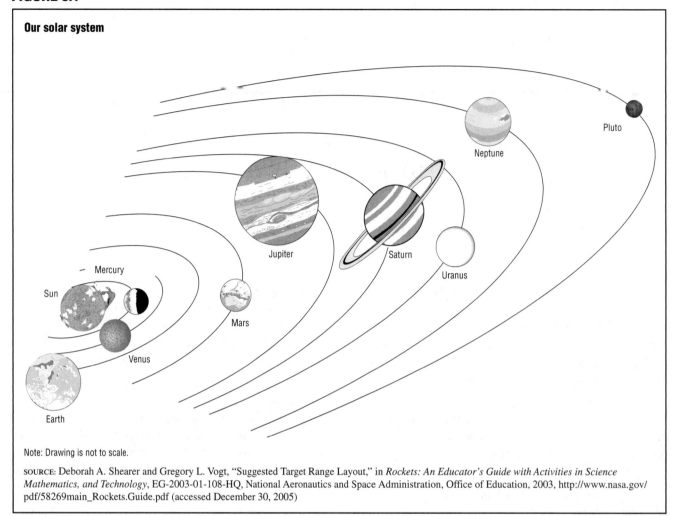

Note: Drawing is not to scale.

SOURCE: Deborah A. Shearer and Gregory L. Vogt, "Suggested Target Range Layout," in *Rockets: An Educator's Guide with Activities in Science Mathematics, and Technology*, EG-2003-01-108-HQ, National Aeronautics and Space Administration, Office of Education, 2003, http://www.nasa.gov/pdf/58269main_Rockets.Guide.pdf (accessed December 30, 2005)

dazzling auroras many times more powerful and bright than the Northern Lights on Earth.

Jupiter's atmosphere is 90% hydrogen. The other 10% is mostly helium, with a tiny bit of methane, water, and ammonia. Its sky is streaked with clouds and often with lightning. A gigantic hurricane-like storm has raged on the planet for hundreds of years, if not longer. It is a cold high-pressure area that is two to three times wider than Earth. The storm is nicknamed Jupiter's Great Red Spot. The red color is probably due to the presence of certain chemicals within the storm.

Scientists believe that Jupiter's surface is not solid, but slushy. The planet has dozens of moons. They are named for the lovers and children of Jupiter or Zeus. Jupiter also has a very light ring of material that orbits the planet.

Galileo is First to Discover Jupiter's Moons

On January 7, 1610, the Italian astronomer Galileo Galilei (1564–1642) was looking through his homemade telescope and discovered four celestial objects near Jupiter. At first he thought they were stars. After watching them for a week Galileo realized that they were satellites in orbit around Jupiter. Two months later Galileo published his findings in *Sidereus Nuncius* (*Starry Messenger*).

That same year the German astronomer Simon Marius (1573–1624) published a book called *Mundus Iovialis* (*The Jovian World*) in which he claimed that he discovered the satellites before Galileo. Marius did not provide any observational data in his book, and Galileo was better respected. The credit was given to Galileo.

In his book Marius proposed the names Io, Europa, Ganymede, and Callisto for the satellites. In Greek mythology these characters were lovers of Zeus. Marius said that fellow astronomer Johannes Kepler (1571–1630) suggested the names to him. Galileo referred to the moons as the Medician stars (to honor the family that ruled his Italian province) and numbered them from one to four. This naming convention was used for two centuries.

Renaming Jupiter's Moons

During the 1800s astronomers decided that a numbering system was too complicated for the moons of planets. More and more of the satellites were being discovered as telescopes improved. It was decided to name moons after literary characters from myths, legends, plays, and poems. Galileo's Medician moons were renamed Io, Europa, Ganymede, and Callisto as Marius had suggested.

More Jupiter Moons

In the years since Galileo's discovery other observers have discovered many smaller moons around Jupiter. The pace of these discoveries accelerated greatly in the late twentieth and early twenty-first centuries as better equipment was developed. In 2003 astronomers at an observatory atop Mauna Kea in Hawaii spotted twenty-three previously unknown moons around Jupiter. In March 2005 the International Astronomical Union (IAU) approved names for ten of these satellites: Hegemone, Mneme, Aoede, Thelxinoe, Arche, Kallichore, Helike, Carpo, Eukelade, and Cyllene. As of 2006 Jupiter's known moons numbered sixty-three.

SATURN

Saturn is the sixth planet from the Sun, and the second largest planet in our solar system. It takes 29.5 Earth years to orbit around the Sun. Saturn's atmosphere is mostly hydrogen, with a little helium and methane. It is a hazy yellow color. The planet is very windy, with wind speeds reaching 1,000 miles per hour.

Saturn is very flat at the poles. The planet is surrounded by many thin rings of orbiting material that circle near its equator. Saturn has dozens of moons. They are named after various characters from Greek and Roman mythology (mainly Saturn's siblings, the titans) and after giants from Gallic, Inuit, and Norse legends.

Galileo Sees Saturn's Handles

In 1610, when Galileo first saw Saturn through his telescope, its rings appeared to him to be two dim stars on either side of the planet. He described these stars as "handles." In 1612 Galileo reported that he could no longer see the dim stars. Much to his amazement, they had disappeared.

In the following years other astronomers saw the strange shapes around Saturn. They were variously described as ears or arms extending from the planet's surface. It would take an improvement in telescopic power before their true nature was revealed.

Huygens Finds a Moon and a Ring

Christiaan Huygens (1629–95) was a Dutch astronomer who became famous for his observations of Saturn. He and his brother Constantyn built new and more powerful telescopes that were greatly admired by astronomers of the time.

On March 25, 1655, Christiaan Huygens discovered a satellite around Saturn. This turned out to be the planet's largest moon. In 1656 Huygens wrote about his discovery in *De Saturni Luna Observatio Nova* (*The Discovery of a Moon of Saturn*). Huygens referred to his discovery as simply Saturn's Moon. Later, it would be called Titan.

Huygens also figured out that the mysterious shapes near Saturn were not stars, arms, or ears, but a ring of material around the planet. Huygens mistakenly thought the ring was one solid object. In 1659 he published his observations in *Systema Saturnium* (*The Saturn System*).

Ring Plane Crossings

Huygens explained that the ring around Saturn was difficult to view, because it is very thin. Every fourteen to fifteen years the Earth moved into the same plane as the ring. If you tried to observe Saturn from Earth at this time, Huygens said, you would be viewing the outer edge of the ring head-on, making it virtually invisible. This explained why Galileo was unable to see the handles around Saturn in 1612. It was a year in which Earth passed through Saturn's ring plane.

There are several planetary alignments that cause Saturn's rings to be invisible to Earth observers. One of these is when Earth passes into the Saturn ring plane. A similar effect occurs when the Sun passes through Saturn's ring plane and when the Sun and Earth are on opposite sides of the ring plane. The next Saturn ring plane passage will occur in August and September of 2009. Throughout history ring crossings have been the best times to discover new moons around Saturn.

More Moons of Saturn

Giovanni Cassini (1625–1712) was born in Italy but lived in France. He was the first director of the Royal Observatory in Paris. During the late 1600s he discovered four more of Saturn's moons. The first two he observed during the ring crossing of 1671–72. The second pair he discovered just prior to the ring crossing of 1685.

Over the next three centuries many more Saturn moons were discovered. In December 2004 astronomers at the Mauna Kea observatory in Hawaii found twelve previously unknown moons around Saturn. The latest discovery occurred in May 2005 when NASA reported that the *Cassini* spacecraft orbiting Saturn had captured an image of a "new" moon.

Cassini and Saturn's Rings

During the 1600s Giovanni Cassini discovered a major gap in the ring around Saturn. This proved that the structure was not one solid object as Huygens had thought. The gap would later be called the Cassini Division. Cassini believed that Saturn's rings were composed of millions of small particles. This view was shared by French astronomer Jean Chapelain (1595–1674). However, it was not generally accepted until the eighteenth century.

FIGURE 8.2

Saturn ring composition

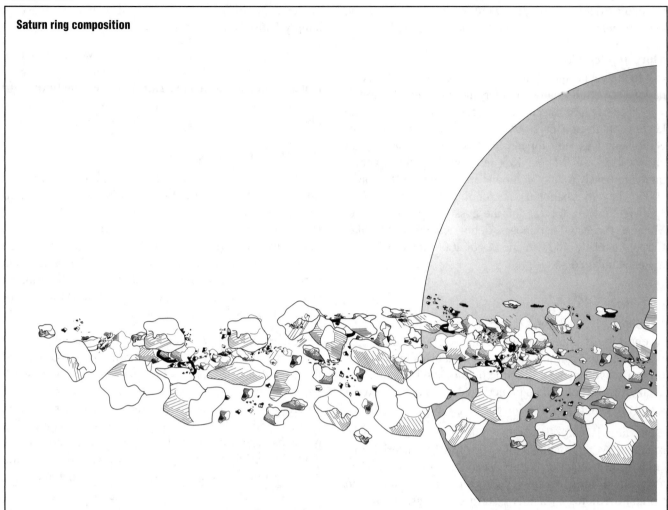

SOURCE: Linda J. Spilker, editor, "Saturn's Flying Snowballs," in *Passage to a Ringed World: The Cassini-Huygens Mission to Saturn and Titan*, National Aeronautics and Space Administration, Jet Propulsion Laboratory, California Institute of Technology, October 1997, http://saturn.jpl.nasa.gov/multimedia/products/pdfs/ptarw.pdf (accessed December 28, 2005)

Modern astronomers believe that the rings are composed of chunks of ice. These chunks range in size from tiny particles to icebergs as large as automobiles. (See Figure 8.2.) Saturn's ring system is actually many ringlets of different sizes nestled within each other with gaps between ring systems. Scientists use letters to designate distinct ring systems around the planet, as shown in Table 8.1.

URANUS

Uranus is the seventh planet from the Sun, with the third largest diameter in the solar system. It was named for the father of the god Saturn in Roman mythology.

Uranus looks featureless through even the most powerful telescopes. Scientists believe the planet is shrouded in clouds that hide it from view. The presence of methane in the upper atmosphere is believed to account for the planet's light blue-green color. It takes eighty-four Earth years for Uranus to orbit around the

TABLE 8.1

The rings of Saturn

Ring	Distance, kilometers*	Width, kilometers
D	66,970	7,500
C	74,500	17,500
B	92,000	25,400
A	122,170	14,610
F	140,180	50
G	170,180	8,000
E	180,000	300,000

*Distance from Saturn to closest edge of ring.

SOURCE: Linda J. Spilker, editor, "The Rings of Saturn," in *Passage to a Ringed World: The Cassini-Huygens Mission to Saturn and Titan*, National Aeronautics and Space Administration, Jet Propulsion Laboratory, California Institute of Technology, October 1997, http://saturn.jpl.nasa.gov/multimedia/products/pdfs/ptarw.pdf (accessed December 28, 2005)

Sun. Uranus is unique in the solar system, because its axis is tilted so far from its orbital plane. The planet lies on its "side" as it orbits with a pole pointed toward the Sun.

As of 2006 Uranus had twenty-seven known satellites. They are named after characters from the plays of William Shakespeare (1564–1616) and from the poem *The Rape of the Lock* by Alexander Pope (1688–1744).

Herschel Discovers Uranus and Two of Its Moons

The astronomer Frederic William Herschel (1738–1822) was born in Germany, but lived and worked in Britain. He dropped his first name and was commonly known as William Herschel. On March 13, 1781, he was searching the sky with his telescope when he discovered Uranus. Herschel wanted to name the planet "Georgium Sidus" in honor of King George III of England. However, the name Uranus was selected from ancient mythology.

A few years later, in 1787, Herschel was the first to spot satellites around the planet. He discovered the two largest moons, named Titania and Oberon.

More Moons

During the mid-1800s two more moons were discovered around Uranus by the British astronomer William Lassell (1799–1880). It was another century before the next moon was found. During the 1980s and 1990s more than a dozen new moons were added to the list. In 2003 four additional moons were discovered by the *Hubble Space Telescope* and astronomers at the Mauna Kea observatory in Hawaii: Margaret, Ferdinand, Mab, and Cupid.

NEPTUNE

Neptune is usually the eighth planet from the Sun. (Occasionally the erratic orbit of Pluto brings that planet in closer than Neptune.) Neptune was named after the mythical Roman god of the sea. The planet is very far from Earth and extremely difficult to observe. The planet has a distinctive bluish hue when viewed through a telescope. It orbits around the Sun once in 165 Earth years.

Neptune has at least thirteen moons. They are named after characters associated with Neptune or Poseidon (his Greek counterpart) or other sea-related individuals in ancient mythology.

Neptune's Controversial Discovery

Neptune's discovery is a twisted tale of mathematics, bureaucrats, and international competition. Following the discovery of Uranus in 1781, astronomers watched that planet for several decades. They were puzzled because its orbit did not follow the path expected. Some astronomers began to suspect that there might be another planet beyond Uranus. The effect of its gravity would explain the irregularities that astronomers saw in the orbit of Uranus.

During the 1840s two scientists used mathematics to figure out where this mystery planet could be located. Their names were Urbain Le Verrier (1811–77) of France and John Couch Adams (1819–92) of Britain. Adams presented his theory to the director of the Cambridge Observatory in England. For some reason the director did not pursue the matter and look for the unknown planet. Le Verrier submitted his theory to the Berlin Observatory. It was directed by Johann Galle (1812–1910). On the night of September 23, 1846, Galle used Le Verrier's notes to locate the planet in the sky.

When Le Verrier and Galle publicized the discovery, the director of the Cambridge Observatory complained that Adams had described the location of the planet months before Le Verrier. It turned into a heated argument between France and England. Astronomers decided to split the credit for the planet's discovery between Adams and Le Verrier. Galle is considered the first to observe the planet. However, a review of Galileo's notes from the 1600s revealed that Galileo actually spotted the planet centuries before, but thought it was a fixed star.

Lassell Discovers a Moon around Neptune

Only weeks after Neptune's discovery its first moon was discovered. In early October 1846 an amateur British astronomer named William Lassell (1799–1880) spotted the moon. It was named Triton after the sea-god son of Poseidon. The name was suggested by the French astronomer Camille Flammarion (1842–1925).

More Moons of Neptune

During the 1940s another moon around Neptune was discovered from a ground-based telescope. In 1989 images from the spacecraft *Voyager 2* revealed six previously unknown moons. The latest discoveries occurred in 2002 and 2003 when five new moons were added to the list, bringing the total to thirteen.

PLUTO

Pluto is the smallest planet in the solar system and usually the farthest planet from the Sun. It is named after the mythical Greek god of the underworld. It takes Pluto 248 Earth years to circle the Sun. Pluto's orbit is very erratic. For twenty of those 248 years the planet orbits closer to the Sun than does Neptune. This last occurred between 1979 and 1999.

Pluto is believed to be a dark and icy world with a surface of frozen nitrogen, methane, and carbon dioxide. It has been observed and photographed only from great distances. No spacecraft have ever been near the planet. In January 2006 NASA launched a robotic probe that should reach Pluto in 2015. It will provide the first-ever detailed images of the distant planet.

Tombaugh Discovers Pluto

Clyde Tombaugh (1906–97) is credited with discovering the planet Pluto. Tombaugh made the discovery on

February 18, 1930, while working at the Lowell observatory in Flagstaff, Arizona. This famous observatory was founded in the 1890s by Percival Lowell. For years Lowell had searched for a planet believed by some astronomers to lie beyond Neptune. Following Lowell's death the observatory continued the search. Tombaugh found Pluto after diligently photographing the sky for many nights and studying the photographs for objects that changed position relative to the fixed stars.

The Naming of Pluto

Lowell's widow wanted to name the planet after her late husband. This was not allowed, because it would have broken the tradition of using names from Greek and Roman mythology. The name Pluto was finally selected from many suggestions made by the public.

Pluto was the Greek god of the underworld, and able to make himself invisible. The name seemed appropriate for the darkest planet in the solar system that had been so difficult to find. Also, the first two letters of the name matched the initials of Percival Lowell. The name Pluto was originally suggested by an eleven-year-old British girl named Venetia Burney.

Christy Discovers Pluto's Moon

Pluto has only one verified moon, Charon (pronounced "Karen"). It is named after a character in Greek mythology who ferried the souls of the dead across the river Styx to the underworld. On June 22, 1978, Charon was discovered by James Christy at the U.S. Naval Observatory in Washington, D.C. Christy was studying photographs of the planet when he noticed an odd shape in some of the images. After comparing photographs he realized that the shape moved over time compared with Pluto and the fixed stars. When the discovery was made public Christy suggested the name that was assigned to the moon. In November 2005 NASA announced that images from the *Hubble Space Telescope* suggested that Pluto has two additional moons orbiting far from the planet. As of December 2005 this finding had not been independently verified.

THE FAR PLANETS IN SCIENCE FICTION

The far planets have not been as popular as the Moon and Mars in science fiction stories. One of the first mentions of Jupiter occurs in an 1894 story called *A Journey in Other Worlds*. In this story Jupiter is similar to a prehistoric Earth. *Skeleton Men of Jupiter* was an unfinished story written by the American writer Edgar Rice Burroughs (1875–1950), who is best known as the creator of Tarzan. It appeared in print during the 1940s. The story referred to a Jupiter-like world called "Sasoom," inhabited by creatures that looked like human skeletons. In Arthur C. Clarke's 1968 novel, *2001: A Space Odyssey*, the exploration of the solar system reaches Saturn,

while in the Stanley Kubrick film of the same name, much of the story takes place near Jupiter.

Advances in telescopes and astronomy made it clear that Jupiter and the other far planets were gaseous worlds without solid surfaces. This made them much less appealing as home worlds for aliens. During the 1990s scientists learned that larger moons in the outer solar system may have thick atmospheres and some organic chemicals in their composition. This increases the chances that life could exist there. These moons became popular home worlds for sea-faring creatures in science fiction stories.

PIONEER

During the early 1970s the United States began a series of interplanetary missions designed to explore the far planets. The first of these missions was aptly named Pioneer.

Pioneer spacecraft were the first to investigate Jupiter and Saturn. The missions were managed by NASA's Ames Research Center in Mountain View, California, for the agency's Office of Space Science. The two spacecraft involved were called *Pioneer 10* and *Pioneer 11*. The total cost of their mission was approximately $350 million.

On each spacecraft was mounted a six-inch by nine-inch metal plaque with greetings from Earth. The plaque included illustrations of a human man and woman, the spacecraft's silhouette, and some mathematical, chemical, and astronomical data represented in binary code symbols. An image of the solar system at the bottom of the plaque shows a Pioneer spacecraft leaving Earth and passing between Jupiter and Saturn on its way out of the solar system. The scientists who designed the plaque hoped that the images and symbols would serve as a viable means of communication, should any intelligent life form happen to encounter the spacecraft.

Pioneer 10

On March 3, 1972, *Pioneer 10* was launched atop an Atlas-Centaur rocket from Cape Kennedy in Florida. It was the first mission ever sent to the outer solar system. Ultimately it became the first human-made object to leave the solar system for interstellar space.

Pioneer 10 was the first spacecraft to travel through the asteroid belt between Mars and Jupiter. Scientists had feared that this would be a dangerous area of space. They learned that the asteroids in the belt are spread far apart and do not pose a significant hazard to spacecraft flying through.

In December 1973 *Pioneer 10* was the first spacecraft to investigate Jupiter. Its closest approach came within 124,000 miles of the planet. *Pioneer 10* carried various instruments to study the solar wind, magnetic fields, cosmic radiation and dust, and hydrogen concentrations in space. Its Jupiter studies focused on the planet's magnetic effects, radio waves, and atmosphere. The atmospheres of Jupiter's satellites (particularly Io) were also investigated.

On June 13, 1983, *Pioneer 10* became the first human-made object to leave the solar system. Over the years the instruments aboard the spacecraft began to fail or were turned off by NASA to conserve power. In 1997 NASA ceased routine tracking of the spacecraft due to budget reasons. The spacecraft was the most distant human-made object in space until February 1998, when it was passed by an even faster spacecraft called *Voyager 1*. NASA last detected a signal from *Pioneer 10* in January 2003. It was approximately 7.6 billion miles away from Earth.

As of 2006 *Pioneer 10* is more than eight billion miles from Earth and on its way out of the solar system. It is headed in the general direction of a star called Aldebaran that forms the eye of the constellation Taurus (The Bull). It will take the spacecraft nearly two million years to reach the star.

Pioneer 11

On April 5, 1973, the *Pioneer 11* spacecraft was launched into space by an Atlas-Centaur rocket. A year and a half later it flew by Jupiter on its way to Saturn. The spacecraft approached within 21,000 miles of Jupiter. It was the first spacecraft to observe the planet's polar regions. It also returned detailed images of the Great Red Spot. Like its sister spacecraft, *Pioneer 11* investigated solar and cosmic phenomena and interplanetary and planetary magnetic fields during its journey.

In September 1979 *Pioneer 11* flew within 13,000 miles of Saturn and returned the first close-up pictures of the planet and its rings. It continued past the planet toward the edge of the solar system. In 1995 routine missions operations were ended, and NASA received its last transmission from the spacecraft. By the end of that year, *Pioneer 11* was approximately four billion miles from Earth. The spacecraft is headed toward the constellation Aquila (The Eagle), which it will pass in approximately four million years.

Pioneer and Plutonium

The Pioneer spacecraft were built with special power systems based on radioisotope thermoelectric generators (RTGs). RTGs generate electricity from the heat released during the natural radioactive decay of a plutonium pellet. Although sending plutonium into space is controversial, NASA has used the power source on all of its missions to the far planets. The planets are too far from the Sun to make solar power a feasible and reliable choice for these spacecraft.

FIGURE 8.3

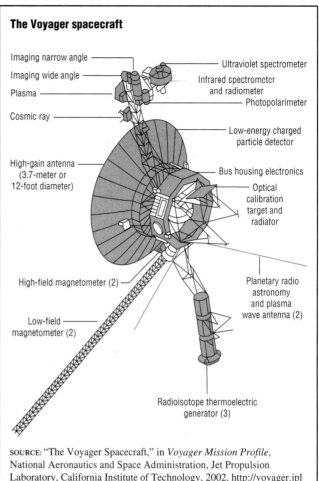

The Voyager spacecraft

SOURCE: "The Voyager Spacecraft," in *Voyager Mission Profile*, National Aeronautics and Space Administration, Jet Propulsion Laboratory, California Institute of Technology, 2002, http://voyager.jpl.nasa.gov/pdf/2073_Voyagerposter_back_AS.pdf (accessed January 31, 2006)

VOYAGER

In 1977 NASA began another bold mission to investigate Jupiter and Saturn. The program was called Voyager and included twin robotic spacecraft named *Voyager 1* and *Voyager 2*. An illustration of a Voyager spacecraft is shown in Figure 8.3.

The various instruments on board were designed to detect and measure the solar wind and other charged particles, cosmic radiation, magnetic field intensities, and plasma waves. The original five-year Voyager mission was so successful that it was extended to include flybys of Uranus and Neptune. The total cost of Voyager's planetary explorations was $865 million.

The Missions

Both spacecraft were launched into space atop Titan rockets. *Voyager 2* was the first to launch, on August 20, 1977. It was followed on September 5, 1977, by *Voyager 1*. Both spacecraft traveled for two years to fly by Jupiter. They made many scientific observations as they passed Jupiter and continued on to Saturn. *Voyager 1* was on a faster trajectory than *Voyager 2* and reached the planet first. *Voyager 2* was directed on to fly by Uranus and Neptune. It was the first spacecraft to do so.

The Voyager spacecraft proved to be so hardy after completing their planetary journeys that they were sent on a new mission called the Voyager Interstellar Mission (VIM). The purpose of the VIM is to use the instruments on the spacecraft to explore the outermost edge of the heliosphere. This is the region of space dominated by energy effects from the Sun.

Planetary Achievements

The Voyager missions were two of the most successful in NASA's history. The spacecraft revealed a number of discoveries about the gas giants in the outer solar system:

- Jupiter, Uranus, and Neptune have faint ring systems.
- Jupiter has a complicated atmosphere in which lightning storms and aurora are common.
- Jupiter's moon Io has active volcanoes.
- Jupiter's moon Europa has a smooth surface composed of water ice.
- The radiation levels experienced during the Jupiter flyby were 1,000 times stronger than what is lethal to humans.
- Saturn's rings comprise thousands of strands (ringlets).
- Saturn's ringlets are not as uniform and separate as expected. Some are kinked or braided together. Additional gaps between rings were discovered.
- Saturn's weather is relatively tame compared to that on Jupiter.
- Saturn's largest moon, Titan, has a dense smoggy atmosphere that contains nitrogen and carbon-containing compounds.
- Saturn's moon Mimas has a massive impact crater.
- Neptune's moon Triton has a thin atmosphere.

The mission also uncovered twenty-three previously unknown moons (three around Jupiter, four around Saturn, ten around Uranus, and six around Neptune). The discovery of water ice on the surface of Jupiter's moon Europa was particularly exciting, because it raised the possibility that there was liquid water underneath.

Voyager Interstellar Mission

In February 1998 *Voyager 1* became the most distant human-made object in space when it reached a distance of 6.5 billion miles from the Sun, surpassing the record of *Pioneer 10*. It continues to travel at a speed of nearly one million miles per day. *Voyager 2* is a little slower than its sister ship.

In a September 2005 update NASA scientists announced that *Voyager 1* had passed the termination shock. (See Figure 8.4.) The termination shock occurs where the solar wind slows considerably due to effects from interstellar wind. Solar wind emanates from the Sun and forms a long "wind sock" that moves with the Sun as it journeys through space. The heliosheath is the outer layer of the heliosphere. The heliopause is the boundary between the heliosphere and interstellar space. It lies more than thirteen billion miles from Earth.

As of October 2005 *Voyager 1* was nearly 9.1 billion miles from Earth. *Voyager 2* was roughly 7.2 billion miles away. Both have been traveling through space for nearly three decades. Some of the redundant instruments on board the spacecraft have been turned off to conserve power. According to NASA, both spacecraft should continue to function until at least 2020. They have enough propellant to last that long and should continue to generate adequate electrical power to run their scientific instruments. *Voyager 1* is expected to pass through the heliopause by 2015. *Voyager 2* should follow about five years later. NASA scientists hope that both spacecraft will be able to transmit data as they cross into the vast unknown of interstellar space.

Messages from Earth

The Voyager spacecraft carry written and recorded messages from Earth, in case they come across any intelligent life. Attached to each spacecraft is a twelve-inch gold-plated copper disk inside a protective aluminum case. The cover of the protective case has symbolic instructions for playing the disc and a diagram of Earth's location in the solar system carved into it. The disks contain recorded greetings in fifty-five different languages and various other sounds, including bits of music and natural and human-made sounds. There are 115 images encoded in analog form on the disks of various Earth scenes. These include pictures of people, objects, and places from around the world. The disks carried printed messages from President Jimmy Carter and the United Nations Secretary General.

GALILEO

NASA's *Galileo* mission was the first to put a spacecraft in orbit around one of the far planets. The $1.4 billion mission to Jupiter included a scientific probe that left the orbiter and plunged into the planet's atmosphere. See Figure 8.5 for a diagram of the spacecraft including the descent probe. The mission was operated by NASA's Jet Propulsion Laboratory in Pasadena, California.

On October 18, 1989, the space shuttle *Atlantis* lifted off from Kennedy Space Center in Florida on mission STS-34 with the *Galileo* spacecraft on board. The shuttle astronauts released *Galileo* in Earth orbit, then the craft

FIGURE 8.4

Locations of Voyager 1 and Voyager 2 as of December 2003

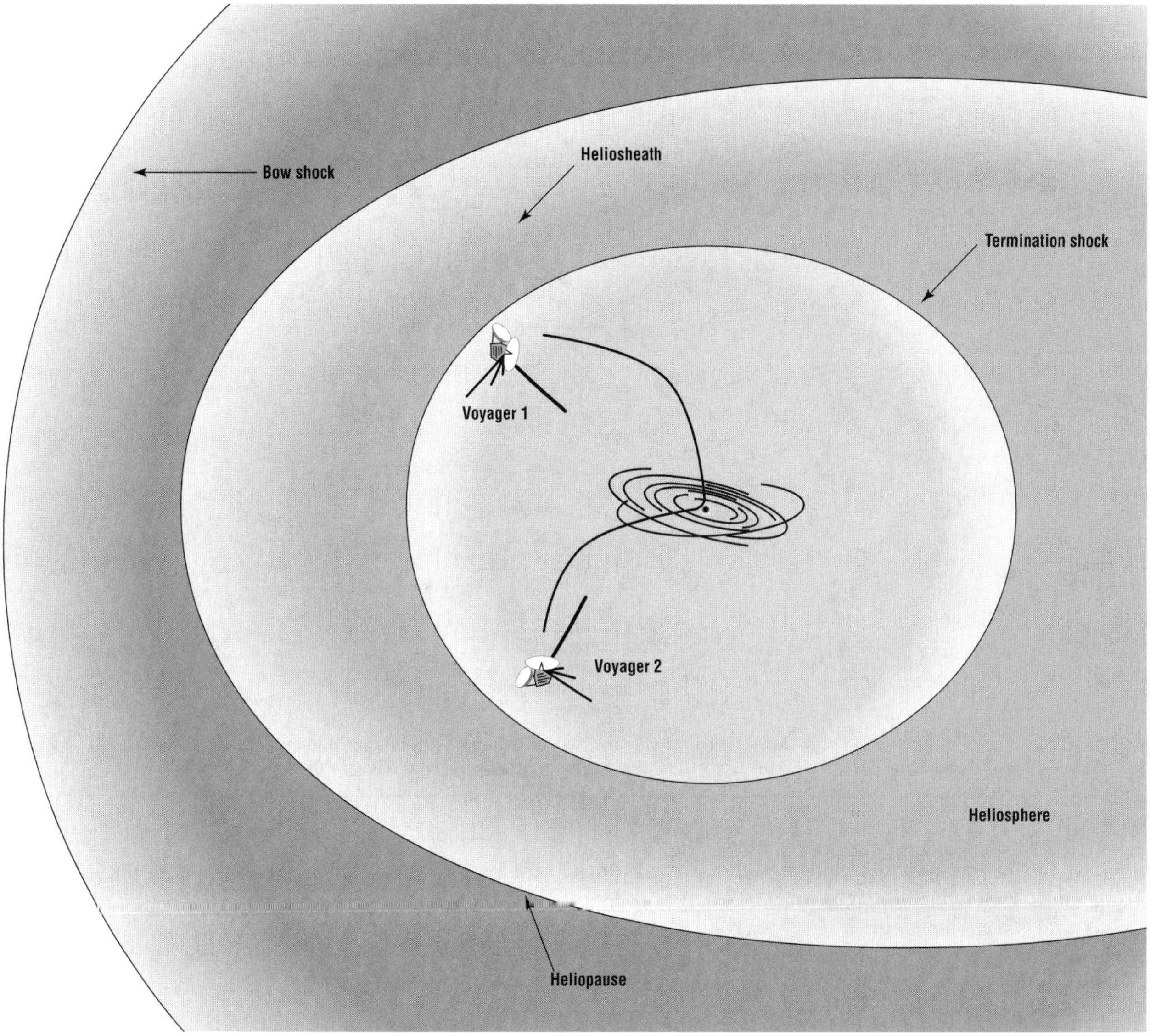

SOURCE: "Voyager Approaches Final Frontier," in *Planetary Photo Journal*, Image No. PIA04927, National Aeronautics and Space Administration, Jet Propulsion Laboratory, California Institute of Technology, December 12, 2003, http://www.jpl.nasa.gov/images/voyager/voyager-interstellar-browse.jpg (accessed December 28, 2005)

used its two-stage inertial upper stage rocket to boost itself toward Venus.

The spacecraft swung by Venus once and Earth twice as part of "gravity assist" maneuvers. These are maneuvers in which a spacecraft flies in close enough to a planet to get a boost from the orbital momentum of a planet traveling around the Sun. NASA compares a gravity assist to throwing a ping pong ball to skim along the top of one of the moving blades of an electric fan. The blades circle the fan's motor at a high rate of speed. The ping pong ball gets close enough to the blade to pick up momentum and

shoot off in a different direction. Using gravity assists during space flight saves on fuel. This is particularly important for long journeys to the outer solar system.

By July 1995 *Galileo* was nearing Jupiter. It released the probe, which was about four feet in diameter and three feet long, and the probe began a five-month plunge toward the planet. On December 7, 1995, the orbiter was in position when the probe began its final descent at more than 105,000 miles per hour. For nearly an hour the heavily protected probe transmitted data about Jupiter's atmosphere, temperature, and weather. It was finally

FIGURE 8.5

The Galileo spacecraft

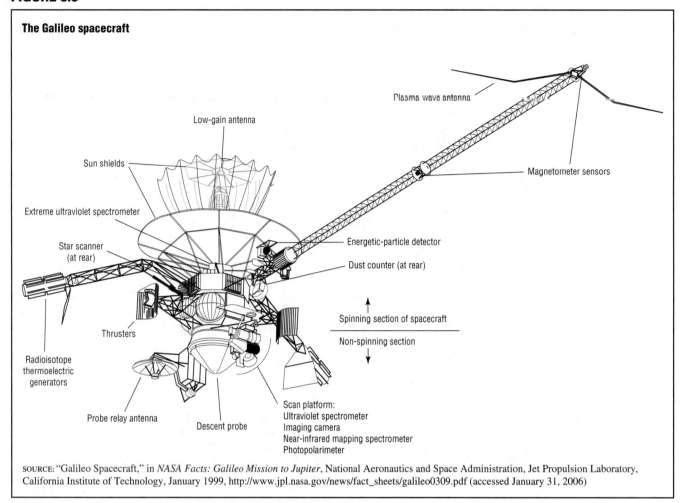

SOURCE: "Galileo Spacecraft," in *NASA Facts: Galileo Mission to Jupiter*, National Aeronautics and Space Administration, Jet Propulsion Laboratory, California Institute of Technology, January 1999, http://www.jpl.nasa.gov/news/fact_sheets/galileo0309.pdf (accessed January 31, 2006)

destroyed by the intense heat and pressure surrounding the planet. It had penetrated 124 miles into the violent atmosphere.

The orbiter spacecraft spent the next eight years in orbit around Jupiter. It conducted numerous flybys of the moons Europa, Ganymede, and Callisto, using its eleven scientific instruments to collect data about radiation, magnetic fields, charged particles, and cosmic dust.

The *Galileo* orbiter was originally designed for a two-year mission. It ended up lasting for fourteen years. In September 2003 NASA scientists destroyed the spacecraft by purposely plunging it into Jupiter's atmosphere. The orbiter was running low on propellant. The scientists feared that it could run out of fuel and crash into one of Jupiter's moons. This could contaminate environments that might possibly contain water and life forms.

The *Galileo* mission was hugely successful. The spacecraft traveled more than 2.8 billion miles during its long journey. It flew by two asteroids, Gaspra and Ida, on its way to the planet and watched Comet Shoemaker-Levy 9 impact Jupiter while it was in orbit there. *Galileo* captured thousands of detailed images of

the planet and its largest moons and collected a wealth of data about these celestial objects.

Major findings attributed to, or confirmed by, the *Galileo* mission include:

- There is an intense radiation belt around Jupiter.

- The surface of the moon Io is constantly being reshaped by heavy volcanic activity.

- There is evidence of liquid water oceans beneath the icy surface of the moon Europa and possibly Callisto.

- The moon Ganymede has its own magnetosphere and probably its own magnetic field.

- Ganymede is heavily cratered from impacts of comets and asteroids and has icy plains, mountains, and basins likely caused by geologic forces.

- Ganymede has a thin ionosphere (electrically charged atmosphere).

- Ganymede, Europa, and Io all appear to have metallic cores.

The possibility of liquid water is considered very strong for the moon Europa and somewhat likely for the

FIGURE 8.6

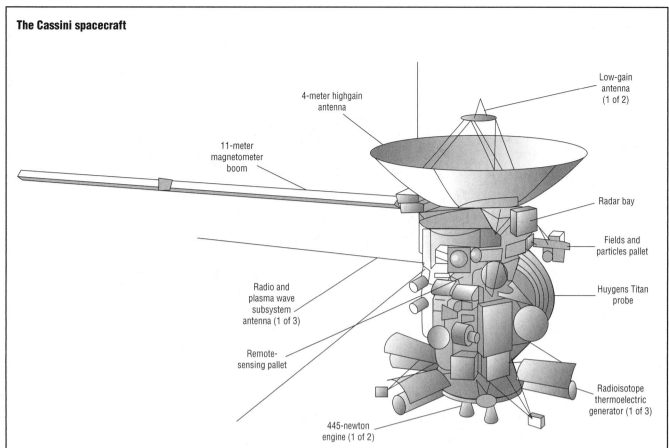

The Cassini spacecraft

4-meter highgain antenna

Low-gain antenna (1 of 2)

11-meter magnetometer boom

Radar bay

Fields and particles pallet

Radio and plasma wave subsystem antenna (1 of 3)

Huygens Titan probe

Remote-sensing pallet

Radioisotope thermoelectric generator (1 of 3)

445-newton engine (1 of 2)

SOURCE: Linda J. Spilker, Editor, "Vehicle of Discovery," in *Passage to a Ringed World: The Cassini-Huygens Mission to Saturn and Titan*, National Aeronautics and Space Administration, Jet Propulsion Laboratory, California Institute of Technology, October 1997, http://saturn.jpl.nasa.gov/multimedia/products/pdfs/ptarw.pdf (accessed January 31, 2006)

moon Callisto. These conclusions are based on analysis of geological and magnetic data collected by *Galileo*.

CASSINI

In 1997 NASA collaborated with the European Space Agency (ESA) and the Italian space agency to launch the *Cassini* mission to Saturn. The spacecraft is depicted in Figure 8.6. It was designed to orbit the planet for four years and release a probe to land on Titan, Saturn's largest moon. Specific mission objectives are to investigate Saturn's magnetosphere and atmosphere, determine the structure and behavior of its rings, and characterize the composition, weather, and geological history of its moons.

On October 15, 1997, the spacecraft was launched atop a Titan IV rocket from the Kennedy Space Center in Florida. Over the next three years it received two gravity assists from Venus and one each from Earth and Jupiter. *Cassini* arrived at Saturn in July 2004, becoming the first spacecraft ever to orbit the planet.

The *Cassini* orbiter is equipped with twelve scientific instruments. It also carried the *Huygens* probe with six instruments of its own. (See Figure 8.7.) The probe,

provided for the mission by the ESA, was released on December 25, 2004, and began its three-week journey to the surface of Titan. It penetrated the thick cloud cover that hides the moon and touched down on January 14, 2005. The probe sampled Titan's atmosphere and provided the first photographs ever of its surface. The orbiter continued to circle Saturn and conducted flybys of Titan and the smaller moons Enceladus, Hyperion, Dione, Rhea, Iapetus, and Phoebe.

In July 2005 NASA compiled a list of the top-ten science highlights of the first year of the *Cassini-Huygens* mission:

- Titan's surprise surface and organic atmosphere—The surface did not include global oceans as scientists expected, but was Earth-like in some ways. There is evidence of volcanoes, erosion, craters, dunes, and dry and wet lake beds. Titan's atmosphere contains organic chemicals, such as benzene and methane. Scientists believe the moon experiences methane showers from clouds sweeping overhead.

- Saturn's complex rings—The rings were found to have "straw-like clumps" several miles long and

FIGURE 8.7

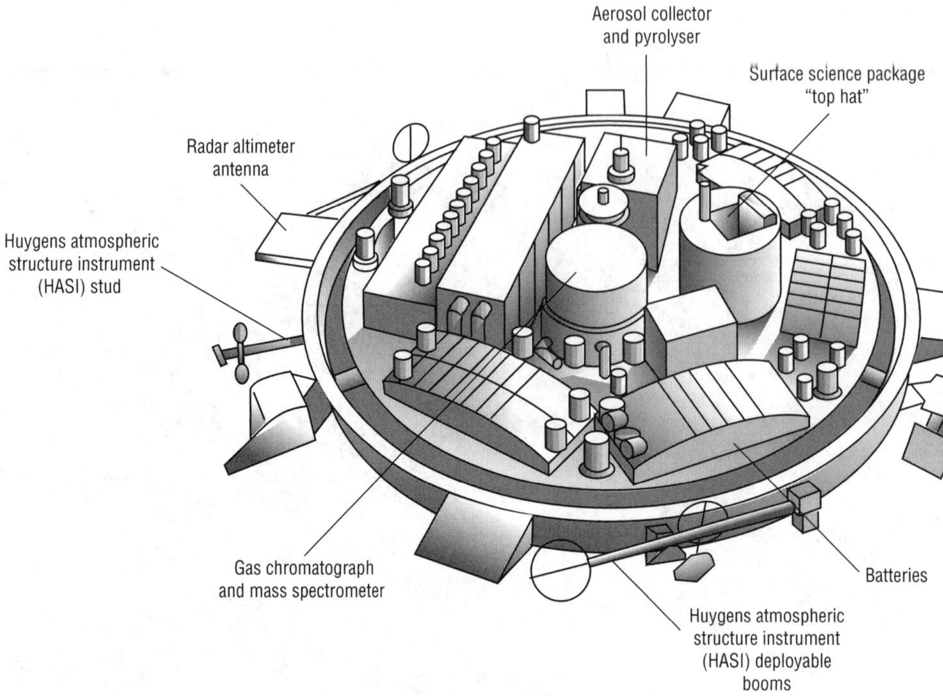

SOURCE: Linda J. Spilker, editor, "Probing Titan's Depths," in *Passage to a Ringed World: The Cassini-Huygens Mission to Saturn and Titan*, National Aeronautics and Space Administration, Jet Propulsion Laboratory, California Institute of Technology, October 1997, http://saturn.jpl.nasa.gov/multimedia/products/pdfs/ptarw.pdf (accessed December 28, 2005)

rotating ring particles. *Cassini* discovered an oxygen atmosphere exists just above the rings.

- First detailed images of Phoebe—*Cassini* found this tiny moon battered and scarred by numerous large crater strikes. There was evidence of water ice and silicate and organic materials on the surface.

- Saturn's colorful and violent atmosphere—Scientists were surprised to find the northern hemisphere of the planet appeared deep blue, rather than hazy yellow like the rest of the world. Also, violent lightning was detected in enormous thunderstorms nearly the size of Earth.

- Enceladus may have an atmosphere—Magnetic field data suggest the tiny icy moon has an atmosphere around it.

- Saturn's inner radiation belt—*Cassini* discovered a previously unknown radiation belt circles the entire planet in between the cloud tops and the edge of the D-ring.

- Dynamic ring-moon relationships—Images revealed unexpected interactions between Saturn's rings and moons, such as particle stealing. Two previously unknown moons were also discovered.

- Saturn's rotational speed—Comparisons of *Cassini* measurements to those made by Voyager spacecraft during the early 1980s suggest that Saturn's internal rotation rate is slowing down.

- Massive mountains on Iapetus—Scientists learned that there is a massive mountain range on the dark side of the moon Iapetus. Some of the mountains exceed twelve miles in height. For comparison, Mount Everest on Earth is approximately 5.5 miles high.

- Cracks in Dione—*Cassini* images reveal that the moon's terrain is creased with giant fractures.

Later in 2005 *Cassini* captured dramatic new images during flybys of the moons Mimas, Tethys, Hyperion, and Dione. As of 2006 the orbiter continues its journey around Saturn conducting detailed studies of the planet, its rings, and moons.

NEW HORIZONS—PLUTO AND THE KUIPER BELT

NASA calls *New Horizons* "the first mission to the last planet." On January 19, 2006, the robotic spacecraft was launched aboard an Atlas V rocket. *New Horizons* is scheduled to reach Pluto in 2015. Following this encounter it will move into the Kuiper Belt and explore there through the year 2020. The Kuiper Belt is a region beyond Pluto believed to contain thousands of "miniature icy worlds."

The spacecraft includes seven scientific instruments designed to assess the geology and atmosphere of Pluto and its primary moon Charon and map their surface compositions. Charon is of particular interest to scientists, because it is believed to be covered by water ice. Following these encounters *New Horizons* will perform flybys of objects in the Kuiper Belt.

New Horizons was conceived in 2001 and is the first mission to be conducted under NASA's New Frontiers program for medium-class planetary missions. The spacecraft will be operated for NASA by the Johns Hopkins University Applied Physics Laboratory in Laurel, Maryland.

FUTURE MISSIONS TO THE FAR PLANETS

NASA's next planned mission to a far planet was supposed to be the Jupiter Icy Moons Orbiter (JIMO). This mission was to focus on the three large moons of Jupiter named Europa, Ganymede, and Callisto. However, in 2004–05 NASA decided to focus its limited funds on crewed missions to Earth's moon and Mars, and the JIMO project was cancelled.

CHAPTER 9
PUBLIC OPINION ABOUT SPACE EXPLORATION

We will build new ships to carry man forward into the universe, to gain a new foothold on the moon, and to prepare for new journeys to worlds beyond our own.

—President George W. Bush, January 14, 2004

How will a country at war and in deficit pay for such things?

—*Harvard Independent Newsmagazine*, February 12, 2004

Humans seem to have an inherent desire to surmount great obstacles and push into new frontiers. There have always been brave people willing to risk their lives on bold and dangerous journeys into uncharted territory. They have climbed Mount Everest, traversed wild jungles, crossed barren deserts, and sailed stormy seas. Successful explorers become popular heroes. Their achievements thrill and delight people who do not have the ability, resources, or courage to go themselves.

The U.S. space program taps into this spirit of adventure. Astronauts became the heroic explorers of the twentieth century. They opened new frontiers and set foot on the Moon. These successes were achieved at a high price. They cost the country human lives and billions of dollars that some critics say could have been spent feeding the poor, healing the sick, and housing the homeless. Was it worth it?

Space exploration is appealing on a psychological level. It is awesome, daring, and closely associated with American can-do optimism and patriotic pride. A robust space program also showcases and strengthens U.S. capabilities in science, engineering, and technology. These are powerful motivations to keep venturing out into space.

On the other side lie the staggering problems facing American society—incurable diseases, crime, poverty, pollution, unemployment, and war. These are expensive problems. People concerned with poor social conditions resent the billions spent in outer space. Within the scientific community many respected researchers would rather see scarce funds devoted to Earth-related research than space science. There are promising scientific and medical frontiers on this planet that still need to be explored.

In a democratic society the public gets to weigh the relative costs and benefits of national goals and decide which ones to pursue. Public opinion polls show that most Americans have an uneasy devotion to the nation's space travel agenda. They love the idea, but hate paying the bill. Sometimes they wonder if money spent on space might be better spent here on Earth. It is a debate that has raged since the earliest days of space exploration and probably always will be a prime issue in space science. As succinctly remarked in the film version of *The Right Stuff*, Tom Wolfe's chronicle of the early days of the American space program: "No bucks, no Buck Rogers."

IS SPACE EXPLORATION IMPORTANT TO SOCIETY?

In November 1999, as the century came to a close, the Gallup Organization asked people to rank eighteen specific events of the twentieth century in order of importance. The resulting list is shown in Table 9.1. Landing a man on the moon ranked seventh in importance. This put it behind major events associated with World War I, World War II, and important social milestones that granted rights to women and minorities. Fifty percent of those asked believed that landing a man on the moon was the most important event of the century.

A second space-travel milestone also made the top eighteen list. Ranked fourteenth was the launching of the Russian Sputnik satellites during the 1950s. Twenty-five percent of those asked rated this one of the most important events of the century. Charles Lindbergh's historic flight across the Atlantic in 1927 also made the list, coming in at number thirteen.

TABLE 9.1

FIGURE 9.1

Public opinion poll on the most important events of the twentieth century, 1999

1. World War II
2. Women gaining the right to vote in 1920
3. Dropping the atomic bomb on Hiroshima in 1945
4. The Nazi Holocaust during World War II
5. Passage of the 1964 Civil Rights Act
6. World War I
7. Landing a man on the moon in 1969
8. The assassination of President Kennedy in 1963
9. The fall of the Berlin Wall in 1989
10. The U.S. Depression in the 1930s
11. The breakup of the Soviet Union in the early 1990s
12. The Vietnam War in the 1960s and early 1970s
13. Charles Lindbergh's transatlantic flight in 1927
14. The launching of the Russian Sputnik satellites in the 1950s
15. The Korean War in the early 1950s
16. The Persian Gulf War in 1991
17. The impeachment of President Bill Clinton in 1998
18. The Watergate scandal involving Richard Nixon in the 1970s

SOURCE: Adapted from Frank Newport, David W. Moore, and Lydia Saad, "The 18 Events Were Then Rank-Ordered Based on the Percentage of Americans Who Placed Each in the Top Category as 'One of the Most Important Events of the Century'," in *The Most Important Events of the Century from the Viewpoint of the People*, The Gallup Organization, December 6, 1999, http://poll.gallup.com/content/default.aspx?ci=3427&pg=1 (accessed February 4, 2006). Copyright © 1999 by The Gallup Organization. Reproduced by permission of The Gallup Organization.

Public opinion poll of teens on the future influence of various technological achievements, 2003

HOW MUCH INFLUENCE DO YOU THINK THESE THINGS WILL HAVE UPON YOUR FUTURE? DO YOU THINK EACH OF THE FOLLOWING WILL HAVE A LOT, SOME, OR NOT MUCH INFLUENCE UPON YOUR FUTURE?

[Percentage of teens aged 13 to 17]

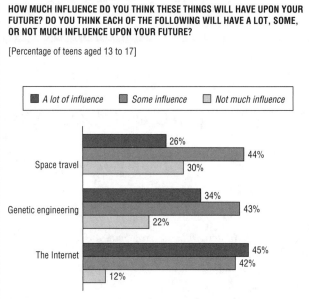

SOURCE: Jennifer Robison, "How Much Influence Do You Think These Things Will Have upon Your Future? Do You Think Each of the Following Will Have A Lot, Some, or Not Much Influence upon Your Future?" in *Teens and the Future: Forecast vs. Fiction*, The Gallup Organization, March 18, 2003, http://poll.gallup.com/content/default.aspx?ci=8008&pg=1 (accessed January 31, 2006). Copyright © 2003 by The Gallup Organization. Reproduced by permission of The Gallup Organization.

In February 2003 Gallup asked 1,200 teenagers across the country to assess three technological phenomena in terms of their potential impact upon the future. The results are shown in Figure 9.1. The Internet received the highest ranking of the three choices. Eighty-seven percent of the teenagers asked believe that the Internet will have at least some influence upon their future. This compares with 77% for genetic engineering and 70% for space travel. Clearly most teenagers believe that space travel is an endeavor of importance to their future, but feel that it may not play as large a role as computer and biological technologies.

A Gallup poll conducted during June and July of 2004 found that a majority of people asked were "interested" in the space program. (See Figure 9.2.) Only 11% reported they were not at all interested. More men (34%) than women (15%) indicated they were "very interested" in the space program. Interest was also higher among respondents aged fifty to sixty-four years old. When provided with five possible reasons for space exploration, people most commonly chose the reason that it is "human nature to explore." This answer garnered 29% of the vote, as shown in Figure 9.3. Another 21% believed that space exploration was primarily performed to help maintain the nation's status as an international leader in space. Nearly as many respondents (18%) thought the main reason was to provide benefits on Earth. Small percentages believed that Americans explore space to ensure national security (12%) or inspire people and motivate children (10%).

SHOULD SPACE TRAVEL BE A NATIONAL PRIORITY?

History shows that space travel was a national priority during the 1960s. President John F. Kennedy and Vice President Lyndon Johnson were convinced that putting a man on the Moon was vital to American political interests during the Cold War. They convinced Congress to devote billions of dollars to the effort. At the time, the public was not very enthusiastic about the idea. According to the Gallup Organization, most polls they conducted during the 1960s showed that less than 50% of Americans considered the endeavor worth the cost.

In 1967 civil rights leader Martin Luther King, Jr., said "Without denying the value of scientific endeavor, there is a striking absurdity in committing billions to reach the moon where no people live, while only a fraction of that amount is appropriated to service the densely populated slums." King's sentiment sums up very well a moral question that has plagued the space program since its inception. Is it right for a nation to spend its money on space travel while there are people suffering on Earth?

NASA would argue that its budget comprises only a tiny fraction of the nation's total spending. Figure 2.1 in

FIGURE 9.2

Public opinion poll on interest in the space program, 2004

No opinion/don't know
4%

Not at all interested
11%

Very interested
24%

Not very interested
18%

Somewhat interested
43%

SOURCE: Adapted from Darren K. Carlson, "Space: Who Cares?" in *Space: To Infinity and Beyond on a Budget*, The Gallup Organization, August 17, 2004, http://poll.gallup.com/content/default.aspx?ci=12727 (accessed December 28, 2005). Copyright © 2004 by The Gallup Organization. Reproduced by permission of The Gallup Organization.

FIGURE 9.3

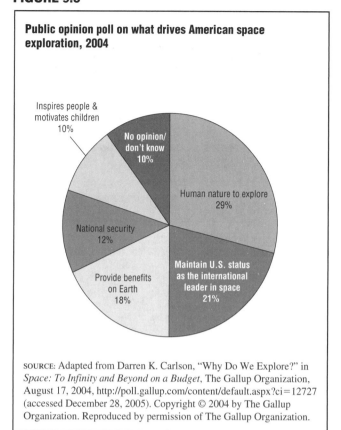

Public opinion poll on what drives American space exploration, 2004

Inspires people & motivates children
10%

No opinion/don't know
10%

Human nature to explore
29%

National security
12%

Maintain U.S. status as the international leader in space
21%

Provide benefits on Earth
18%

SOURCE: Adapted from Darren K. Carlson, "Why Do We Explore?" in *Space: To Infinity and Beyond on a Budget*, The Gallup Organization, August 17, 2004, http://poll.gallup.com/content/default.aspx?ci=12727 (accessed December 28, 2005). Copyright © 2004 by The Gallup Organization. Reproduced by permission of The Gallup Organization.

Chapter 2 shows that in 2005 NASA received less than 1% of the federal budget and has been near this level since the end of the Apollo program in 1972.

In June and July 2004 Gallup asked 1,000 adults their thoughts about government spending on space exploration. Poll participants were reminded that NASA's budget request for the next year would be less than 1% of the federal budget and would amount to an average of approximately $55 per taxpayer. In response, 37% said that NASA's budget should remain at its current level, while 26% thought it should be increased. Another 23% thought NASA's budget should be decreased, and 13% wanted no NASA funding at all. (See Figure 9.4.)

Gallup has been asking this same poll question since 1984. The percentage of people wanting to increase NASA's budget has varied between 10% and 27% over time. Consistently, the largest group of people (37% to 51%) has advocated maintaining the agency's budget at its existing level.

There are many programs competing for funding in the federal budget. During a 2003 survey, Gallup asked poll participants if money should be taken away from the space program and devoted to other programs instead. The answers varied widely, depending on the program against which space travel was paired.

The largest number of respondents (74%) would transfer money from the space program to increase funding

for healthcare. National defense was also a pressing priority, with 60% of those asked willing to take money away from the space program for national defense. Only 38% of poll participants wanted to increase funding for the nation's welfare program at the expense of the space program.

SHOULD SPACE TRAVEL BE A SCIENCE PRIORITY?

In the 1960s television show *Star Trek,* space was called "the final frontier." While this may be true from a philosophical viewpoint, it does not apply as well to the realm of science. Geneticists, oceanographers, geologists, and biologists maintain that there are still many scientific and medical frontiers to be explored on Earth.

Since 1998 marine biologist Sylvia Earle has been an explorer-in-residence at the National Geographic Society. That same year *Time* magazine named her a "hero for the planet." In 2004 the Associated Press asked Earle about NASA's discovery that water once existed on Mars (*Mars Critics Wonder if Billions Aren't Better Spent Elsewhere*, Environmental News Network, March 9, 2004). Earle stated that "the resources going into the investigation of our own planet and its oceans are trivial compared to investment looking for water elsewhere in the universe. Real oceans need scientific attention more

FIGURE 9.4

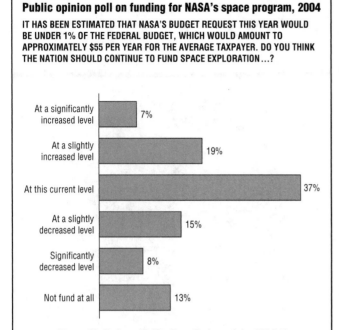

Public opinion poll on funding for NASA's space program, 2004

IT HAS BEEN ESTIMATED THAT NASA'S BUDGET REQUEST THIS YEAR WOULD BE UNDER 1% OF THE FEDERAL BUDGET, WHICH WOULD AMOUNT TO APPROXIMATELY $55 PER YEAR FOR THE AVERAGE TAXPAYER. DO YOU THINK THE NATION SHOULD CONTINUE TO FUND SPACE EXPLORATION...?

At a significantly increased level	7%
At a slightly increased level	19%
At this current level	37%
At a slightly decreased level	15%
Significantly decreased level	8%
Not fund at all	13%

SOURCE: Darren K. Carlson, "It Has Been Estimated that NASA's Budget Request This Year Would be Under 1% of the Federal Budget, Which Would Amount to Approximately $55 Per Year for the Average Taxpayer. Do You Think the Nation Should Continue to Fund Space Exploration?" in *Space: To Infinity and Beyond on a Budget*, The Gallup Organization, August 17, 2004, http://poll.gallup.com/content/default.aspx?ci=12727 (accessed December 28, 2005). Copyright © 2004 by The Gallup Organization. Reproduced by permission of The Gallup Organization.

than the dried-up remnants on Mars." She said that she does not want to cut funding for space science, but noted that "we have better maps of Mars than our own ocean floor. That's just not right."

Amitai Etzioni is a sociologist at George Washington University in Washington, D.C., and a long-time critic of the U.S. space program. He believes that the scientific community should focus more attention on Earth's oceans because of their potential to yield new energy and food sources or medical breakthroughs that would benefit humanity. He also criticizes the money spent looking for water on Mars and asks, "What difference does it make to anyone's life? Will it grow any more food? Cure a disease? This doesn't even broaden our horizons." Etzioni believes that any crewed space missions should be financed by private investors, not with taxpayers' dollars.

CREWED VERSUS ROBOTIC MISSIONS?

Etzioni wrote the 1964 book *The Moon-Doggle* that questioned the scientific value of putting astronauts on the moon and criticized NASA for favoring expensive manned missions over cheaper, more productive robotic

missions. This complaint has been a common one in the scientific community from the 1960s onward.

It is very expensive to send explorers into space, particularly human ones. Robotic spacecraft can accomplish more for less money, but they lack the glamour of human explorers. Machines do not give television interviews from space or get ticker tape parades when they come back. Astronauts do. Human explorers inspire young people to be astronauts and encourage voters and politicians to keep funding space travel. NASA knows that machines simply do not reap the same public relations benefits as human astronauts.

Astronomers and physicists fought throughout the 1970s and 1980s for large, sophisticated observatories to be put in space to gather data about solar and galactic phenomena. Time and again funding for their programs was slashed, because NASA needed more money for the space shuttle and *International Space Station (ISS)* programs. NASA's Great Observatories (including the *Hubble Space Telescope*) eventually did make it into orbit.

Although smaller and weaker than what scientists originally wanted, these observatories are considered some of space science's greatest triumphs. The *Hubble Space Telescope* alone has captured thousands of images of celestial objects and greatly advanced human understanding about the origins and workings of the universe.

In 2004, however, NASA announced it would let *Hubble* fall out of orbit before the end of its useful life. The observatory needs an altitude boost that only a space shuttle mission can give it, but NASA is reluctant to risk astronaut lives for such a purpose. Since the 2003 *Columbia* disaster the Agency has been hypersensitive about shuttle safety issues. Also, NASA has switched its focus to President George W. Bush's new space travel mandate. This plan calls for devoting shuttle missions to finishing the *ISS* as soon as possible and then retiring the shuttle fleet. Bush wants NASA to concentrate on developing new spacecraft for long-distance flights to the Moon and Mars.

The *Hubble* telescope decision met with disapproval from many astronomers and space scientists who were once again disappointed to see human missions given priority over robotic ones. In April 2005 a new NASA Administrator, Mike Griffin, indicated the agency is rethinking its decision to cancel the servicing mission. He instructed NASA engineers to begin preparing for such a mission in case the funding and opportunity for it become available. However, there was no money included for an *HST* servicing mission in NASA's 2006 budget request.

Two of the most outspoken critics of human space-flight are physicists James Van Allen and Robert Park. Van Allen is an astrophysicist credited with discovering the Van Allen radiation belts around Earth. Park is a physics professor at the University of Maryland. Both were very vocal in 2004 in their opposition to the plan presented by President George W. Bush to send astronauts to the Moon and to Mars.

In an interview with radio show *Democracy Now!* Allen explained his opinion of human spaceflight: "I'm a critic of it in terms of the yield of either scientific results or any results from the human space flight program that's been very meager" ("Bush's New Space Program Criticized over Costs & Nuclear Fears," January 15, 2004). In an article for the journal *New Atlantis*, Park also advocated robotic space travel over astronauts ("The Virtual Astronaut," winter 2004, http://www.thenewatlantis.com/archive/4/park.htm). He speculates that Christopher Columbus would have sent out a drone (an unmanned vessel) to search for the new world, if he had the technology. Park maintains that "the great adventure worthy of the twenty-first century is to explore where no human can ever set foot."

Following the *Challenger* and *Columbia* shuttle disasters in 1986 and 2003, respectively, Gallup asked whether the United States should concentrate on unmanned missions or also include manned missions. (See Figure 9.5.) In both polls a significant majority of the respondents expressed support for manned missions. The percentage actually increased from 67% in 1986 to 73% in 2003. Obviously Americans want human explorers to venture into space.

AMERICA RATES NASA'S PERFORMANCE

During 2005 Gallup asked Americans to rate the job being done by NASA as excellent, good, fair, or poor. Polls were conducted before and after the successful mission of the space shuttle *Discovery* on its first "return to flight." The results are shown in Table 9.2. Prior to the *Discovery* mission 53% of those asked rated NASA's performance as good or excellent. This percentage increased to 60% after the mission.

Gallup has asked this same question about NASA's performance in many polls conducted since 1990. As shown in Table 9.2, there have been a number of peaks and valleys in the percentage of people rating the agency as "excellent" or "good." The highest approval (76%) was recorded in a poll conducted November 22–24, 1998. This was only two weeks after John Glenn's flight aboard the space shuttle *Discovery*. The next three polls saw NASA's rating slip dramatically, reaching 50% in September 2003. This was a few months after the *Columbia* shuttle disaster. NASA's image has improved slightly since that time.

FIGURE 9.5

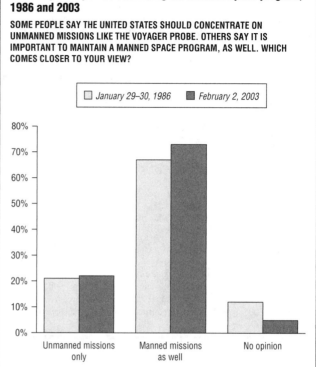

Public opinion poll on continuing the manned space program, 1986 and 2003

SOME PEOPLE SAY THE UNITED STATES SHOULD CONCENTRATE ON UNMANNED MISSIONS LIKE THE VOYAGER PROBE. OTHERS SAY IT IS IMPORTANT TO MAINTAIN A MANNED SPACE PROGRAM, AS WELL. WHICH COMES CLOSER TO YOUR VIEW?

SOURCE: Adapted from Frank Newport, "Some People Say the United States Should Concentrate on Unmanned Missions like the Voyager Probe. Others Say It is Important to Maintain a Manned Space Program, as Well. Which Comes Closer to Your View?" in *Americans Want Space Shuttle Program to Go On*, The Gallup Organization, February 3, 2003, http://poll.gallup.com/content/default.aspx?ci=7708&pg=1 (accessed January 31, 2006). Copyright © 2003 by The Gallup Organization. Reproduced by permission of The Gallup Organization.

NASA's most horrific failures occurred in 1986 and 2003, when space shuttles were lost in accidents. Seven astronauts died each time. Days after each disaster Gallup assessed the public's confidence in NASA's ability to avoid similar accidents in the future. Following the 1986 loss of the space shuttle *Challenger*, 79% of respondents expressed confidence that another shuttle loss could be avoided. When the *Columbia* shuttle was destroyed during reentry in 2003 this confidence proved to be misplaced. Interestingly enough, the public's confidence level actually increased to 82% after the second accident. The numbers suggest that Americans remain optimistic about NASA's competency in regards to the space shuttle program.

In August 2003 Gallup surveyed 1,003 adults regarding their expectations about the risks associated with the space shuttle program. Most respondents (43%) believed that a fatal crash every 100 missions was an "acceptable price to pay" to advance America's space exploration goals. In reality NASA's shuttle program has experienced two crashes during 114 missions. This is an average of one fatal crash every fifty-seven missions.

TABLE 9.2

Public opinion poll on the performance of NASA, selected years 1990–2005

HOW WOULD YOU RATE THE JOB BEING DONE BY NASA—THE U.S. SPACE AGENCY? WOULD YOU SAY IT IS DOING AN EXCELLENT, GOOD, ONLY FAIR, OR POOR JOB?

	Excellent	Good	Only fair	Poor	No opinion
	%	%	%	%	%
2005 Aug 5–7	16	44	29	8	3
2005 Jun 24–26	11	42	34	6	7
2003 Sep 8–10	12	38	36	10	4
1999 Dec 9–12	13	40	31	12	4
1999 Jul 13–4	20	44	20	5	11
1998 Nov 20–22	26	50	17	4	3
1998 Jan 30–Feb 1	21	46	21	4	8
1994 Jul 15–17	14	43	29	6	8
1993 Dec 17–19	18	43	30	7	2
1993 Sept 13–15	7	36	35	11	11
1991 May 2–5	16	48	24	6	6
1990 July 19–22	10	36	34	15	5

SOURCE: Frank Newport, "How Would You Rate the Job Being Done by NASA—the U.S. Space Agency? Would You Say It is Doing an Excellent, Good, Only Fair, or Poor Job?" in *Space Shuttle Program a "Go" for Americans,* The Gallup Organization, August 10, 2005, http://poll.gallup.com/content/default.aspx?ci=17761&pg=1 (accessed February 4, 2006). Copyright © 2005 by The Gallup Organization. Reproduced by permission of The Gallup Organization.

Seventeen percent of those asked expressed the belief that a successful space shuttle program should experience no fatal crashes at all. The vast majority (75%) appear to accept the loss of human lives as a regrettable, but expected, price to pay to advance the nation's space program.

This viewpoint also appeared in another Gallup poll conducted the day after the *Columbia* disaster. The vast majority of those asked (94%) said they were upset about the accident. However, 71% of the poll participants said that a second fatal shuttle accident was not unexpected. More than one-quarter of those surveyed (28%) were surprised that another shuttle had been lost during their lifetime.

Recent polls show support may be waning for the space shuttle program. Soon after the *Challenger* and *Columbia* space shuttle disasters the Gallup Organization polled 462 adults about their opinions on shuttle missions. More than 80% of the people asked in each poll thought the space shuttle program should continue. The same question was asked in June 2005 (before the first return-to-flight mission). Nearly three-quarters of those asked (74%) said the space shuttle program should continue, while 21% said the program should not continue.

A CBS News poll conducted in August 2005 during the shuttle return-to-flight mission found that support for the space shuttle program was down compared to years past ("Poll: Public up in Air on Shuttle," August 3, 2005, http://www.cbsnews.com/stories/2005/08/03/opinion/polls/

main713808.shtml). The 2005 poll indicated that 59% of respondents thought the space shuttle program was worth continuing. This value was down from 75% in 2003 and 72% in 1999.

In August 2005 during *Discovery*'s return-to-flight mission, Gallup pollsters asked Americans how confident they were that NASA could make the space shuttle safe to fly on future missions. The results indicated that a large majority (83%) were "very confident" or "somewhat confident." Another 13% were "not too confident" and 3% were "not at all confident."

PUBLIC OPINION ABOUT FUTURE SPACE PROGRAMS TO THE MOON AND MARS

In January 2004 President George W. Bush proposed a new agenda for the nation's space program that focuses on sending astronauts to the Moon and Mars. During the summer of 2004 Gallup polled Americans about their level of support for this new space exploration plan. The pollsters described the plan in general terms and noted that it was to be assumed that NASA's budget would not exceed 1% of the total federal budget. As shown in Figure 9.6 a majority of those asked (68%) supported the plan. Another 24% were opposed to the plan, while 6% expressed a neutral opinion.

However, a Gallup poll conducted later indicated that Americans provide somewhat different viewpoints when this question is phrased in a different manner. In June 2005 Gallup asked people if they favored or opposed the United States "setting aside money" for a project to land an astronaut on Mars. The pollsters found that a majority of those asked (58%) opposed the idea, while 40% were in favor of it.

NASA administrator Michael Griffin publicly criticized the wording of this poll question. In an interview televised on the *Meet the Press* television show Griffin said that the question should have been phrased to ask Americans how best to spend the budget that NASA was going to be allocated in future years. Griffin believes that, given a choice between shuttle missions in low-Earth orbit and more adventuresome plans, Americans will choose the bolder undertaking.

PRACTICAL BENEFITS OF SPACE TRAVEL

Although it is widely acknowledged that space travel has psychological and scientific benefits to society, it is more difficult to point to everyday products that have directly resulted from the nation's space program. Certainly satellites have brought about great changes in telecommunications, navigation, military operations, and weather prediction. All of these developments do affect American lives. The technologies associated with space exploration have advanced the fields of robotics, computer programming,

FIGURE 9.6

Public opinion poll on support for President Bush's new plan for space exploration, 2004

THIS YEAR, A NEW PLAN OR GOAL FOR SPACE EXPLORATION WAS ANNOUNCED. THE PLAN INCLUDES A STEPPING-STONE APPROACH TO RETURN THE SPACE SHUTTLE TO FLIGHT, COMPLETE ASSEMBLY OF THE SPACE STATION, BUILD A REPLACEMENT FOR THE SHUTTLE, GO BACK TO THE MOON, AND THEN ON TO MARS AND BEYOND. IF NASA'S BUDGET DID NOT EXCEED 1% OF THE FEDERAL BUDGET, TO WHAT EXTENT WOULD YOU SUPPORT OR OPPOSE THIS NEW PLAN FOR SPACE EXPLORATION? WOULD YOU STRONGLY SUPPORT IT, SUPPORT IT, OPPOSE IT, OR STRONGLY OPPOSE IT?

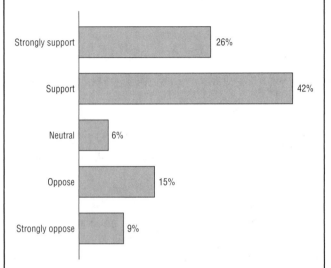

SOURCE: Darren K. Carlson, "This Year, a New Plan or Goal for Space Exploration was Announced. The Plan Includes a Stepping-Stone Approach to Return the Space Shuttle to Flight, Complete Assembly of the Space Station, Build a Replacement for the Shuttle, Go Back to the Moon, and Then on to Mars and Beyond. If NASA's Budget Did Not Exceed 1% of the Federal Budget, to What Extent Would You Support or Oppose This New Plan for Space Exploration? Would You Strongly Support It, Support It, Oppose It, or Strongly Oppose It?" in *Space: To Infinity and Beyond on a Budget*, The Gallup Organization, August 17, 2004, http://poll.gallup.com/content/default.aspx?ci=12727 (accessed December 28, 2005). Copyright © 2004 by The Gallup Organization. Reproduced by permission of The Gallup Organization.

and cryogenics (the physics of extremely cold temperatures). In addition, improvements based on NASA technologies have been incorporated into such diverse products as memory foam mattresses, medical imaging devices, eyeglass lenses, golf balls, baby food, pacemakers, and life rafts.

One of the mandates of the 1958 National Aeronautics and Space Administration Act is that the agency (and its contractors) must publicize any new developments significant to commercial industry. NASA accomplishes this through four publications: the newsletter *Technology Innovation*; a monthly magazine for engineers, managers, and scientists called *NASA Tech Briefs* that briefly describes new technologies; *Technical Support Packages*, which describe in detail the technologies presented in *NASA Tech Briefs*; and *Spinoff*, an annual publication describing successfully commercialized NASA technology.

In 2003 NASA released a booklet called *NASA Hits: Rewards from Space—How NASA Improves Our Quality of Life*. The booklet describes many practical benefits associated with NASA's work in space flight, space science, Earth science, and aeronautical research and development, including:

- Communications satellite technology
- Medical monitoring systems used in intensive care units
- The Hazard Analysis and Critical Control Point (HACCP) system for ensuring food safety
- The NASTRAN software system for computerized design
- Space-based beacon locators used in satellite-based search and rescue systems
- Use of thin grooves in concrete airport runways and highways to improve drainage and reduce hydroplaning
- Advances in hydroponics (growing crops using water rather than soil to support plants)
- Improved hurricane forecasting and wildfire tracking using Earth-observing satellites
- Developments in microelectromechanical systems (extremely small devices and sensors about the diameter of a human hair)
- Combustion research that has improved the performance of jet engines
- Suspension techniques used by animal researchers
- A new light source now used to improve chemotherapy treatment for cancer patients
- Needle-based biopsies used in breast cancer diagnosis
- Bioreactors (devices used to turn cell cultures into functional tissue)
- Lifeshears (a hand-held shearing tool used by rescue workers to free people trapped in cars or underneath rubble)

The booklet also discusses patents and Nobel prizes associated with NASA-funded research and development.

In 1988 NASA and the private organization, The Space Foundation, established the Space Technology Hall of Fame. Its stated purpose is "to honor the innovators who have transformed space technology into commercial products, to increase public awareness of the benefits of space technology and to encourage further innovation." Each year a handful of space-based technologies are selected for induction into the Hall of Fame. Inductees are honored at an annual conference held in

TABLE 9.3

Space Technology Hall of Fame inductees, 2004–05

Year	Technology	Description
2005	InnerVue™ Diagnostic Scope System	Uses space image enhancement technology and a disposable micro-invasive endoscope to enable doctors to see clearly inside joints with minimum patient discomfort.
	NanoCeram® Superfilters	Water filters comprised of nanometer-size particles. Far exceed current filtration systems and can handle extremely difficult treatment requirements.
	Outlast Technologies, Inc. Smart Fabric Technology™	Contains micro-materials that can absorb, store and release heat. Used in consumer products such as active wear. Derived from research on materials to protect astronauts in space.
	Portable Hyperspectral Imaging Systems	Portable device for hyperspectral imagery (energy measurement). Has diagnostic applications in the bio-medical, forensics, counter terrorism, food safety, and Earth imaging markets.
2004	LADARVision 4000 (LASIK eye surgery)	Uses a laser and eye-tracking device to reshape the cornea with extreme precision. Based on technology used to assist spacecraft in delicate docking maneuvers.
	MedStar Medical/Health Monitoring System	Miniature physiological monitoring device that can collect and analyze numerous signals in real time. Used to monitor astronauts on the ISS.
	Multi-Junction (MJ) Space Solar Cells	High-efficiency solar-cell technology that reduces costs for the life cycle of space missions, telecommunication, weather forecasting and other services.
	Precision Global Positioning System (GPS) Software System	Sophisticated system that delivers information enabling real-time positioning accurate to within a few inches. Based on software developed at JPL to determine the location of satellite orbits.

SOURCE: Adapted from "2004 Space Technology Hall of Fame," in *Technology Innovation*, vol. 11, no. 4, National Aeronautics and Space Administration, Fall 2004, http://nctn.hq.nasa.gov/innovation/innovation114/0-content.html and "2005 Space Technology Hall of Fame," in *Technology Innovation*, vol. 12, no. 1, National Aeronautics and Space Administration, 2005, http://ipp.nasa.gov/innovation/innovation115/0-content.html (accessed December 29, 2005)

Colorado Springs, Colorado, called the National Space Symposium. Table 9.3 lists technologies honored in recent years. Previous Hall of Fame winners familiar to consumers include satellite radio technology and DirecTV satellite systems.

The Public Speaks Out

Polls were conducted on the tenth, twenty-fifth, and thirtieth anniversaries of the *Apollo 11* moon landing to quiz the public regarding the benefits of the nation's space program.

In each poll the participants were asked whether they believe the space program has benefited the country enough to justify its costs. A 1979 poll conducted by NBC News and the Associated Press found that only 41% of respondents considered the benefits worth the costs. A majority (53%) thought the expense was not worth what was accomplished. In a 1994 Gallup poll Americans were evenly split on the issue, with 47% taking each side. By 1999 the space program had earned a bit more respect. Gallup reported that 55% of those asked believed the space program's benefits justified its cost, while 40% did not.

In June and July 2004 Gallup asked Americans whether they agreed or disagreed with the statement "the quality of our daily lives has benefited from the knowledge and technology that have come from our nation's space program." As shown in Figure 9.7 more than two-thirds of those asked (68%) agreed with the statement, while 16% disagreed. Another 16% were neutral on the subject.

NASA WOOS THE AMERICAN PUBLIC

NASA employs a number of public relations tools designed to interest and excite people about space travel. Since its inception the agency has recognized that public support is crucial to fostering a successful long-term space program.

Television

Throughout the Space Age NASA has used television as a publicity tool to try to spark greater interest in the space program. Television turned out to be one of the greatest public relations tools of the Apollo program. In 1968 the *Apollo 7* astronauts conducted the first live television interview from space. All of the remaining Apollo flights carried television cameras. The worldwide television audience for the *Apollo 11* moon landing was estimated at half a billion people.

In July 1999 the Gallup Organization polled Americans about their memories of the first manned lunar landing by *Apollo 11*. The survey found that 76% of people aged thirty-five and up claimed to have watched the event on television as it happened.

NASA's Web Site

NASA operates one of the most colorful and elaborate Web sites of any federal agency (http://www.nasa.gov/home/). It includes thousands of mission photographs and millions of documents related to the nation's space endeavors. The Web site provides detailed information about NASA facilities, programs, and missions. There are a variety of multimedia features, including interactive displays, video and audio downloads, and spectacular images of Earth and space captured by NASA spacecraft. Enormous

FIGURE 9.7

Public opinion poll on benefits from the space program, 2004

Neutral
16%

Disagree with
statement
16%

Agree with statement
68%

SOURCE: Adapted from Darren K. Carlson, "The Quality of Our Daily Lives Has Benefited from the Knowledge and Technology that Have Come from Our Nation's Space Program," in *Space: To Infinity and Beyond on a Budget*, The Gallup Organization, August 17, 2004, http://poll.gallup.com/content/default.aspx?ci=12727 (accessed December 28, 2005). Copyright © 2004 by The Gallup Organization. Reproduced by permission of The Gallup Organization.

FIGURE 9.8

Interpreting spacecraft sighting data

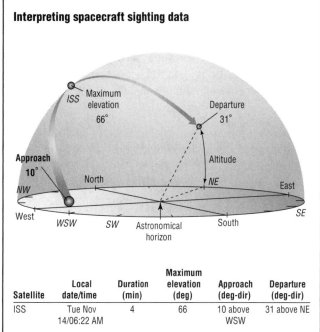

Satellite	Local date/time	Duration (min)	Maximum elevation (deg)	Approach (deg-dir)	Departure (deg-dir)
ISS	Tue Nov 14/06:22 AM	4	66	10 above WSW	31 above NE

SOURCE: "Interpreting the Data," in *Sighting Opportunities*, National Aeronautics and Space Administration, April 17, 2003, http://spaceflight.nasa.gov/realdata/sightings/help.html (accessed January 31, 2006)

historical archives that include documentation dating back to the earliest days of space travel are available online.

According to NASA, the Web site is visited millions of times each day. The number of "hits" increases dramatically during highly publicized missions. For example, NASA reported receiving 6.53 billion hits between January 4, 2004, and February 19, 2004. This period of time coincides with the highly successful landings of the Mars Exploration rovers on Mars.

SIGHTING OPPORTUNITIES. One of the ways that NASA tries to engage public interest in space travel is by posting sighting opportunities for its satellites, particularly the *International Space Station (ISS)* and any ongoing shuttle missions. The NASA Web site instructs people how and where to look in the nighttime sky to see the spacecraft as it is passing overhead. Figure 9.8 shows a set of instructions for viewing the *ISS* at a particular location, assuming that skies are clear enough.

The listing identifies the exact date and time at which the *ISS* should become visible to observers on the ground and how long it will remain visible. It also gives information about the station's location in the sky based on direction (north, south, east, or west) and angle of elevation compared to the horizon. A spacecraft flying directly overhead would be at 90° maximum elevation.

In the example shown the *ISS* is to first appear in the west-southwest (WSW) direction approximately 10°

above the horizon. It will then climb to a maximum elevation of 66° above the horizon and travel out of sight heading toward the northeast (NE). It should disappear from view about 31° above the horizon. This progression is illustrated in the diagram at the top of Figure 9.8.

NASA says that a spacecraft looks like "a steady white pinpoint of light moving slowly across the sky." Viewers are urged to observe spacecraft with the naked eye or through binoculars. The speed at which spacecraft move makes telescope viewing impractical.

NASA's Web site provides links to sighting data for hundreds of cities around the world. People at locations not listed can use an applet (a small application program) called SkyWatch to enter their latitude and longitude and receive viewing information for numerous orbiting satellites.

NTV

NASA operates its own television network called NASA TV or NTV. NTV broadcasts via satellite and cable and is streamed over the Internet. It features live coverage of NASA activities and missions, video of events for the news media, and educational programming for teachers and students.

The show *NASA Education Hour* plays at 8:00 AM EST every weekday morning and is rebroadcast at regular intervals throughout the day and night. Hour-long coverage of the *ISS* mission is presented live at 11:00 AM EST daily.

Ham Radio

Ham radios vary in signal strength and capability. The strongest stations can reach operators on the other side of the world by bouncing signals off the upper atmosphere or using satellites.

In November 1983 astronaut Owen Garriott (1930–) carried a small ham radio with him aboard the space shuttle *Columbia* (STS-9). During his spare time he used the radio to contact fellow ham operators around the world. This was the first of more than twenty-four shuttle missions that carried ham radio equipment so astronauts could communicate with their families and other ham operators worldwide and perform interviews for school children. The program was called the Space Amateur Radio Experiment (SAREX). The Soviet space agency operated a similar ham radio program for cosmonauts aboard the *Mir* space station.

In September 2000 the crew of the space shuttle *Atlantis* (STS-106) carried a ham radio to the *International Space Station (ISS)* for use by Expedition crews. The SAREX program was given the new name of Amateur Radio on the International Space Station (ARISS). Under the ARISS program *ISS* crewmembers can communicate with ham radio operators all over the world.

Educational Programs

NASA's strategic plan says that one of the agency's primary goals is to "inspire the next generation of explorers." In order to accomplish this goal NASA operates an extensive student education program designed to encourage young people to pursue studies in science, mathematics, technology, and engineering and careers in aeronautics and space science. The program's proposed budget for 2006 was $190 million. NASA prides itself on its educational programs and the partnerships it establishes with schools, museums, libraries, and science centers around the nation to reach as many young people as possible.

NASA EXPLORER SCHOOLS. The NASA Explorer School program was started in 2003 for educators teaching grades four through nine. Each year NASA selects fifty schools for the program and enters into a three-year partnership agreement with them. The schools are eligible for grants and summer training courses for science and mathematics teachers. The courses are provided at NASA centers around the country and present new teaching resources and tools to better educate students. According to a May 2005 press release, 87% of all NASA Explorer Schools are located in high-poverty areas and 76% are in predominantly minority communities ("NASA Selects Florida and Puerto Rico Schools for

Explorer Program," May 25, 2005, http://www.nasa.gov/centers/kennedy/news/releases/2005/43-05.html).

TEACHER RESOURCE CENTERS. Teacher Resource Centers (TRCs) are offices maintained at NASA facilities around the country. TRCs serve as libraries that loan educational materials including lesson plans, audio and video tapes, slides, and miscellaneous print publications to teachers.

ASTRONAUT INTERVIEWS VIA AMATEUR RADIO. One of the most innovative ways that students can interact with astronauts in orbit is via ARISS, which is sponsored by NASA in conjunction with the American Radio Relay League and the Radio Amateur Satellite Corporation. Volunteers set up ham radio stations at schools so that students can interview *ISS* crewmembers.

CURRICULUM DEVELOPMENT. In 2004 NASA announced a new partnership with Pearson Scott Foresman (PSF), a leading publisher of educational products for elementary schools. PSF will draw upon publications in the NASA archives to create new science textbooks and other learning materials for the classroom. According to NASA, the goal of the program is to "spark student imagination, encourage interest in space exploration, and enhance elementary science curricula."

Art Program

In 1962 NASA administrator James Webb established the NASA Art Program to encourage and collect works of art about aeronautics and space. As of 2005 the NASA art collection includes more than 800 works of art in a variety of media including paintings, drawings, poems, and songs. NASA has donated more than 2,000 of its art works (including a number by Norman Rockwell) to the National Air and Space Museum in Washington, D.C. Other famous artists that have participated in the program include Annie Leibovitz, William Wegman, Andy Warhol, and Jamie Wyeth.

More than 200 artists have provided art works to the program. Many of the pieces are displayed at art galleries and museums around the country. NASA centers, particularly the Kennedy Space Center in Florida, also display the art works in their visitor areas.

Astronauts

Astronauts have always been NASA's greatest public relations agents. The early astronauts became instant heroes during the 1950s and 1960s. They were flooded with fan mail and held up by the media as sterling role models of what was great and daring about America. After the first Moon landing in 1969 public interest in the space program began to fade. The astronauts of later decades were still admired and respected, but not treated to the same level of hero worship as their predecessors.

During the early 1980s NASA decided to include a new type of astronaut on space shuttle flights to catch the public's attention. The agency began the Educator-in-Space program. NASA hoped that sending a teacher into space would excite the nation's schoolchildren and foster goodwill toward the space program. Teacher Christa McAuliffe (1948–86) was selected and trained for a mission aboard the space shuttle *Challenger*. Sadly, she was killed with the other crew members in 1986 when the shuttle exploded soon after liftoff.

NASA's public relations experiment turned into a nightmare. The catastrophe brought harsh criticism of the agency. The shuttle program was found to have serious management and safety problems. The loss seemed even more poignant to the public because a teacher, an everyday kind of person, had been one of the victims. NASA decided that space shuttle travel was not routine enough to risk the lives of private citizens as good-will ambassadors.

In 1998 NASA relented somewhat and allowed former Mercury astronaut John Glenn to ride aboard the space shuttle *Discovery* on mission STS-92. At the time Glenn was a seventy-seven-year-old senator from Ohio. NASA said the mission would reveal new knowledge about the effects of weightlessness and bone loss in older people. Critics complained that it was nothing more than a publicity stunt. Whatever the motivation, the event did greatly improve NASA's image. The public was entranced by the idea of an old hero traveling back into space.

Tourist Attractions

Many NASA facilities have become popular tourist attractions. This is particularly true for centers associated with the Apollo and space shuttle programs. Most NASA facilities operate their own visitor centers for which admission is free. Johnson Space Center (Houston, Texas), Kennedy Space Center (Cape Canaveral, Florida), and Marshall Space Flight Center (Huntsville, Alabama) have privately operated tourist centers that charge a fee for admittance.

Contests and Gimmicks

One relatively new way that NASA engages the public in space travel is by holding spacecraft-naming contests. During the 1990s NASA held contests that chose the names for the Mars Pathfinder mission's *Sojourner* rover and the *Magellan* spacecraft.

In 1998 the agency asked people to suggest names for an x-ray telescope to be launched as part of the Great Observatories Program. Each entry had to be supported by a short essay justifying why the name was appropriate. More than 6,000 people entered the contest, representing every state in the country and sixty-one other nations.

Two winning essays were selected. Both suggested the name Chandra, in honor of late Indian-American Nobel laureate, Subrahmanyan Chandrasekhar (1910–95). The winners were a high-school student from Laclede, Idaho, and a high-school teacher from Camarillo, California.

During 2001 a similar contest was held to name an infrared telescope intended for the Great Observatories Program. More than 7,000 entries were received from people around the world. NASA chose the name Spitzer in honor of the late American physicist Dr. Lyman Spitzer (1914–97). The winning essay came from a Canadian astronomy enthusiast.

In 2002 NASA held a contest for children to name the planned Mars Rover craft. The contest was held in partnership with the Planetary Society and the Lego Company. A nine-year-old girl from Scottsdale, Arizona, wrote the winning essay, which suggested the names *Spirit* and *Opportunity*. Hers was one of nearly 10,000 entries in the contest.

Another public relations device used by NASA is to ask people to submit their names for inclusion on CDs or DVDs carried by spacecraft. Numerous NASA missions conducted since the 1990s have included electronic disks carrying the names of millions of people. In 1999 the *Mars Polar Lander* carried a CD containing the names of one million schoolchildren from around the world. Unfortunately, the spacecraft was lost before it landed on Mars.

The highly successful Mars Explorer rovers *Spirit* and *Opportunity* carried mini-DVDs including the names of more than 3.5 million people. In 2003 and early 2004 hundreds of thousands of people registered their names for inclusion on a CD that would travel aboard the *Deep Impact* spacecraft to comet Tempel 1.

PUBLIC KNOWLEDGE ABOUT SPACE TRAVEL

In July 1999 the Gallup Organization conducted an extensive poll on space issues to commemorate the thirtieth anniversary of the *Apollo 11* moon landing. One of the questions concerned the number of astronauts who have actually walked on the Moon. Figure 9.9 shows that only 5% of those asked correctly answered that twelve different men have walked on the Moon. A vast majority (77%) of the respondents guessed too low. Another 11% guessed too high, and 7% had no opinion.

As shown in Table 9.4, only 50% of the people asked in July 1999 correctly named Neil Armstrong as the first person to walk on the Moon. Gallup reports that young people aged eighteen to twenty-nine were most likely to give the correct answer, despite the fact that the event occurred before they were born.

FIGURE 9.9

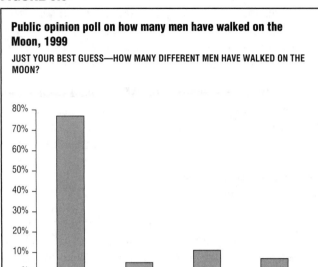

Public opinion poll on how many men have walked on the Moon, 1999

JUST YOUR BEST GUESS—HOW MANY DIFFERENT MEN HAVE WALKED ON THE MOON?

SOURCE: Adapted from Frank Newport, "Just Your Best Guess—How Many Different Men Have Walked on the Moon?" in *Landing a Man on the Moon: The Public's View*, The Gallup Organization, July 20, 1999, http://poll.gallup.com/content/default.aspx?ci=3712&pg=1 (accessed January 31, 2006). Copyright © 1999 by The Gallup Organization. Reproduced by permission of The Gallup Organization.

TABLE 9.4

Public opinion poll on the identity of the first person to walk on the Moon, 1989 and 1999

DO YOU HAPPEN TO KNOW WHO WAS THE FIRST PERSON TO WALK ON THE MOON?

	99 Jul 13–14	89 Jul 6–9
Neil Armstrong	50%	39%
John Glenn	13	
Alan Shepard	4	
Buzz Aldrin	2	
Other	3	
No opinion	28	61**
	100%	100%

**Incorrect; don't know.

SOURCE: Frank Newport, "Do You Happen to Know Who Was the First Person to Walk on the Moon?" in *Landing a Man on the Moon: The Public's View*, The Gallup Organization, July 20, 1999, http://poll.gallup.com/content/default.aspx?ci=3712&pg=1 (accessed January 31, 2006). Copyright © 1999 by The Gallup Organization. Reproduced by permission of The Gallup Organization.

Other astronauts receiving votes included John Glenn (13%), Alan Shepard (4%), and Buzz Aldrin (2%). More than a quarter of the people asked (28%) could not come up with an answer at all.

Gallup asked the same question back in July 1989 on the twentieth anniversary of the first human lunar landing. At that time only 39% of respondents gave the correct answer. The other 61% either did not know the answer or gave an incorrect answer.

Over a three-day period in early August 2003 Gallup pollsters asked 534 adults whether there were any U.S. astronauts in space at that time or not. Exactly half of the people contended that there were no U.S. astronauts in space at that time. Another 35% believed that U.S. astronauts were in space, while 15% had no opinion or did not know.

In reality there was one U.S. astronaut in space at the time. NASA Science Officer Ed Lu was aboard the *ISS*. The *ISS* has been continually inhabited by at least one American astronaut since 2000.

PERSONAL DESIRE TO EXPLORE SPACE

In June 1965 the Gallup Organization asked Americans if they would personally go to the Moon if given the chance. Only 13% said yes. Thirty-four years later Gallup asked the same question and found that 27% of respondents would go to the Moon.

Enthusiasm was greater for riding aboard the space shuttle. Three Gallup polls conducted between 1986 and 2003 found consistently that 30% to 40% of those asked wanted to be a passenger on a space shuttle flight.

In March 1986 nearly 40% of the respondents wanted to ride aboard the space shuttle. This value is surprising because the poll was conducted less than two months after the *Challenger* disaster, in which seven astronauts were killed. The 2003 poll occurred only a week after the *Columbia* shuttle was lost during reentry over the western United States. Again, seven astronauts died. Even in the face of that tragedy, 31% of respondents indicated a willingness to fly aboard a space shuttle sometime in the future.

The 2003 poll showed the desire to take a shuttle flight varied greatly by gender and age. Fifty-five percent of all male respondents under the age of fifty were eager to take the trip. Men older than fifty were less enthusiastic; 31% of them expressed a desire to go. Women respondents were even cooler about the idea. Only 21% of women younger than fifty and 13% of women older than fifty were enthused about taking a space shuttle flight.

In early 2004 Gallup polled teenagers aged thirteen to seventeen years about their desire to visit the Moon and Mars. As shown in Figure 9.10 a majority of those asked (59%) said they would like to go to the Moon if given the chance. Support was slightly weaker for going to Mars; 48% of the teens wanted to be the first person to go to Mars. Gallup found that the desire for space travel varied greatly by gender. Nearly three-quarters of the boys (74%) wanted to go to the Moon, compared to only 43% of the girls. Similarly, 64% of the boys wanted to be the first person to go to Mars, but only 31% of the girls expressed this desire.

FIGURE 9.10

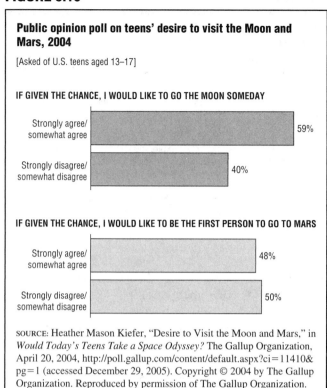

Public opinion poll on teens' desire to visit the Moon and Mars, 2004

[Asked of U.S. teens aged 13–17]

IF GIVEN THE CHANCE, I WOULD LIKE TO GO THE MOON SOMEDAY

Strongly agree/somewhat agree — 59%

Strongly disagree/somewhat disagree — 40%

IF GIVEN THE CHANCE, I WOULD LIKE TO BE THE FIRST PERSON TO GO TO MARS

Strongly agree/somewhat agree — 48%

Strongly disagree/somewhat disagree — 50%

SOURCE: Heather Mason Kiefer, "Desire to Visit the Moon and Mars," in *Would Today's Teens Take a Space Odyssey?* The Gallup Organization, April 20, 2004, http://poll.gallup.com/content/default.aspx?ci=11410&pg=1 (accessed December 29, 2005). Copyright © 2004 by The Gallup Organization. Reproduced by permission of The Gallup Organization.

FIGURE 9.11

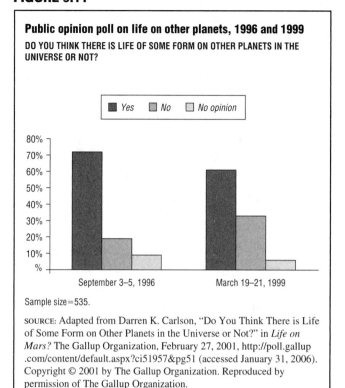

Public opinion poll on life on other planets, 1996 and 1999

DO YOU THINK THERE IS LIFE OF SOME FORM ON OTHER PLANETS IN THE UNIVERSE OR NOT?

Sample size=535.

SOURCE: Adapted from Darren K. Carlson, "Do You Think There is Life of Some Form on Other Planets in the Universe or Not?" in *Life on Mars?* The Gallup Organization, February 27, 2001, http://poll.gallup.com/content/default.aspx?ci51957&pg51 (accessed January 31, 2006). Copyright © 2001 by The Gallup Organization. Reproduced by permission of The Gallup Organization.

OPINIONS ABOUT SPACE TOPICS

Extraterrestrial Life

In 1996 and 1999 the Gallup Organization surveyed people about the possible existence of extraterrestrial (not Earth-related) life. The results are shown in Figure 9.11. Each time a fairly strong majority of poll participants expressed the opinion that some form of life does exist on other planets in the universe. Belief in extraterrestrial life declined somewhat between 1996 and 1999, dropping from 72% to 61%.

According to several Gallup polls conducted between 1973 and 1999, Americans are less convinced in the possibility of extraterrestrial human life. (See Figure 9.12.) Only 38% to 51% of poll participants agreed that there could be people somewhat like us living elsewhere in the universe. The latest poll, taken in 1999, reflects the greatest skepticism for the idea of extraterrestrial people. For the first time a majority (54%) of those asked did not believe that such people exist.

A Moon Hoax?

One of the most offbeat conspiracy theories of the Space Age is that the U.S. government faked the Apollo moon landings. In 2001 the Fox television network broadcast a show called *Conspiracy Theory: Did We Land on the Moon?* Guests on the show claimed that the Apollo program never actually put a man on the

FIGURE 9.12

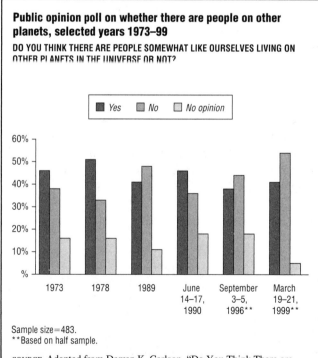

Public opinion poll on whether there are people on other planets, selected years 1973–99

DO YOU THINK THERE ARE PEOPLE SOMEWHAT LIKE OURSELVES LIVING ON OTHER PLANETS IN THE UNIVERSE OR NOT?

Sample size=483.
**Based on half sample.

SOURCE: Adapted from Darren K. Carlson, "Do You Think There are People Somewhat Like Ourselves Living on Other Planets in the Universe or Not?" in *Life on Mars?* The Gallup Organization, February 27, 2001, http://poll.gallup.com/content/default.aspx?ci=1957&pg=1 (accessed January 31, 2006). Copyright © 2001 by The Gallup Organization. Reproduced by permission of The Gallup Organization.

Moon but faked the lunar landing for television cameras. The theory continues to be supported on various Web sites on the Internet.

Advocates of the hoax theory rely on several key points to support their position. Chief among these are:

- NASA's moon photographs do not show stars in the background behind the astronauts.

- The American flag supposedly planted on the Moon by *Apollo 11* astronauts is rippling in a breeze, yet there is no atmosphere on the Moon.

- Humans could not have survived exposure to the intense radiation of the Van Allen belts lying between Earth and the Moon.

In general, NASA ignores the hoax claims and does not address them publicly. The NASA Web site does include one article of rebuttal titled "The Great Moon Hoax" (http://science.nasa.gov/headlines/y2001/ast23feb_2.htm). In it Dr. Tony Phillips addresses questions about moon photographs and the rippling flag. He points out that the exposure on the moon cameras had to be adjusted to tone down the dazzling brightness of the astronauts' sunlit spacesuits. This caused the background stars to be too faint to appear in the photographs. The rippling flag is explained by the wire inserts built into the fabric and by the twisting motion the astronauts used to push the flagpole into the lunar ground.

Phillips notes that Moon rocks are the best evidence that astronauts visited the Moon. There are more than 800 pounds of these rocks, and they have been investigated by researchers all over the world. Moon rocks differ greatly in mineral and water content from any rocks found on Earth. They also contain isotopes created by long-term exposure to high-energy cosmic rays on the lunar surface.

Another NASA Web site addresses the hoax issue concerning the Van Allen Radiation Belts, which are regions of highly energized ionized particles trapped within the geomagnetic fields surrounding Earth (http://imagine.gsfc.nasa.gov/docs/ask_astro/answers/

TABLE 9.5

Public opinion poll on whether the government faked the Apollo Moon landing, 1995 and 1999

THINKING ABOUT THE SPACE EXPLORATION, DO YOU THINK THE GOVERNMENT STAGED OR FAKED THE APOLLO MOON LANDING, OR DON'T YOU FEEL THAT WAY?

	Yes, staged	No	No opinion
1999 Jul 13–14	6%	89	5
1995 Jul 19–20**	6%	83	11

SOURCE: "Thinking about the Space Exploration, Do You Think the Government Staged or Faked the Apollo Moon Landing, or Don't You Feel That Way?" in *Did Men Really Land on the Moon?* The Gallup Organization, February 15, 2001, http://poll.gallup.com/content/default.aspx?ci=1993&pg=1 (accessed January 31, 2006). Copyright © 2001 by The Gallup Organization. Reproduced by permission of The Gallup Organization.

970630a.html). Astronomer Laura Whitlock of the Laboratory for High-Energy astrophysics says that early NASA researchers were also worried about the radiation belts. Scientists at the Oak Ridge National Laboratory (ORNL) in Oak Ridge, Tennessee, devised experiments in which bacteria and blood cells were sent aboard unmanned probes into space and returned to Earth. Animal experiments were also performed. ORNL used the resulting data to design special radiation shields for the Apollo spacecraft. The shields utilized materials left over from nuclear testing performed during the 1950s. Whitlock also notes that the Apollo spacecraft traveled so fast that the astronauts were exposed to Van Allen radiation for only a short time.

In a survey conducted in July 1999, Gallup asked poll participants their view about a possible moon-landing hoax. As shown in Table 9.5 the vast majority of those interviewed (89%) did not believe that the government staged the Apollo moon landing. Only 6% agreed that the landing was a hoax. Another 5% had no opinion. The results closely match those of a poll taken in 1995 by Time, CNN, and Yankelovich Partners, Inc. That poll also found that 6% of the people asked believed the moon landing was staged. Most people (83%) did not.

IMPORTANT NAMES AND ADDRESSES

Amateur Radio on the International Space Station
ARRL Headquarters
225 Main St.
Newington, CT 06111-1494
(860) 594-0200
FAX: (860) 594-0259
E-mail: hq@arrl.org
URL: http://www.rac.ca/ariss/

American Institute of Aeronautics and Astronautics
1801 Alexander Bell Dr., Suite 500
Reston, VA 20191-4344
(703) 264-7500
1-800-639-2422
FAX: (703) 264-7551
E-mail: custserv@aiaa.org
URL: http://www.aiaa.org/

Ames Research Center
Moffett Field, CA 94035-1000
(650) 604-5000
URL: http://www.nasa.gov/centers/ames/

Arecibo Observatory
HCO 3 Box 53995
Arecibo, PR 00612
(787) 878-2612
FAX: (787) 878-1861
E-mail: jacevedo@naic.edu
URL: http://www.naic.edu/

Canadian Space Agency
John H. Chapman Space Centre
6767 Route de l'Aéroport
Longueuil, Quebec, Canada J3Y 8Y9
(450) 926-4800
FAX: (450) 926-4352
E-mail: julie.simard@space.gc.ca
URL: http://www.space.gc.ca/asc/eng/

China Aerospace Science and Technology Corporation
E-mail: info@spaceproducts.com.cn
URL: http://www.spacechina.com/espace/

Columbia Accident Investigation Board
2900 South Quincy St., Suite 800
Arlington, VA 22206
URL: http://caib.nasa.gov/

Department of Energy Los Alamos National Laboratory
Community Relations
1619 Central Ave., MS A117
Los Alamos, NM 87545
(505) 665-4400
1-888-841-8256
FAX: (505) 665-4411
E-mail: community@lanl.gov
URL: http://www.lanl.gov/

Dryden Flight Research Center
P.O. Box 273
Edwards, CA 93523-0273
(661) 276 3311
URL: http://www.nasa.gov/centers/dryden/

European Space Agency
8-10 rue Mario Nikis
75738 Paris, France Cedex 15
(33) 1-53-69-76-54
FAX: (33) 1-53-69-76-60
E-mail: contactesa@esa.int
URL: http://www.esa.int/esaCP/index.html

Fédération Aéronautique Internationale
Avenue Mon-Repos 24
Lausanne, Switzerland CH-1005
(41) 1-21-345-1070
URL: http://www.fai.org/

Goddard Space Flight Center
Mail Code 130, Office of Public Affairs
Greenbelt, MD 20771-0001
(301) 286-2000
FAX: (301) 286-1781
E-mail: gsfcpao@pop100.gsfc.nasa.gov
URL: http://www.nasa.gov/centers/goddard/

Government Accountability Office
441 G St. NW
Washington, DC 20548

(202) 512-3000
E-mail: webmaster@gao.gov
URL: http://www.gao.gov/

International Astronomical Union
98bis, bd Arago
75014 Paris, France
(33) 1-43-25-83-58
FAX: (33) 1-43-25-26-16
E-mail: iau@iap.fr
URL: http://www.iau.org/

Japan Aerospace Exploration Agency
7-44-1 Jindaiji Higashi-machi, Chofu-shi
Tokyo, Japan 182-8522
(81) 3-6266-6400
URL: http://www.jaxa.jp/index_e.html

Jet Propulsion Laboratory
4800 Oak Grove Dr.
Pasadena, CA 91109-8099
(818) 354-4321
URL: http://www.jpl.nasa.gov/

John H. Glenn Research Center
Lewis Field
21000 Brookpark Rd.
Cleveland, OH 44135
(216) 433-4000
URL: http://www.nasa.gov/centers/glenn/

Keck Observatory
65-1120 Mamalahoa Highway
Kamuela, HI 96743
(808) 885-7887
URL: http://www.keckobservatory.org/

Kennedy Space Center
Kennedy Space Center, FL 32899
(321) 867-5000
URL: http://www.nasa.gov/centers/kennedy/

Langley Research Center
100 NASA Rd.
Hampton, VA 23681-2199
(757) 864-1000
URL: http://www.nasa.gov/centers/langley/

Lowell Observatory
1400 West Mars Hill Rd.
Flagstaff, AZ 86001
(928) 774-3358
E-mail: steele@lowell.edu
URL: http://www.lowell.edu/

Lyndon B. Johnson Space Center
2101 NASA Pkwy.
Houston, TX 77058
(281) 483-0123
URL: http://www.nasa.gov/centers/johnson/

Marshall Space Flight Center
One Tranquility Base
Huntsville, AL 35807
(256) 837-3400
URL: http://www.nasa.gov/centers/marshall/

NASA Headquarters
Suite 1M32
Washington, DC 20546-0001
(202) 358-0001
FAX: (202) 358-3469
E-mail: public-inquiries@hq.nasa.gov
URL: http://www.nasa.gov/home/

National Oceanic and Atmospheric Administration
14th Street & Constitution Ave. NW,
Room 6217
Washington, DC 20230
(202) 482-6090
FAX: (202) 482-3154
URL: http://www.noaa.gov/

National Science Foundation
4201 Wilson Blvd.
Arlington, VA 22230
(703) 292-5111
E-mail: info@nsf.gov
URL: http://www.nsf.gov/

Office for Outer Space Affairs
United Nations Office at Vienna
A-1400 Vienna, Austria
(43) 1-260-60-4950
FAX: (43) 1-260-60-5830

E-mail: oosa@unvienna.org
URL: http://www.unoosa.org/

The Planetary Society
65 North Catalina Ave.
Pasadena, CA 91106-2301
(626) 793-5100
FAX: (626) 793-5528
E-mail: tps@planetary.org
URL: http://www.planetary.org/

Scaled Composites, LLC
1624 Flight Line
Mojave, CA 93501
(661) 824-4541
FAX: (661) 824-4174
E-mail: info@scaled.com
URL: http://www.scaled.com/

Space Adventures, Ltd.
8000 Towers Crescent Dr., Suite 1000
Vienna, VA 22182
(703) 524-7172
1-888-85-SPACE
1-888-857-7223
FAX: (703) 524-7176
E-mail: info@spaceadventures.com
URL: http://www.spaceadventures.com/

Space Environment Center
W/NP9
325 Broadway
Boulder, CO 80305
E-mail: bpoppe@sec.noaa.gov
URL: http://www.sec.noaa.gov/

Space Telescope Science Institute
3700 San Martin Dr.
Johns Hopkins University Homewood Campus
Baltimore, MD 21218
(410) 338-4700
E-mail: help@stsci.edu
URL: http://www.stsci.edu/resources

Stennis Space Center
Stennis Space Center, MS 39529-2000
(228) 688-3341
E-mail: pao@ssc.nasa.gov
URL: http://www.nasa.gov/centers/stennis/

United Space Alliance
1150 Gemini
Houston, TX 77058
(281) 212-6200
E-mail:
communications@usahq.unitedspace
alliance.com
URL: http://www.unitedspacealliance.com/

United States Geological Survey Astrogeological Program
2255 North Gemini Dr.
Flagstaff, AZ 86001
1-888-ASK-USGS
1-888-275-8747
E-mail: info@astrogeology.usgs.gov
URL: http://astrogeology.usgs.gov/

United States Strategic Command
901 SAC Blvd., Suite 1A1
Offutt Air Force Base, NE 68113-6020
(402) 294-4130
URL: http://www.stratcom.mil/

Virgin Galactic
(609) 897-7865
E-mail: Jackie.mcquillan@virgin.co.uk
URL: http://www.virgingalactic.com

Wallops Flight Facility
Wallops Island, VA 23337
(757) 824-1579
E-mail: Keith.A.Koehler@nasa.gov
URL: http://www.wff.nasa.gov/

White Sands Test Facility
12600 NASA Rd.
Las Cruces, NM 88012
(505) 524-5771
E-mail: RMelton@wstf.nasa.gov
URL: http://www.wstf.nasa.gov/

The X Prize Foundation
320 Wilshire Blvd., Suite 303
Santa Monica, CA 90401
(310) 587-33551-866-XPRIZE-8
FAX: (310) 393-4207
E-mail: press@xprize.org
URL: http://www.xprizefoundation.com/

RESOURCES

The National Aeronautics and Space Administration (NASA) is the premier resource for information about the space activities of the United States. NASA headquarters operates a very informative Web site at www.nasa.gov. There are links to all of the facilities operated by NASA around the country. A great number of these NASA Web sites were consulted for this book. In addition, some specific NASA publications proved to be very useful. These include *Basics of Space Flight, Space Transportation System, National Aeronautics and Space Administration 2006 Strategic Plan, Fiscal Year 2007 Budget Estimates: Agency Summary, NASA FY 2007 Budget, Fiscal Year 2005 Performance and Accountability Report, The Vision for Space Exploration, Columbia Accident Investigation Board Report, Final Report of the Return to Flight Task Group, NASA's Implementation Plan for Space Shuttle Return to Flight and Beyond, How We'll Get Back to the Moon, Rockets: An Educator's Guide with Activities in Science, Mathematics, and Technology, Suited for Spacewalking, Passage to a Ringed World, Statement of Michael D. Griffin, Administrator, National Aeronautics and Space Administration, before the Committee on Science, House of Representatives: November 3, 2005, NASA Hits*, and *NASA Workforce.*

NASA's History Office maintains an extensive collection of historical documents at its Web site (www.hq.nasa.gov/office/pao/History). The collection includes the complete text of books written for NASA about space activities of previous decades. Online books valuable to this project include *This New Ocean: A History of Project Mercury, On the Shoulders of Titans: A History of Project Gemini, Apollo by the Numbers: A Statistical Reference, Chariots for Apollo: A History of Manned Lunar Spacecraft*, and *The Human Factor: Biomedicine in the Manned Space Program to 1980.*

NASA publishes a *Press Kit* for each mission conducted for the Space Shuttle and International Space Station Programs and major missions of the robotic space program. These press kits contain key information on mission objectives, spacecraft design, crewmembers, and science experiments. Another important NASA series is called *NASA Facts*. This series provides data about missions and space science and biographies of historical figures key to the nation's space program. The quarterly NASA newsletter *Discovery Dispatch* describes the progress of ongoing robotic missions. The series *Technology Innovation* discusses businesses and technologies related to NASA's Innovative Partnerships Program.

The National Oceanic and Atmospheric Administration's Space Environment Center has published a series of educational brochures called *Space Environment Topics*. Informative brochures in the series include "Solar Maximum," "Navigation," "Aurora," "The Ionosphere," "Radio Wave Propagation," "Satellite Anomalies," "NOAA Scales Help Public Understand Space Weather," and "Satellites and Space Weather." Another excellent publication from the Space Environment Center is *Solar Physics and Terrestrial Effects: A Curriculum Guide for Teachers, Grades 7–12.*

The Congressional Research Service (CRS) is the public policy research arm of the U.S. Congress. CRS publications useful to this book were *China's Space Program: An Overview* (RS21641), *Space Stations* (IB93017), and *U.S. Space Programs: Civilian, Military, and Commercial* (IB92011).

The Government Accountability Office (GAO) is the investigative arm of the U.S. Congress. GAO publications used in this book include *Space Transportation: Challenges Facing NASA's Space Launch Initiative* (GAO-irse02-irse1020), *Space Station: Impact of the Grounding of the Shuttle Fleet* (GAO-03-1107), *Shuttle Fleet's Safe Return to Flight Is Key to Space Station Progress* (GAO-04-201T), *Space Shuttle: Further Improvements Needed in NASA's Modernization Efforts* (GAO-04-203), *Space Shuttle: Actions Needed to Better Position NASA to Sustain Its Workforce through Retirement* (GAO-05-230),

Defense Space Activities: Management Guidance and Performance Measures Needed to Develop Personnel (GAO-05-833), and *National Aeronautics and Space Administration: Long-Standing Financial Management Challenges Threaten the Agency's Ability to Manage Its Programs* (GAO-06-216T).

Other government agencies with online publications about space activities include the U.S. Army at Redstone Arsenal, the U.S. Air Force Space Command, the U.S. Naval Observatory, the National Science Foundation, the National Academy of Sciences, the U.S. Department of Energy Los Alamos National Laboratory, and the U.S. Geological Survey Astrogeology Research Program.

Information about international space programs and missions was obtained from the Web sites of the European Space Agency, Rosaviakosmos (the Russian space agency), the Japanese Aerospace Exploration Agency, the Canadian Space Agency, and China Aerospace Science and Technology Corporation. The United Nations Office for Outer Space Affairs operates an excellent Web site describing international space law and treaties.

Private companies and organizations engaged in space-related enterprises and educational programs include the Planetary Society, the X Prize Foundation, the American Radio Relay League, Scaled Composites, LLC, Mojave Aerospace Ventures, LLC, and Space Adventures, LLC. Breaking news about space activities is available from online news services, including *Spaceflight Now*, *Environmental News Network*, *National Geographic News*, and *Space.com*.

The Smithsonian Institution operates the National Air and Space Museum in Washington, D.C. The museum's Web site is an excellent resource for information about the history of airplane flight and space flight (http://www.nasm.si.edu). Many thanks to the Gallup Organization for their polls and surveys about American attitudes regarding programs and missions conducted throughout the Space Age.

INDEX

*Page references in italics refer to
photographs. References with the letter* t
*following them indicate the presence of a
table. The letter* f *indicates a figure. If more
than one table or figure appears on a particular page, the exact item number for the
table or figure being referenced is provided.*

A

AAP (Apollo Applications Program), 82

Abort Once Around, 69

Abort-to-Orbit procedure, 69

ACE mission, 119

Adams, John Couch, 145

Adolescents
  public opinion on space tourism, 166,
    167(*f* 9.10)
  public opinion on technological
    achievements, 156*f*

Advisory Committee for Aeronautics
  (ACA). *See* National Advisory
  Committee for Aeronautics (NACA)

Aelita: Queen of Mars (movie), 125

AFSPC (Air Force Space
  Command), 50

"Agreement between the United States of
  America and the Russian Federation
  Concerning Cooperation in the
  Exploration and Use of Outer Space
  for Peaceful Purposes," 84

Air Force, U.S.
  DynaSoar concept, 62
  Manned Orbiting Laboratory, 82
  manned space flight plans, 46–47
  military satellite programs, 46–47
  rocket science, 44, 45
  space chimp colony, 46
  space weapons, 49

Air Force Space Command (AFSPC), 50

Aldrin, Edwin (Buzz), 27, 238

ALH84001, 128–129

Allen, Paul G., 16–17, 58

Amateur Satellite Corporation
  (AMSAT), 16

American Institute of Aeronautics and
  Astronautics, 57

American Interplanetary Society, 57

American Rocket Society (ARS), 57

Ames Research Center (ARC), 32

AMSAT (Amateur Satellite
  Corporation), 16

Anderson, Michael, 20

Andromeda galaxy, 114

Ansari, Anousheh and Amir, 17

Ansari X Prize, 17–18, 58–59

Antarctica, 129

Anti-Ballistic Missile Treaty, 47

Anti-satellite weapons (ASATs), 49

Anti-ship missiles, 55

Apollo 1 mission, 10, 20, 26

Apollo 11 mission, 27, 162,
  165–166, 166*t*

Apollo 13 mission, 27

Apollo 17 mission, 27–28

Apollo 13 (movie), 27

Apollo Applications Program (AAP), 82

Apollo program
  Apollo-*Soyuz* rendezvous and docking
    test project, 13, 53, 53*f*
  chronology of manned missions, 28*t*
  launch configuration for lunar landing
    mission, 26*f*
  overview, 10–11, 25–28

Apollo-Soyuz rendezvous and docking
  test project, 13, 53, 53*f*

Application satellites, 15–16

ARC (Ames Research Center), 32

Arecibo Observatory, 108

Arianespace, 18

Armstrong, Neil, 11, 27, 28, 165

Army, U.S.
  manned space flight plans, 46
  rocket science, 44, 45

ARS (American Rocket Society), 57

Art program, 164

ASATs (anti-satellite weapons), 49

ASPERA-3, 105

Asteroids, 56, 106, 129

Astronauts
  International Space Station, 94
  "Mercury Seven," 12
  NASA, 37–39
  public relations, 164–165
  selection process timeline, 38*t*
  space shuttle, 67–68
  training, 34, 38, 97–98

Astronomers. *See* Specific names

ATK Thiokol Propulsion, 57

Atmosphere of Mars, 125–126

Aurora Exploration Programme, 55

Auroras, 116, 117

Autour de la lune (Verne), 2, 3

Aviation history, 3–4

B

Baikonur cosmodrome, 97

Ballistic Defense Missile Organization
  (BDMO), 47

Ballistic missile defense system, 47–48, 48*t*

Ballistic missiles, 7, 51

BDMO (Ballistic Defense Missile
  Organization), 47

Beagle 2, 132–133

Bean, Alan, 28

Belyayev, Pavel, 51

Berkeley Online Infrastructure for
  Network Computing (BOINC), 57

Berkner, Lloyd, 7

Big Bang, 107

Binnie, Brian, 17, 59

Biosatellite program, 29, 44

BIS (British Interplanetary Society), 57

"Blueberries," 137

Bluford, Guion, 38

Boeing Corporation, 18

BOINC (Berkeley Online Infrastructure for Network Computing), 57
Bone loss, 96, 139
Bradbury, Ray, 125
Branson, Sir Richard, 59
"The Brick Moon" (Hale), 3
British Interplanetary Society (BIS), 57
Brooks, Overton, 46
Brown, David, 20, 72
Budget issues. *See* Costs
Bulganin, Nikolai, 22–23
Buran program, 53
Burroughs, Edgar Rice, 125
Bush, George H. W., 47, 84
Bush, George W.
  International Space Station, 81, 92, 94
  manned space exploration, 159
  Mars travel, 123
  Missile Defense Agency, 47
  NASA commercial partnerships, 37
  public opinion on space program of, 160, 161*f*
  space program, 14, 40, 61, 74
  space weapons policy, 49
Bykovsky, Valeri, 51

C

CAIB (*Columbia* Accident Investigation Board), 72–73, 74, 74*t*
Canadarm, 67, 91
Canadian Space Agency (CSA) Headquarters, 97
Carpenter, M. Scott, 12
Cassini, Giovanni, 143
Cassini program, 15, 31, 55, 151–152, 151*f*
Casualties
  Apollo 1, 10, 26
  Challenger, 13, 29, 71
  chronology, 19–20
  Columbia, 14, 30, 72
  Soviet space program, 52
  Soyuz 11, 52, 83
CAV (Common Aero Vehicle), 49
Central Intelligence Agency (CIA), 46
Cernan, Eugene A., 28
CEV (Crew Exploration Vehicle), 40
CGRO (Compton Gamma-Ray Observatory), 111
Chaffee, Roger B., 10, 20
Challenger disaster
  Educator-in-Space program, 165
  effect on military satellite programs, 47
  investigation, 29
  overview, 20
  private citizens on shuttle flights policy, 38
  public opinion, 159, 160
  Scobee, Francis (Dick), 71

space shuttle fleet grounding, 13
Chandra X-Ray Observatory (CXRO), 111–113, 112*f*, 113*t*
Chandrasekhar, Subrahmanyan, 111, 165
Chang, Iris, 55
Chapelain, Jean, 143
Chawla, Kalpana, 20, 72
Chinese space program, 55–56
Christy, James, 146
Chronology
  Apollo program missions, 28*t*
  casualties, 19–20
  Gemini program missions, 25*t*
  International Space Station missions, 95*t*
  Mars expeditions, 127*t*
  Mercury program missions, 24*t*
  Salyut space stations, 83*t*
  space shuttle-*Mir* timeline, 86*t*–87*t*
  space shuttle missions, 75*t*–77*t*
CIA (Central Intelligence Agency), 46
Civil servants, 35–36
Clark, Laurel, 20, 72
Clarke, Arthur C., 14, 146
Classical mythology, 1–2
Clinton, Bill
  International Space Station, 91
  Russian/U.S. space collaboration, 84
  space station program, 88
CMEs (coronal mass ejections), 118
Cold war
  space race, 7–12
  Star Wars program, 47
Collins, Michael, 27
Columbia Accident Investigation Board (CAIB), 72–73, 74, 74*t*
Columbia disaster
  International Space Station, effect on the, 14, 93
  investigation, 72–73
  NASA, effect on, 30
  overview, 20
  public opinion, 159, 160
Comets, 105
Commander, space shuttle, 67
Commercial enterprises, 18, 36–37, 58
Commercial Space Act of 1998, 18
Commercial Space Launch Act, 18
Commercial Space Launch Amendments Act, 19
Common Aero Vehicle (CAV), 49
Communications, 35, 36*f*
Communications satellites
  commercial launch services, 58
  NASA, 15
Compton Gamma-Ray Observatory (CGRO), 111, 111*f*
Conrad, Charles (Pete), Jr., 28
Conspiracy Theory: Did We Land on the Moon? (television show), 167–168

Constellation Systems program, 40
Contests, 165
Contingency abort, 69
CONTOUR, 103
Contractors, NASA, 36–37
Cooper, L. Gordon (Gordo), Jr., 12
Copernicus, Nicolaus, 2
Corona missions, 46
Coronal mass ejections (CMEs), 118
Cosmonauts, 51
Cosmos-1 spacecraft, 58
Costs
  Apollo program, 28
  Challenger disaster, 71
  Defense Department space budget, 44, 44*f*
  Explorer missions, 102
  Freedom space station, 88
  International Space Station, 91–92
  Mars Exploration Rovers, 138
  Missile Defense Agency funding, 48*f*
  NASA budget, 22*f*, 39, 39*t*
  NASA/commercial partnerships, 36–37
  public opinion on NASA's budget, 158*f* 156–157
  space shuttle program, 62
  space shuttle program shutdown, 93
Crew Exploration Vehicle (CEV), 40
Crew module, space shuttle, 67–68, 69*f*
Cuban Missile Crisis, 12
Curriculum development, 164

D

Da Vinci, Leonardo, 2
Dawn mission, 106
DDMS (Manned Space Flight Support Office), 50
De la terre à la lune (Verne), 2
Deep Impact program, 105
Deep Space Network (DSN), 35, 36*f*
Definitions, 107*t*
Destiny module, 90, 95, 96*f*
Détente, 12
DeWinne, Frank, 92
DFRC (Dryden Flight Research Center), 32
Diamandis, Peter, 17, 18, 19, 58
Direct flight technique, 25
Discovery program, NASA, 102–106, 104*f*, 129
Discovery space shuttle, 30, 31, 72–73
Dobrovolskiy, Georgiy, 20, 52
Dogs in space, 8, 20, 52
Dryden Flight Research Center (DFRC), 32
DSN (Deep Space Network), 35, 36*f*
Duke, Charles M., Jr., 28
DynaSoar program, 45, 62

E

Earle, Sylvia, 157
Earth orbit flight, 25
Earth orbiters, NASA, 31–32, 102–106
Education
    Educator Astronaut program, 38
    Educator-in-Space program, 165
    NASA, 164
Eisenhower, Dwight D., 22–23
Electromagnetic spectrum, 106–107, 107*f*
ELVs (Expendable Launch Vehicles), 46, 47
Emergency flight options, space shuttle, 68–69
Employees
    NASA, 35–36
    space shuttle program, 70
ESA (European Space Agency). *See* European Space Agency (ESA)
Etzioni, Amitai, 24, 158
European Space Agency (ESA)
    Cassini program, 151–152
    Mars Express, 132–133
    NASA collaboration with, 31–32
    overview, 54–55
    solar exploration missions, 119–121
    XMM-Newton observatory, 115
Exosolar planet search, 58
Expendable Launch Vehicles (ELVs), 46, 47
Experimental Satellite Series (XSS), 49
Exploration Systems Mission Directorate, NASA, 39
Explorer 1, 8, 8*f*, 45
Explorer program, NASA, 31, 102, 103*t*
External rocket boosters, 62
Extrasolar planets, 113–114
Extraterrestrial life, 124–125, 167, 167(*f* 9.11), 167(*f* 9.12)
Extravehicular activity. *See* Space walks

F

FAA (Federal Aviation Administration), 19
Facilities, NASA, 32–35, 34*f*
FAI (Fédération Aéronautique Internationale), 3
Far planets
    Cassini program, 15, 31, 55, 151–152, 151*f*
    future missions, 153
    Galileo program, 15, 31, 148–151, 150*f*
    New Horizons program, 152–153
    Pioneer program, 146–147
    Voyager program, 30–31, 147–148, 147*f*, 149*f*
    See also specific planets
Federal Aviation Administration (FAA), 19

Fédération Aéronautique Internationale (FAI), 3
Fei Junlong, 55
Feoktistov, Konstantin, 51
Fire aboard *Mir*, 84
The First Men on the Moon (Wells), 3
Firsts
    African American in space, 38
    American in space, 23
    American space program casualties, 26
    American to complete an Earth orbit, 30
    American woman in space, 38
    artificial satellite, 22, 45, 51
    Chinese manned space flight, 55
    Chinese satellite, 55
    commercial manned space flight, 56, 58
    hot-air balloons, 2
    human to die in space flight, 52
    intercontinental ballistic missile, 51
    Japanese citizen to fly in space, 83
    Japanese satellite, 56
    live television broadcast from a manned spacecraft, 26
    manned space flight, 9, 51
    microsatellite, 49
    Moon landing, 27
    safe landing of a spacecraft on another planet, 30
    Soviet space program firsts, 52
    space shuttle flight, 73
    space station, 29
    space telescope, 108
    space tourist, 58
    space walk, 24, 51–52
    spacecraft landing on an asteroid, 103
    supersonic flight, 22
    sustained flight of powered aircraft, 3
    television transmission from the moon, 25
    U.S. satellite, 45, 102
    woman in space, 51
Flammarion, Camille, 124
Flight system, Mars Exploration Rovers, 136*f*
Flight techniques, 25
Flight trajectories, Mars Exploration Rovers, 133, 135*f*
FOOT (Foot/Ground Reaction Forces during Space flight), 96
Fox television network, 167–168
Freedom space station, 88
Friedman, Louis, 57
From Earth to the Moon (Verne), 2
Future plans
    Chinese space program, 56
    International Space Station, 94
    Japanese space program, 56

Mars exploration, 139
NASA, 39–40, 42*f*
solar missions, 121
space shuttle program, 74, 77

G

Gagarin, Yuri, 9, 23, 51, 52
Gagarin Cosmonaut Training Center, 97
Galilei, Galileo
    Jupiter's moons, 142
    Mars, 123
    Saturn, 143
    sunspots, 116
    telescopes, 2, 107–108
Galileo program, 15, 31, 148–151, 150*f*
Galle, Johann, 145
Garriott, Owen, 164
Gehman Board. *See Columbia* Accident Investigation Board (CAIB)
Gemini program, 24, 25*t*
Genesis, 104–105
Geology, 125, 141–142
Geomagnetic storms, 118, 119*f*
Germany
    rocket-powered weapons, 5–6
    Verein für Raumschiffahrt, 56–57
Gilruth, Robert, 24
Glenn, John, 12, 23, 30, 159, 165
Glenn Research Center (GRC), 32
Glennan, T. Keith, 23
Global Positioning System (GPS), 16
Global Surveyor mission, 129
Goddard, Robert, 5, 57
Goddard Space Flight Center (GSFC), 32–34
Gorbachev, Mikhail, 47
Government contractors, 36
GPS (Global Positioning System), 16
Grantees, NASA, 36–37
Gravity, 2
GRC (Glenn Research Center), 32
"The Great Moon Hoax" (Phillips), 168
Great Observatories Program, 106, 165
    See also specific space observatories
Greek mythology, 1–2
Griffin, Mike, 39, 94, 110, 158, 160
Grissom, Virgil (Gus), 10, 12, 20
Ground-based telescopes, 107–108
Group for Investigation of Reactive Motion, 51, 57
Gruppa Isutcheniya Reaktivnovo Dvisheniya, 51, 57
GSFC (Goddard Space Flight Center), 32–34
Guggenheim Foundation, 5

H

H-II rocket, 56
Hale, Edward Everett, 3

Hall, Asaph, 124
Ham radio, 15–16, 164
Hayabusa program, 15, 56
Heavy-lift vehicles, 40, 41*f*
Heppenheimer, T. A., 62
Herschel, Frederic William, 123–124, 145
History
    astronomy, 2
    rocket science, 4–6
    science fiction, 2–3
    space race, 7–12
    telescopes, 107–108
    U.S. military space programs, 43–45
Hoax theory of moon landings, 167–168, 168*t*
Holmes, D. Brainerd, 25
Hot-air balloons, 2
Hubble, Edwin, 108
Hubble Space Telescope, 77, 108–111, 109*f*, 110*t*, 158
Hurricane Katrina, 74
Husband, Rick, 20, 72
Huygens, Christiaan, 55, 123, 143
Huygens probe, 55, 151–152, 152*f*
Hypervelocity rods, 49

I

ICBMs (intercontinental ballistic missiles), 7, 47–48, 51
ICSU (International Council of Scientific Unions), 7, 116
IGY (International Geophysical Year), 7, 44
Infrared radiation detection, 113–114
Innovative Partnerships Program (IPP), 37, 37*f*
INTEGRAL (International Gamma-Ray Astrophysics Laboratory), 32
Intercontinental ballistic missiles (ICBMs), 7, 47–48, 51
Intergovernmental Agreement on Space Station Cooperation, 89, 93
International collaboration
    Cassini mission, 31
    Earth and Sun orbiters, 31–32
International Council of Scientific Unions (ICSU), 7, 116
International Gamma-Ray Astrophysics Laboratory (INTEGRAL), 32
International Geophysical Year (IGY), 7, 44
International Solar-Terrestrial Physics (ISTP) program, 31, 119–120
International space agencies, 50–56, 51*t*
International Space Station
    assembly, 89–91, 93*t*
    costs, 91–92
    crew training, 96–97
    crew training locations, 98*f*

crews, 91*t*
downsizing, 94
early visions, 81–82
expeditions, 90–93, 91*t*, 94, 95*t*
grounding of the space shuttle fleet, impact of the, 74
NASA/Rosaviakosmos collaboration, 30
partner nations, 89*t*
plan, 89
rescue operations, 68
Rosaviakosmos, 54
science, 94–96
space shuttle program, 67, 72, 93–94
space tourism, 19, 56, 91, 92
spacesuits, 97–98, 99*f*
United States/Russian cooperation, 14
Internet, 162–163
Interplanetary exploration
    European Space Agency participation, 54–55
    NASA programs, 30–31
    robotic space explorers, 14–15
    space shuttle deployment, 74
Interstellar exploration, 148
Investigations
    Challenger disaster, 71
    Columbia disaster, 72
IPP (Innovative Partnerships Program), 37, 37*f*
Iran Nonproliferation Act, 74
Irwin, James B., 28
Is Mars Habitable? (Wallace), 124
ISS Management and Cost Evaluation Task Force, 91–92
ISTP (International Solar-Terrestrial Physics) program, 31, 119–120
Itokawa, Hideo, 56

J

James Webb Space Telescope (JWST), 115
Jansky, Karl, 108
Japan
    Hayabusa mission, 15
    International Solar-Terrestrial Physics (ISTP) Science Initiative, 31
    space program, 56
Jarvis, Gregory, 20, 71
Jet Propulsion Laboratory (JPL), 35
Johnson, Lyndon B., 24, 29, 46, 156
Johnson Space Center (JSC), 34, 38, 69, 96
Joint Declaration on Cooperation in Space, 89
The Jovian World (Marius), 142
JPL (Jet Propulsion Laboratory), 35
JSC. *See* Johnson Space Center (JSC)
Jupiter
    Galileo program, 15, 31, 148–151

geology and atmosphere, 141–142
Jupiter Icy Moons Orbiter (JIMO) program, 153
moons, 142–143
Pioneer program, 146–147
Ulysses mission, 120–121
Voyager program, 148
Jupiter Icy Moons Orbiter (JIMO) program, 153
JWST (James Webb Space Telescope), 115

K

Kelvin scale, 106
Kennedy, John F.
    Cuban Missile Crisis, 12
    lunar mission, 9, 156
    NASA, 21, 23
    space program dangers, 20
Kennedy Space Center (KSC), 34, 63–64, 69, 96–97
Kepler, Johannes, 2
Khrushchev, Nikita, 12
Kibo laboratory, 56
King, Martin Luther, Jr., 156
Knight, William (Pete), 59
Komarov, Vladimir, 20, 51, 52
Korolev, Sergei, 7, 50–52, 57
KSC (Kennedy Space Center). *See* Kennedy Space Center (KSC)
Kuiper Belt, 152–153

L

Lagrange points, 115–116
Laika, 8
Landing
    Mars Exploration Rovers, 133–134, 137*f*
    space shuttle, 68
Lang, Fritz, 5
Langley Research Center (LRC), 34
Large Space Telescope (LST), 108
Lassell, William, 145
Lasser, David, 57
Launch configuration for lunar landing mission, 26*f*
Launch facilities, space shuttle, 63–64, 64*f*
Launch vehicles, 40, 41*f*, 56, 113, 114*f*
Launches
    Mars Exploration Rovers, 133, 134*f*
    NASA tracking, 19
Le Verrier, Urbain, 145
Legislation
    Commercial Space Act of 1998, 18
    Commercial Space Launch Act, 18
    Commercial Space Launch Amendments Act, 19
    Iran Nonproliferation Act, 74
    National Aeronautics and Space Act, 21–22, 23

Lego toy company, 58
Leibovitz, Annie, 164
LEMS (Lower extremity monitoring suit), 96, 97f
LEO (low Earth orbit), 13–14
Leonardo module, 90–91
Leonov, Aleksei, 51, 52
Light years, 107
Lindbergh, Charles, 4, 5, 155
Long March rockets, 55
Lovell, James, 27
Low Earth orbit (LEO), 13–14
Lowell, Percival, 124, 130–131, 146
Lowell Observatory, 124, 146
Lower extremity monitoring suit (LEMS), 96, 97f
LRC (Langley Research Center), 34
LST (Large Space Telescope), 108
Lu, Ed, 166
Lucian of Samosata, 2
Lucid, Shannon, 84
Luna program (Soviet Union), 51
Lunar launch vehicles, 40, 41f
Lunar missions
    future, 40
    moon landing as hoax, 167–168, 168t
    NASA, 23–28
    public knowledge, 165–166, 166f, 166t
    Soviet space program, 52–53
    television, 162
    See also Apollo program
Lunar orbit technique, 25
Lunar Prospector, 103

M

Magellan mission, 15, 31
Magnetosphere, 116, 117f
Man in Space Soonest (MISS) project, 45
Manarov, Musa, 83
Manned Earth Reconnaissance Project, 46
Manned Orbiting Laboratory (MOL), 82
Manned space flight
    first, 9
    future, 40, 42f
    Johnson Space Center, 34
    Manned Space Flight Support Office, 50
    Mars missions, 139
    NASA's Mercury project, 23–24
    private, 16–18
    public opinion, 158–159, 159f
    Soviet Union, 51
    U.S. military programs, 46
    vs. robotic exploration, 28–29
Manned Space Flight Support Office (DDMS), 50
Manning, Laurence, 57

Mapping
    application satellites, 15
    gamma-rays, 108, 111
    lunar, 9–10
    Mars, 31, 129, 131
    x-rays, 112
Maps
    Deep Space Network communications complexes, 36f
    International Space Station crew training locations, 98f
    NASA human space flight program locations, 70f
    NASA sites, 34f
    space shuttle launch facilities, 64f
MARIE (Mars Radiation Environment Experiment), 131
Mariner program, 30, 127–128
Marius, Simon, 142
Mars
    ALH84001, 128–129
    ASPERA-3 mission, 105
    chronology of Mars expeditions, 127t
    future exploration, 40, 139
    Global Surveyor mission, 129
    manned missions, 139
    Mars Exploration Rovers, 133–138, 134f, 135f, 136f, 137f, 138f
    Mars Express, 55, 105, 132–133
    Mars Reconnaissance Orbiter, 138–139, 139f
    naming planetary features, 135–137
    NASA explorer programs, 30–31
    Odyssey mission, 130–131, 132f
    Pathfinder mission, 129–130, 130f
    perihelic oppositions, 126, 131–132
    Red Rover Goes to Mars program, 57–58
    robotic space explorers, 14–15
    telescopic views, 123–124
    Viking program, 128
Mars Climate Orbiter, 126
Mars Exploration Rovers (MERs), 133–138, 134f, 135f, 136f, 137f, 138f, 165
Mars Express, 55, 105, 132–133
Mars in opposition, 126, 131–132
Mars Observer, 126
Mars Polar Lander, 126–127
Mars Radiation Environment Experiment (MARIE), 131
Mars Reconnaissance Orbiter, 138–139, 139f
Mars Science Laboratory, 139
Marshall Space Flight Center (MSFC), 34–35, 70
The Martian Chronicles (Bradbury), 125
Mauna Kea observatories, 108
McAuliffe, Christa, 20, 38, 71, 165
McCool, William, 20, 72
McNair, Ronald, 20, 71

MDA (Missile Defense Agency), 47–48, 48f
Méliès, George, 3
Melvill, Mike, 16, 17, 59
Mercury (planet), 106
Mercury program, 12, 23–24, 24t
MERs (Mars Exploration Rovers). See Mars Exploration Rovers (MERs)
Messages from Earth, 148
Messenger program, 31, 105–106
Messier 31 galaxy, 114
Meteorites, 128–129
Microgravity science glovebox (MSG), 94–96, 96f
Military space program, Russian, 54
Military space programs, U.S.
    Department of Defense program budget, 43–44, 44f
    military and intelligence satellites, 46–47
    missile defense system, 47–48, 48t
    NASA takeover of, 45–46
    rocket science, 44–45
    space weapons, 49
    Star Wars program, 47–48
    unmanned satellite, 45
    U.S. Strategic Command, 49–50
Mir space station
    chronology, 86t–87t
    history, 83–84, 88
    mishaps, 84, 88
    Rosaviakosmos/NASA cooperation, 14, 54
    shuttle-Mir missions, 29–30, 84, 85f
    space tourism, 18–19, 58
Mishaps
    Mars missions, 126–127
    Mir, 84, 88
MISS (Man in Space Soonest) project, 45
Missile Defense Agency (MDA), 47–48, 48f
Missile defense system, 47–48, 48t
Mission specialists, 67
Missions of Opportunity (MOs), 102, 105
Mitchell, Edgar D., 28
MOL (Manned Orbiting Laboratory), 82
Montgolfier, Joseph and Etienne, 2
Moon
    Apollo missions, 10–11
    mapping, 9–10
The Moon-Doggle (Etzioni), 24, 158
Moon landing
    as hoax, 167–168, 168t
    public knowledge, 165–166, 166f, 166t
    television, 162
Moon rocks, 27–28, 168
Moons
    Cassini program, 151–152
    Galileo program, 150–151
    Jupiter, 142–143

Jupiter Icy Moons Orbiter (JIMO) program, 153
Neptune, 145
New Horizons program, 153
Pluto, 146
Saturn, 143
Uranus, 145
Voyager program, 148
Moons, Martian, 126
MOs (Missions of Opportunity), 102, 105
Movies, 14, 27, 125, 131
MPLMs (multi-purpose logistics modules), 67, 90–91
MSFC (Marshall Space Flight Center), 34–35, 70
MSG (microgravity science glovebox), 94–96, 96*f*
Multi-purpose logistics modules (MPLMs), 67, 90–91
Mundus Iovialis (Marius), 142
Murray, Bruce, 57
Muscle deterioration, 96
My Favorite Martian (television show), 125
Mythology, 1–2

N

N-1 rockets, 52
NACA (National Advisory Committee for Aeronautics), 3–4
Naming
    contests, 165
    Mars planetary features, 135–137
    Pluto, 146
NASA
    Apollo program, 25–28, 28*t*
    Apollo-*Soyuz* rendezvous and docking test project, 53, 53*f*
    application satellites, 15
    astronauts, 37–39, 38*t*
    budget, 22*f*, 39, 39*t*
    Cassini program, 151–152, 151*f*
    Chandra X-Ray Observatory (CXRO), 111–113, 112*f*, 113*t*
    Compton Gamma-Ray Observatory (CGRO), 111
    culture of, 27, 30, 71
    Discovery program, 102–106, 104*f*, 129
    early space telescopes, 108
    Explorer program, 102, 103*t*
    facilities, 32–35, 34*f*
    formation, 8–9, 21–22
    future goals, 39–40, 42*f*
    Galileo program, 148–151, 150*f*
    Gemini program, 24, 25*t*
    Global Surveyor mission, 129
    Great Observatories, 106
    hoax theory of moon landings, 168
    Hubble Space Telescope, 108–111, 109*f*, 110*t*

human space flight program locations, 70*f*
International Space Station, future plans for the, 94
James Webb Space Telescope (JWST), 115
Jupiter Icy Moons Orbiter (JIMO) program, 153
lunar launch vehicles, 40, 41*f*
lunar missions, 9–11
manned *vs.* robotic space exploration, 28–29
Mariner program, 127–128
Mars Climate Orbiter, 126
Mars Exploration Rovers, 133–138, 134*f*, 135*f*, 136*f*, 137*f*, 138*f*
Mars Observer, 126
Mars Polar Lander, 126–127
Mars Reconnaissance Orbiter, 138–139, 139*f*
Mercury program, 11–12, 23–24, 24*t*
military satellite deployment, 46–47
New Horizons program, 152–153
Odyssey mission, 130–131, 132*f*
organization, 32, 33*f*, 71
Pathfinder mission, 129–130, 130*f*
Pioneer program, 146–147
public opinion on NASA's budget, 156–157, 158*f*
public opinion on the benefits *vs.* costs of the space program, 162, 163(*f* 9.7)
public opinion on the performance of, 159–160, 160*t*
public relations, 163–165
robotic space explorers, 14–15, 30–32
Rogers Commission findings, 71
Rosaviakosmos, collaboration with, 14, 29–30, 54
Rosaviakosmos, reimbursement to, 74
Russian/U.S. space collaboration, 84
science goals, 101
Search for Extra-Terrestrial Intelligence (SETI) project, 57
Skylab, 29
solar exploration missions, 119–121, 120*t*
space casualties, 20
space launches, tracking of, 19
space shuttle program, 13, 14, 29–30
space station programs, 88–89
Space Technology Hall of Fame, 161–162, 162i*t*
space tourism policy, 91
Spitzer Space Telescope, 113–114, 114*f*, 115*f*
takeover of military space programs, 45–46
technological contributions, 161–162
Viking program, 128
Voyager program, 147–148, 147*f*, 149*f*
workforce, 35–36

NASA Explorer School program, 164
NASA Hits: Rewards from Space—How NASA Improves Our Quality of Life (NASA), 161
National Academy of Public Administration, 94
National Advisory Committee for Aeronautics (NACA), 3–4, 6, 22–23
National Aeronautics and Space Act, 21–22, 23
National Oceanic and Atmospheric Administration (NOAA), 118
National Research Council, 94
National Science Foundation, 108, 129
National Space Science Data Center (NSSDC), 33
Navigation satellites, 16, 16*f*
Navy, U.S.
    manned space flight plans, 46
    rocket science, 44, 45
Near Field Infrared Experiment (NFIRE), 49
NEAR mission, 103
Neptune, 145, 148
New Horizons program, 152–153
Newton, Sir Isaac, 2
NFIRE (Near Field Infrared Experiment), 49
Nie Haisheng, 55
Nikolayev, Andriyan, 51
Nixon, Richard, 61, 62
NOAA (National Oceanic and Atmospheric Administration), 118
Noordung, Hermann, 81
North American Aerospace Defense Command (NORAD), 49
NSSDC (National Space Science Data Center), 33
NTV, 163–164

O

OAO-1 (Orbital Astronomical Observatory), 108
Oberg, James E., 52
Oberth, Hermann, 4–5, 56–57, 81, 108
Observatories. *See* Space observatories
OBSS (Orbiter Boom Sensor System), 67
Oceans as a science priority, 157–158
Odyssey mission, 130–131, 132*f*
Office of Space Science, NASA, 101
O'Keefe, Sean, 20, 77, 92, 110
Olsen, Greg, 94
Olympus Mons, 125
OMS (orbital maneuvering system), 65–66
Onizuka, Ellison, 20, 71
Opportunity rover, 133–138
Orbit
    Mars, 126
    Newton's theory, 2
    space shuttle, 65–68

Orbital Astronomical Observatory (OAO-1), 108
Orbital maneuvering system (OMS), 65–66
Orbiter Boom Sensor System (OBSS), 67
Orbiters, space shuttle
  boom sensor system, 73*f*
  components, 63*f*
  design and development, 62–63
  flights per, 79*f*
  orbital maneuvering system (OMS), 65–66
  payload bay, 68*f*
  reaction control system, 65–66
  safety zone, 50*f*
  views, dimensions and weight, and minimum ground clearances, 66*f*
Organizational structure, NASA, 32, 33*f*
Orteig Prize, 4, 17
OSCAR project, 15–16
Osumi satellite, 56
Outer space particle samples, 15
Outer Space Treaty, 12–13, 49

P

Park, Robert, 159
Particle samples, 15
Pathfinder mission, 130*f*, 130*f* 129–130
Patsayev, Viktor, 20, 52
Payload bays, 67, 68*f*
Payload specialists, 67
Pendray, G. Edward, 57
Perihelic oppositions, 126, 131–132
Phillips, Tony, 168
Phoenix Mars Scout, 139
Physical effects of space travel, 96, 139
Pickering, W. H., 125
Pilots, space shuttle, 67
Pioneer program, 146–147
Pirs, 92
The Planetary Society, 57–58
PLSM (primary life support module), 98
Pluto, 145–146, 152–153
Plutonium in space, 147
Polyakov, Valeri, 54, 83
Popovich, Pavel, 51
Post–World War II era, 44
Primary life support module (PLSM), 98
Primates
  Air Force space chimp colony, 46
  Biosatellite program, 29, 44
  casualties, 19–20
Principles Regarding Processes and Criteria for Selection, Assignment, Training and Certification of ISS (Expedition and Visiting) Crewmembers, 92
Private manned space flight, 16–18, 58–59, 60*t*

Private space organizations
  American Institute of Aeronautics and Astronautics, 57
  early European, 56–57
  The Planetary Society, 57–58
Prizes
  Ansari X Prize, 17–18, 58–59
  Orteig Prize, 4, 17
Project Adam, 46
Project MER, 46
Project Vanguard, 44–45
Proton K rocket, 89, 90
Public knowledge about space travel, 165–166, 166*f*, 166*t*
Public opinion
  benefits *vs.* costs of the space program, 162, 163(*f* 9.7)
  Bush's space program, 160, 161*f*
  extraterrestrial life, 167, 167(*f* 9.11), 167(*f* 9.12)
  on the future influence of technological achievements, 156*f*
  importance of space exploration, 155–156, 156*t*
  interest in the space program, 157(*f* 9.2)
  moon landing as hoax, 167–168, 168*t*
  NASA's budget, 156–157, 158*f*
  NASA's performance, 159–160, 160*t*
  reasons for space exploration, 157(*f* 9.3)
  space tourism, 166, 167(*f* 9.10)
  space travel knowledge, 165–166, 166*f*, 166*t*
Public relations, NASA, 162–165

R

Radiation in space, 102, 106, 107, 168
Radio telescopes, 108
Radioisotope thermoelectric generators (RTGs), 147
Ramon, Ilan, 72
Ranger probes, 25
RCS (reaction control system), 65–66
Reaction control system (RCS), 65–66
Reaction Motors, Inc., 57
Reagan, Ronald
  space shuttle program, 70–71
  space station, 88
  Star Wars program, 47–48
  U.S. Space Command, 49
Reconnaissance satellites, 46
Red Rover Goes to Mars program, 57–58
Redstone Arsenal, 44, 45
Reentry, space shuttle, 68
A Renewed Spirit of Discovery: The President's Vision for U.S. Space Exploration (Bush), 40
Rescue operations, 68
Research, NASA, 35–36

Resnik, Judith, 20, 71
Return-to-flight mission, 159–160
  See also Discovery space shuttle
Return to Flight (RTF) Task Group, 72–73
Return to Launch Site (RTLS) abort, 68
Riccioli, Giovanni, 9
Ride, Sally, 38
The Right Stuff (Wolfe), 12
Rings of Saturn, 143–144, 144*f*, 144*t*
Robotic space exploration
  Cassini program, 151–152, 151*f*
  Chandra X-Ray Observatory, 111–113, 112*f*, 113*t*
  Compton Gamma-Ray Observatory, 111, 111*f*
  Discovery program, 102–106, 104*f*
  Explorer program, 102, 103*t*
  Galileo program, 148–151, 150*f*
  Global Surveyor mission, 129
  Hayabusa asteroid sampler, 15, 56
  Hubble Space Telescope, 108–111, 109*f*, 110*t*
  James Webb Space Telescope (JWST), 115
  Jupiter Icy Moons Orbiter (JIMO) program, 153
  Mariner program, 127–128
  Mars Exploration Rovers, 133–138, 134*f*, 135*f*, 136*f*, 137*f*, 138*f*
  Mars Express, 132–133
  Mars Observer, Climate Orbiter, and Polar Lander, 126–127
  Mars Reconnaissance Orbiter, 138–139, 139*f*
  NASA, 14–15, 30–32
  NASA's Great Observatories, 106
  NASA's science goals, 101
  New Horizons program, 152–153
  Odyssey mission, 130–131, 132*f*
  Pathfinder mission, 129–130, 130*f*
  Pioneer program, 146–147
  public opinion, 158–159
  solar exploration missions, 119–120, 120*t*
  Spitzer Space Telescope, 113–114, 114*f*, 115*f*
  Viking program, 128
  Voyager program, 147–148, 147*f*, 149*f*
Rocket boosters, 10, 62
Rocket propulsion systems, 35
Rocket science
  Chinese space program, 55
  Goddard's rocket design, 6*f*
  history, 4–6, 4*f*
  Marshall Space Flight Center, 34–35
  Post–World War II era, 44
  private space organizations, 56–57
Rockwell, Norman, 164
Roddenberry, Gene, 14

Rogers Commission, 71
Roman, Ilan, 20
Rosaviakosmos
  grounding of the space shuttle fleet,
    impact of the, 74
  International Space Station modules,
    89–90
  low Earth orbit (LEO) missions, 13–14
  NASA, collaboration with, 14,
    29–30, 54
  space tourism, 19, 91
'Round the Moon (Verne), 2–3
Rovers
  Mars, 57–58, 133–138
  Sojourner rover, 129–130, 131*f*
RTF (Return to Flight) Task Group,
  72–73
RTGs (radioisotope thermoelectric
  generators), 147
RTLS (Return to Launch Site) abort, 68
Russia
  low Earth orbit (LEO) missions, 13–14
  Mir space station, 84, 85*t*, 86*t*–87*t*, 88
  Rosaviakosmos/NASA collaboration,
    29–30
  space tourism, 18–19, 56
Rutan, Burt, 59
RXJ1242-11 galaxy, 112

S

SAFER units, 98
Safety
  International Space Station, 93
  NASA culture, 30, 71
Sagan, Carl, 57, 129
Salyut program, 53, 83, 83*t*
Sampling spacecraft
  Genesis, 15, 104–105
  Hayabusa asteroid sampler, 15, 56
  SOHO, 119
  Stardust, 104
SAREX (Space Amateur Radio
  Experiment), 164
*Satellite Carrying Amateur Radio
  (OSCAR)*, 15–16
Satellites
  commercial launch enterprises, 58
  commercial services, 18
  Explorer I, 8*f*
  Japanese space program, 56
  Project Vanguard, 44–45
  space shuttle missions, 61–62, 67,
    73–74
  Sputnik, 7–8, 45, 45*f*
  U.S. military, 46–47
  See also Robotic space exploration
Saturn
  Cassini program, 55, 151–152
  overview, 143, 144*f*, 144*t*

Voyager program, 148
Saturn rocket, 10
Scaled composites, 58, 59
Schiaparelli, Giovanni, 124
Schirra, Walter, Jr., 12
Schmitt, Harrison (Jack), 27, 28
Schwabe, Heinrich, 116
Science
  Chandra X-Ray Observatory (CXRO),
    111–113, 112*f*, 113*t*
  *Compton Gamma-Ray Observatory
    (CGRO)*, 111
  Discovery program, 102–106
  electromagnetic spectrum, 106–107,
    107*f*
  Explorer program, 102, 103*t*
  Hubble Space Telescope, 108–111,
    109*f*, 110*t*
  International Space Station, 94–96
  Mars, 125–126
  NASA goals, 101
  NASA programs, 31–32
  NASA's Great Observatories, 106
  Science Mission Directorate,
    NASA, 39
  solar exploration missions, 119–121,
    120*t*
  space shuttle program, 74
  space travel as a science priority,
    157–158
  Spitzer Space Telescope, 113–114, 114*f*,
    115*f*
  See also Rocket science
Science fiction
  early, 2–3
  far planets, 146
  L3 point, 116
  Mars, 125
  space-age, 14
Scobee, Francis (Dick), 20
Scott, David R., 28
Scout rocket, 10
SDI (Space Defense Initiative), 47–48
SDIO (Strategic Defense Initiative
  Organization), 47
Sea Launch Company, LLC, 58
Search for Extra-Terrestrial Intelligence
  (SETI) project, 57
SEC (Space Environment Center), 118
SETI (Search for Extra-Terrestrial
  Intelligence) project, 57
Shargin, Yuri, 94
Sharman, Helen, 83
Shenzhou program, 55
Shepard, Alan B., Jr., 9, 12, 23, 28
Shuttleworth, Mark, 92
Sighting opportunities, 163, 163(*f* 9.8)
Silkworm missiles, 55
Skylab, 13, 29, 82–83, 82*t*
Slayton, Donald (Deke), 12

Smith, Michael, 20, 71
Smithsonian Institution, 5
Society for Spaceship Travel. *See Verein
  für Raumschiffahrt (VfR)*
SOHO observatory, 119–120
Sojourner rover, 129–130, 131*f*
Solar exploration missions, 119–121, 120*t*
Solar flares, 116–117
Solar prominences, 117
Solar system, 141, 142*f*
Solar wind, 116
Soviet space program
  anti-ballistic missile treaties, 47
  Apollo-*Soyuz* rendezvous and docking
    test project, 53, 53*f*
  détente, 12–13
  firsts, 52
  *Gruppa Isutcheniya Reaktivnovo
    Dvisheniya*, 57
  Korolev, Sergei, 50–52
  lunar missions, 11, 25
  Mars missions, 126
  Mir space station, 54, 83–84, 85*t*,
    86*t*–87*t*, 88
  Salyut program, 53, 83, 83*t*
  space casualties, 20
  space race, 7–12, 21, 22–23
  Voskhod, 51–52
  Vostok, 51
Soyuz program, 20, 52, 53, 53*f*, 93
Space Adventures, Ltd., 19, 58
Space Amateur Radio Experiment
  (SAREX), 164
Space biology program. *See* Biosatellite
  program
Space Control Center, AFSPC, 50
Space Defense Initiative (SDI), 47–48
Space Environment Center (SEC), 118
Space Flight Operations Contract, 69, 70
The Space Foundation, 161
Space observatories
  Chandra X-Ray Observatory (CXRO),
    111–113, 112*f*, 113*t*
  *Compton Gamma-Ray Observatory
    (CGRO)*, 111, 111*f*
  early space telescopes, 108
  Great Observatories, 106
  Hubble Space Telescope, 108–111,
    109*f*, 110*t*
  James Webb Space Telescope (JWST),
    115
  solar exploration missions, 119–121,
    120*t*
  Spitzer Space Telescope, 113–114, 114*f*,
    115*f*
Space race
  NASA, formation of, 21–23
  Soviet space program, 51–54
Space shuttle program
  accomplishments, 73–74

boom sensor system, 73*f*
budget, 39
Challenger disaster, 71
Columbia Accident Investigation Board
    recommendations, 74*t*
Columbia disaster, 72
crew module layout, 69*f*
design and development, 62–65,
    62*f*, 63*f*
development, 13
emergency flight options, 68–69
flight profile, 65–69, 67
flights per orbiter, 79*f*
flights per year, 78*f*
future of, 40
Hubble Space Telescope servicing
    mission, 110–111
launch, 65
military satellite deployment, 46–47
Mir-shuttle missions, 84, 85*f*,
    86*t*–87*t*
mission profile, 67*f*
missions, 70–73, 75*t*–77*t*
orbit, 65–68
orbiter, 66*f*
orbiter safety zone, 50*f*
organization, 69–70
overview, 29–30
payload bay, 68*f*
physical specifications, 65*t*
public opinion, 159–160
public opinion on space tourism, 166
Return to Flight (RTF) Task Group,
    72–73
shutdown, 93
as a transport business, 61
*Space Station: Impact of the Grounding of
the Shuttle Fleet* (General Accounting
Office), 93
Space stations
    early visions, 81–82
    Freedom space station, 88
    Skylab, 13, 29, 82–83, 82*t*
    Soviet and Russian, 83–84, 83*t*, 85*t*,
        86*t*–87*t*, 88
    Station Alpha, 88–89
    See also International Space Station
Space Systems Division, U.S. Air
    Force, 46
Space Technology Hall of Fame,
    161–162, 162*t*
Space tourism
    commercial enterprises, 58
    International Space Station, 91, 92
    Mir space station, 83–84
    public opinion, 166, 167(*f* 9.10)
    Russian missions, 18–19
    shuttle flights, 38
Space walks

International Space Station, 97–98
    Soviet space program, 51–52
    spacesuits, 97–98
    White, Edward, 24
Space weather, 118–119
The Spaceship Company, 59
SpaceShipOne, 16–18, 58–59, 60*t*
Spacesuits, 97–98, 99*f*
Spirit rover, 133–138
Spitzer, Lyman, 108
Spitzer Space Telescope, 113–114,
    114*f*, 115*f*
Sputnik, 7–8, 45, 45*f*
Spy satellites, 46
SSC (Stennis Space Center), 35
SSD (Space Systems Division), U.S. Air
    Force, 46
Star Trek (television show), 14
Star Wars (movie), 14
Star Wars program, 47–48
Stardust mission, 15, 104, 105*f*
Station Alpha, 88–89
Statistics
    Department of Defense program
        budget, 44*f*
    Missile Defense Agency funding, 48*f*
    moon landing as hoax, 168*t*
    NASA budget, 22*f*
    public opinion on Bush's space
        program, 161*f*
    public opinion on extraterrestrial life,
        167(*f* 9.11), 167(*f* 9.12)
    public opinion on interest in the space
        program, 157(*f* 9.2)
    public opinion on NASA's budget,
        156–157, 158*f*
    public opinion on NASA's
        performance, 160*t*
    public opinion on reasons for space
        exploration, 157(*f* 9.3)
    public opinion on space tourism,
        167(*f* 9.10)
    public opinion on space travel
        knowledge, 166*f*, 166*t*
    public opinion on the benefits *vs.* costs
        of the space program, 163(*f* 9.7)
    public opinion on the importance of
        space exploration, 156*t*
Stellar tidal disruption (STD), 112
Stennis Space Center (SSC), 35
StratCom (U.S. Strategic Command),
    49–50
Strategic Defense Initiative Organization
    (SDIO), 47
Studies, reports, and surveys
    *Space Station: Impact of the Grounding
    of the Shuttle Fleet*, 93
    *Technical Realities: An Analysis of the
    2004 Deployment of a U.S. National
    Missile Defense System* (Union of
    Concerned Scientists), 48

Sun-Earth connection, 101, 102, 115–116
Sun orbiters, 31–32
Sunspots, 116, 118*f*
Superboosters, 52
Surveyor missions, 25

T

Taikonauts, 55
TAL (Transatlantic Abort Landing), 68
Teacher Resource Centers (TRCs), 164
Teachers, 38
*Technical Realities: An Analysis of the 2004
Deployment of a U.S. National Missile
Defense System* (Union of Concerned
Scientists), 48
Technological achievements, 156*f*, 161–162
Telescopes, 107–108
    See also Space observatories
Television
    Apollo 11 mission, 27
    *Conspiracy Theory: Did We Land on the
    Moon?*, 167–168
    Mars, 125
    NASA, 162
    NTV, 163–164
    Star Trek, 14
Tereshkova, Valentina, 51, 52
Termination shock, 148
Test pilots, 6–7
Theophrastus, 116
Time and telescopes, 107
Tito, Dennis, 91
Titov, Gherman, 9, 51, 52
Titov, Vladimir, 83
Tombaugh, Clyde, 145–146
Tourism, 165
Transatlantic Abort Landing (TAL), 68
Transatlantic flight, 4
TRCs (Teacher Resource Centers), 164
Treaties
    "Agreement between the United States
        of America and the Russian
        Federation Concerning
        Cooperation in the Exploration
        and Use of Outer Space for
        Peaceful Purposes," 84
    Anti-Ballistic Missile Treaty, 47
    Intergovernmental Agreement on
        Space Station Cooperation, 89, 93
    Joint Declaration on Cooperation in
        Space, 89
    Outer Space Treaty, 12–13, 49
Treaty on Principles Governing the
    Activities of States in the Exploration
    and Use of Outer Space, including the
    Moon and Other Celestial Bodies. *See*
    Outer Space Treaty
Trip to the Moon (Méliès), 3
Tsien Hsue-shen, 55
Tsiolkovsky, Konstantin, 4

2001: A Space Odyssey (Clarke), 146
2001: A Space Odyssey (movie), 14, 131

U

Ulysses program, 31–32, 120–121, 121*f*
Union of Concerned Scientists, 48
United Nations, 12–13
United Space Alliance, 69, 70
United States
    aviation history, 3–4
    détente, 12–13
    Freedom space station, 88
    low Earth orbit (LEO) missions, 13–14
    robotic space explorers, 14–15
    space casualties, 20
    space race, 7–12
    X series planes, 6–7
Unity module, 89–90, 89*f*, 90*f*
Uranus, 144–145, 148
U.S. Department of Defense
    Manned Space Flight Support
        Office, 50
    space budget, 44, 44*f*
    space shuttle launch sites, 63–64
U.S. Strategic Command (StratCom),
    49–50

V

V weapons, 6
Valles Marineris, 125
Van Allen, James, 45, 159
Van Allen radiation belts, 102, 168
Vandenberg Air Force Base, 63–64
Venus
    Magellan mission, 15, 31
    robotic space explorers, 15
    Venus Express, 55
Verein für Raumschiffahrt (VfR), 5, 56–57
Verne, Jules, 2–3

Very Large Array, 108
Very Long Baseline Array, 108
VfR *(Verein für Raumschiffahrt)*, 5, 56–57
Viking program, 30, 128
Virgin Galactic, 59
Volkov, Vladislav, 20, 52
Von Braun, Wernher
    Hermann Oberth, influence of, 81
    Marshall Space Flight Center, 34–35
    rocket program, 23
    rocket science, 5–6, 44
    Saturn V rocket, 26
    space race, 8–9
    Verein für Raumschiffahrt (VfR), 57
Voskhod, 51–52
Vostok, 51
Le voyage dans la lune (Méliès), 3
Voyager program, 30–31, 147–148, 147*f*,
    149*f*
Vulcan, 116

W

Wages and earnings
    astronauts, 39
    NASA employees, 36
Wallace, Alfred Russel, 124
Wallops Flight Facility, 32
The War of the Worlds (Wells),
    3, 125
Warhol, Andy, 164
Water on Mars, 137–138
Weapons
    anti-ship missiles, 55
    ballistic missiles, 7
    military space programs, 49
    Von Braun, Wernher, 5–6
Web site, NASA's, 162–163
Webb, James, 164
Wegman, William, 164

Weightlessness, 9
Weiner, Time, 49
Wells, H. G., 3, 125
White, Edward, 10, 20, 24
White, Robert, 59
White Knight, 17, 18, 59
White Sands Test Facility
    (WSTF), 35
Whitlock, Laura, 168
Wiesner, Jerome, 46
Wind mission, 120
Wolfe, Tom, 12
Women
    American astronauts, 38
    first woman in space, 51, 52
World War II, 5
Wright, Orville and Wilbur, 3
WSTF (White Sands Test Facility), 35
Wyeth, Jamie, 164

X

X Prize Foundation, 17–18
X series planes, 6–7
XMM-Newton observatory, 115
XSS (Experimental Satellite Series), 49

Y

Yang Liwei, 55
Yeager, Charles (Chuck), 6, 22
Yegorov, Boris, 51
Yeltsin, Boris, 84
Young, John W., 28

Z

Zarya module, 89, 89*f*, 90*f*
Zimmerman, Robert, 88
Zvezda module, 90, 93
Zvezdny Gorodok, 51, 97